Praise for *Natural Beekeeping*

"*Natural Beekeeping* is a wonderful book, beautifully written and illustrated, about how one can have healthy hives of bees without using synthetic pesticides, antibiotics, or artificial diets. Ross Conrad explains in fine detail that the key ingredients of organic beekeeping are disease-resistant stock, favorable apiary sites, and good colony management, plus a reverence and respect for the bees."

—Thomas D. Seeley, Cornell University,
author of *Honeybee Democracy* and *The Wisdom of the Hive*

"Ross Conrad keeps bees the way bees should be kept. . . . His natural approach to caring for these magnificent pollinators makes sure they will continue their tasks so we can continue to rely on them. Natural beekeeping is not only the best way to keep bees, it is the *only* way we should keep bees."

—Kim Flottum, editor of *Bee Culture: The Magazine of American Beekeeping* and author of *The Backyard Beekeeper*

"Ross Conrad buzzes and brainstorms where other angels have feared to tread. He makes organic apiculture seem not only possible, but necessary."

—From the foreword by Gary Paul Nabhan

"Ross Conrad has created an intimate guide to beekeeping that clearly details holistic methods aimed at hive health. The honeybee faces many affronts to its immune system, [making] it all the more imperative that local farmers and gardeners learn ways to assist the honeybee. Ross has laid out the ground rules; the rest of us need to heed the buzz."

—Michael Phillips, author of *The Holistic Orchard* and *The Apple Grower*

"*Natural Beekeeping* describes opportunities for the seasoned professional to modify existing operations, increase profits, and eliminate the use of chemical treatments. Beginners will need no other book to guide them. Whether you are an experienced apiculturist looking for ideas to develop an integrated pest-management approach or someone who wants to sell honey at a premium price, this is the book you've been waiting for."

—*Pest Inspections* blog

NATURAL BEEKEEPING

Organic Approaches to Modern Apiculture

REVISED AND EXPANDED EDITION

ROSS CONRAD

CHELSEA GREEN PUBLISHING
WHITE RIVER JUNCTION, VERMONT

Project Manager: Patricia Stone
Project Editor: Benjamin Watson
Copy Editor: Fern Marshall Bradley
Proofreader: Eric Raetz
Indexer: Lee Lawton
Designer: Melissa Jacobson

Printed in the United States of America.
First printing February, 2013.
10 9 8 7 6 5 4 3 2 1   13 14 15 16 17

Our Commitment to Green Publishing
Chelsea Green sees publishing as a tool for cultural change and
ecological stewardship. We strive to align our book manufacturing
practices with our editorial mission and to reduce the impact of our
business enterprise in the environment. We print our books and
catalogs on chlorine-free recycled paper, using vegetable-based
inks whenever possible. This book may cost slightly more because
it was printed on paper that contains recycled fiber, and we hope
you'll agree that it's worth it. Chelsea Green is a member of the
Green Press Initiative (www.greenpressinitiative.org), a nonprofit
coalition of publishers, manufacturers, and authors working to protect the world's endangered forests and conserve natural resources. *Natural
Beekeeping* was printed on FSC®-certified paper supplied by QuadGraphics that contains at least 10% postconsumer recycled fiber.

Library of Congress Cataloging-in-Publication Data
Conrad, Ross, 1963-
  Natural beekeeping : organic approaches to modern apiculture / Ross Conrad.—2nd ed.
     p. cm.
  Includes bibliographical references and index.
  ISBN 978-1-60358-362-6 (pbk.)—ISBN 978-1-60358-363-3 (ebook)
1. Bee culture. 2. Organic farming. I. Title.

SF523.C75 2013
638'.1—dc23
                        2012044566

Chelsea Green Publishing
85 North Main Street, Suite 120
White River Junction, VT 05001
(802) 295-6300
www.chelseagreen.com

Chelsea Green Publishing is committed to preserving
ancient forests and natural resources. We elected to print
this title on paper containing at least 10% postconsumer
recycled paper, processed chlorine-free. As a result, for
this printing, we have saved:

19 Trees (40' tall and 6-8" diameter)
8,931 Gallons of Wastewater
9 million BTUs Total Energy
598 Pounds of Solid Waste
1,647 Pounds of Greenhouse Gases

Chelsea Green Publishing made this paper choice because
we are a member of the Green Press Initiative, a nonprofit
program dedicated to supporting authors, publishers,
and suppliers in their efforts to reduce their use of fiber
obtained from endangered forests. For more information,
visit www.greenpressinitiative.org.

Environmental impact estimates were made using the
Environmental Defense Paper Calculator. For more
information visit: www.papercalculator.org.

MIX
Paper from
responsible sources
FSC® C084269

Disclaimer
The use of the hive-management techniques described in this book is strictly your responsibility. Anything that is used with the intent to kill
unwanted organisms (pests or weeds) is considered a pesticide by the United States Environmental Protection Agency (EPA) and therefore falls
under its authority and jurisdiction. EPA regulations prohibit the sale of honey and other bee products taken from hives treated with unapproved
pesticides, no matter how safe or nontoxic they may be. It is your responsibility to see to it that any hive products you offer for sale to the public
do not violate this requirement.

Dedicated to you, the Reader.
May the honey bee bring as much joy and
wonder into your life as it has to mine.

CONTENTS

FOREWORD TO THE FIRST EDITION

In *Natural Beekeeping,* Ross Conrad takes us on a journey, from his initially casual, coincidental involvement with honey bees as part-time wage work to his multifaceted engagement with these colonial creatures as collaborators in a struggle to keep the world we live in productive, safe, and healthy. He begins, as many of us do, with barely an inkling of how the lives of domesticated bees are intertwined with our own. Gradually, as he accumulates skills and as his passion for these creatures deepens, he attempts to solve problems that honey bee colonies now face, ones that natural and agricultural scientists have somewhat neglected.

Conrad uses honey bees as the proverbial "canaries in the coal mine" to provide a sense of the health and natural wealth remaining in the world around us, a world that has been damaged by the chemical and physical fragmentation of habitats that both honey bees and humans require to survive. He ingeniously finds ways to deal with the parasites and pesticides that trip up honey bees as they go out, day after day, to do their sacred work of pollination. Ultimately, because Ross Conrad buzzes and brainstorms where other angels have feared to tread, he makes organic apiculture seem not only possible but necessary.

In many ways, this is the worst of times for the honey bee in America, as the species *Apis mellifera* is facing unprecedented threats. There are fewer honey bees to be seen today in the wild and agrarian habitats of this continent than were witnessed by any human generation during the last 150 years.

This is because of the inadvertent introductions of tracheal and varroa mites during the 1980s, the arrival of Africanized bees in our southern states (and the killing off of hives genetically contaminated by them), the prevalent and persistent use of agricultural chemicals, and, most recently, the so-called colony collapse disorder (CCD) that is decimating many managed honey bee populations. The dramatic die-offs of honey bee colonies that are occurring around the time this book is being released have not yet been fully explained by scientists, but there is little doubt that honey bee exposure to pesticides and other agricultural chemicals has weakened individual resistance and disrupted their complex of immune responses.

In short, the insights embedded in *Natural Beekeeping* would be interesting and welcome to beekeepers living at any point in history, but they are all the more vital, all the more urgent, at this peculiar moment in time. As I read Conrad's litany of the perils that the formerly ubiquitous, highly adaptable, unflappably resilient domestic honey bee faces, I couldn't help but worry even more about the tens of thousands of other truly wild pollinator species in America, whose work sometimes complements and sometimes competes with that of the domesticated colonists. Wild bees, butterflies, bats, and birds are getting slam-dunked by many of the same insults that affect the honey bee, and many are even more susceptible to pesticides, due to their body weights and behaviors. Most of the honey-bee-friendly agricultural practices that Conrad champions are, by and large, pollinator-friendly across the board. There are, of course, best practices that specifically aid the health of managed colonies of honey bees in ways that do not pertain to other wild pollinators, but they do pertain to us, in the sense that we consume bee products for food and medicine.

"Do least harm" is a tenet shared not only among Native Americans, Buddhists, and Jainists, but among many others of diverse ethnicities and faiths as well. Ross Conrad has walked a path of learning how to do least possible harm to the honey bee, and emulating him may teach the rest of us

how to live with this precautionary principle and make it evident in our own actions.

GARY PAUL NABHAN
FEBRUARY 2007

Gary Paul Nabhan is the founder of the Forgotten Pollinators campaign, the Migratory Pollinators Project, and the Renewing America's Food Traditions (RAFT) consortium. He is coauthor or editor of *The Forgotten Pollinators, Conserving Migratory Pollinators and Nectar Corridors in Western North America,* and *Coming Home to Eat.*

PREFACE

It's hard to believe, but it has been over five years since the first edition of *Natural Beekeeping* was published. The book was released on the heels of colony collapse disorder's first widespread impacts, and much has changed since then. There has been phenomenal growth in the number of new people getting into beekeeping, and our knowledge and understanding of the honey bee and its pests and pathogens has grown tremendously, thanks to increased funding for research. There are scores of new products on the market based upon these new understandings. Beekeeping management has continued to evolve at a brisk pace, with beekeepers both new and old experimenting with novel approaches and philosophies. Interest in urban beekeeping has exploded, and top bar hives have become commonplace.

In this revised and expanded edition, I've made minor changes throughout the book based on recent research and new information gained from my continued learning and experience. The primary changes from the original text appear in chapters 2, 3, 5, 8, 9, and 12. I've added a new chapter on marketing (Chapter 11). "Organics and the Evolution of Beekeeping" (Chapter 11 in the first edition) is now Chapter 12. As in the first edition, I have included numerous personal anecdotes and philosophical discussions with the goal of writing a book that not only includes serious issues and technical details, but is injected with humor and is, I hope, interesting to read: more like a novel than a textbook. I hope you find that I have succeeded.

PREFACE TO THE FIRST EDITION

Read not to contradict nor to believe, but to weigh and consider.
—FRANCIS BACON

I did not develop my passionate affection for the honey bee early in life. In fact, other than the occasional sting from stepping on a bee barefooted, or enjoying some honey in my tea, my interaction with the bee world was extremely limited. This changed during the summer of 1989. After moving to Vermont in the winter of '88, I had landed a job at a radio station and was finding it difficult to make ends meet. The rate of pay at my small-town station was quite different from what I had been used to in Manhattan, and as a result, I picked up part-time work from time to time to help pay the bills. That was the summer I met Charlie Mraz (1905–1999).

Charles was in his eighties and still heated his home with a woodstove, as is the fashion throughout the state of Vermont. As a result of his advanced years, Charlie hired me to assist him in stacking the four cords of firewood he had purchased for the upcoming winter. After the work was completed, I asked Charlie if he had anything else I could help him with. He said he didn't, but that he would speak with his son Bill, who might have some work over at the honey house for me.

Charlie had moved to Vermont in the late 1920s and had started a beekeeping business called Champlain Valley Apiaries. He had gone on to build up the company into the largest beekeeping outfit in Vermont that kept all its bees within the state. Although Charles was still very involved in the business, some sixty years after its formation, Bill had taken over most of the day-to-day activities of running the company. I helped harvest and extract the excess honey that the bees produced for Champlain Valley Apiaries that summer. After this experience, I started my own disc jockey service and refocused my attention on the broadcast/music industry.

During the winter of 1990–91, I found myself working as the DJ at a nightclub near the Killington ski resort. I was working about six hours every night and had my days available to pursue any interest or activity that happened to strike my fancy. In fact, this was the first time since I had been a toddler that I had my days totally to myself. As a result I found myself doing a lot of thinking, reflecting, and soul-searching. I began to seriously evaluate my life. I also began to read more and was drawn to a book written by a Native American Indian elder named Sun Bear. Sun Bear had founded the Bear Tribe, which sought to teach Native wisdom to anyone who was willing to learn about the old ways that had been handed down to

him by his parents and grandparents. This book resonated with what I was feeling at the time, and, as a result, I decided to spend a month with the Bear Tribe at their teaching center known as Vision Mountain near Spokane, Washington.

It was from Sun Bear and the Bear Tribe that I received a new appreciation for nature. I began to see how little I had respected and cared for the natural world around me, and I took a vow, to myself, to bring more healing to the Earth and, by extension, to those around me. While I was with the Bear Tribe, I also had the fortunate opportunity to participate in several American Indian ceremonies. One of these ceremonies was a vision quest. This ancient mystical journey is undertaken as an odyssey of self-knowledge and self-fulfillment—a spiritual journey into the wilderness and the soul. Traditionally boys undertook the vision quest as they made the transition from childhood to manhood. As part of my quest, I sat alone for four days and four nights, fasting, praying, and seeking vision on a mountain where the Bear Tribe's learning center was located. It was during this time, on my second day on the mountain, that a single bee came to visit me. I looked down at one point, and there on my big toe sat a honey bee. How it got there, I do not know. I didn't see it arrive. Nor had I heard it arrive or felt its presence on my skin.

Native wisdom teaches that we can learn from everything in nature, so I spoke to the bee and let it know that I was open to learning whatever it had to teach me. The bee then spent about five minutes flying around me, landing occasionally on various parts of my body, before flying off on its merry way. I didn't think anything more of the bee's visit until the next day while sitting in the same spot, when a bee flew by. It came up to my ear and hovered there for a moment, buzzing, and then it flew over my head and buzzed in my other ear before taking straight off to my right, as if it had been on a beeline and my head was in its way. The bee's flight path would have looked like one of those fake arrows that comedians put on their heads to make it look like they have been shot. Whether it was the same bee that visited me the previous day I do not know for sure, but I suspect that it was. It almost seemed as if the bee were trying to speak to me and say, "Now don't forget what I told you yesterday."

I returned to Vermont following my time with the Bear Tribe. I didn't think much more about honey bees until about six months after my return, when Bill Mraz stopped by to see me. This was the first time I had seen him in quite some time. He told me that his primary beekeeping assistant had been badly injured in a car accident and would be unavailable for the remainder of the season. He wondered if I was interested in filling the position on a full-time basis. Due mainly to my experience with that bee during my vision quest, I accepted the offer, and spent the next six years receiving a whole new education.

Not only did I find the honey bees fascinating, but I also became intrigued by the work that Charlie was doing with apitherapy. I was astounded by how effective the various products of the beehive were in restoring health and vitality to individuals with diseases or other health conditions. Charlie had drawers full of letters from people with debilitating diseases who had found relief by using honey, pollen, propolis, royal jelly, and/or bee venom therapeutically. A typical letter told of a person so crippled she had been confined to a wheelchair for years. Doctors were unable to do much more than relieve her pain and make her comfortable. After apitherapy treatments, she was up and walking around as if she had never been sick in the first place. These stories fascinated me and led me to research how apitherapy worked and why. Curiosity got a firm hold of me, and I began looking into other forms of alternative healing such as homeopathy, herbs, natural foods, and dietary supplements.

All these experiences have nurtured a connection between the honey bee and myself that is deeper and more meaningful than I can put into words. The bees have taught me so much over the years and have brought so many wonderful people into my world. It is my hope that I will always have honey bees around me, until the end of my days.

Since the honey bee exerts such a powerfully beneficial influence on the natural world around us, it seems logical to assume that our own efforts to help the honey bee thrive can indirectly benefit all of nature. As a result, we beekeepers are, for the most part, a collegial lot; we exist in a kind of friendly competition with one another. We certainly do not have a situation where there are too many beekeepers in the United States. In fact, I believe that we need more beekeepers in this world, and those beekeepers need to be successful in apiculture. This motivates me to do whatever I can to assist my fellow apiculturists, which includes answering questions, offering advice, and even occasionally providing on-site inspections and evaluations of hives. The most common questions I encounter revolve around how I manage to keep my bees healthy and achieve relatively low winter losses compared to the conventional beekeepers all around our area, who typically experience much larger losses. By sharing what I've learned with less experienced folks, I hope to aid them in keeping their bees alive and healthy. I also hope that, through the good work we accomplish with our honey bees, we can all help to leave the world in better condition than we've found it. In these pages, I share some of the insights I have gained, along with specific "how to" information on keeping bees in an ecological way, in the hope that others will find something of use that may be applied to their work in apiculture. This is my primary motivation for writing this book. As with beekeeping in general, it certainly isn't for the money.

During my own efforts over the years to obtain information on natural, nontoxic approaches to beekeeping, I have noticed that there are very few sources that list, all in one place, a critical mass of information on the numerous organic techniques that can be utilized by beekeepers. It is my hope that the information contained in this book will concentrate a significant number of natural and sustainable solutions in a practical way to give both the commercial and the hobby beekeeper a variety of ideas and suggestions for reducing chemical usage and to provide a number of alternative options in the effort to keep bees healthy without toxic chemotherapy controls. The more options and knowledge we as beekeepers have at our disposal, the greater the opportunity for flexibility—something that is sorely needed today to help us adapt successfully to changes and events that challenge our beekeeping operations.

I'll be the first to tell you that I do not have all the answers. In fact, no small part of my enjoyment of the craft of beekeeping stems from the creativity, adventure, and sense of discovery that is inherent in the process of attempting to maintain healthy hives in a natural and sustainable way. From the very beginning, I have always refused to use toxic chemicals in my hives. My stubbornness caused me to lose many hives in the early days, but by persevering I have proven to myself that it is possible in this day and age to keep honey bees without resorting to the use of dangerous synthetic chemical compounds. I have attempted to present my experiences and what I have heard or read in a manner that will be of the most benefit to others in regard to natural and organic apiculture. As a result, I have taken pains to indicate which techniques I have directly tried and experienced and which I have only heard or read about in passing.

I see the activity of beekeeping in much the same way that I view activities like gardening and raising children. There is no single correct approach that applies to everyone. Each one of us who participates in the craft of apiculture will develop approaches and techniques that are unique to our particular needs, styles, and situations. As with gardening and child rearing, how individuals approach beekeeping will depend on their resources, knowledge, purpose, level of confidence, and philosophies for becoming involved in the activity. A full-time migratory beekeeper, for example, will manage hives quite differently from a commercial honey producer. And both of these approaches to beekeeping will differ greatly from those of a person who maintains a hive simply as a hobby in order to provide themselves with a source of honey bees for apitherapy. No matter our reasons for keeping bees, it is beneficial to seek out beekeeping information from a variety of sources and to choose those forms of hive management that best fit our personal situation, goals, and finances.

Thanks to the continual development of innovative technologies and management techniques, new methods seem to surface every year that fit the natural and organic philosophy, and improvements are gradually being made to many of the older ways of going about things. Some people may choose to incorporate just a few of these ideas into their hive management routines and simply reduce the number of chemical treatments they need to apply on a yearly basis. This would lead to an integrated pest management (IPM) approach, which dictates that treatments are applied not simply routinely but only according to need. Others may choose to be more aggressive in implementing nontoxic beekeeping and manage their hives in an entirely natural manner, as I endeavor to do. Either way, significant cost savings can be gained from reduced chemical, drug, and feed supplement expenditures. More important, greater peace of mind will be obtained, by easing concerns about toxic chemical contamination and exposure. The final outcome—healthier bees and robust hives—is everyone's ultimate prize.

ACKNOWLEDGMENTS

I would like to thank the following people, who—each in their own unique way—assisted, supported, and/or encouraged this book along to its fruition:

Jeff Cunningham, Nicole Dehne, Peter Grant, Geri Hens, Dee Lusby, Larry Plesent, Suky Shoen, Kirk Webster, Hilde Whalley, and Joshua White.

Special thanks to John Barstow, Ben Watson, and all the good folks at Chelsea Green Publishing for their belief in the value of this book and their willingness to take a chance on an ambitious yet inexperienced book author.

Extra special thanks to John Cleary, Mark Conrad, C. Wallon Hazen Denney, Alice Eckles, James Gabriel, and Sandy Lincoln for their comments on various sections of the original draft manuscript.

And finally, my eternal gratitude goes out to Michelle McCauley and Kevin Kite, whose unwavering support has been instrumental in helping me get this revised and expanded edition of *Natural Beekeeping* accomplished.

CHAPTER 1

Why Organic Beekeeping?

We must be the change we wish to see in the world.

—MAHATMA GANDHI

• THE HIVE AS TEACHER •

We can learn so much from the honey bee, who also goes by the Latin name of *Apis mellifera*. Of all the insects, the honey bee seems to lend itself most perfectly to anthropomorphism. For example, the relationship between the bee and the plant kingdom is a powerful and intricate orchestration of interdependence and cooperation. To live its day to-day life, the bee need only collect nectar and pollen from the flowers in bloom. These gifts from the plant kingdom, along with some water, plant resins that the bees use to make propolis, fresh air, and sunshine, are all the bees need from the world around them to survive and prosper within their colony. Thus, unless it feels threatened and is forced to defend itself or its hive, the bee is the only creature in the animal kingdom, that I am aware of, that does not kill or injure any other being as it goes through its regular life cycle. *Apis mellifera* damages not so much as a leaf. In fact, honey bees take what they need in such a way that the world around them is improved. By pollinating blossoms during nectar- and pollen-foraging activities, the honey bee contributes directly to the abundance found on Earth. This industrious little creature even transforms the nectar it collects from sugar water into deliciously sweet and health-promoting honey.

As a result, beekeeping is a wonderful way to give back to the world and help make it a better place, while at the same time receiving many incredible gifts. It is a pleasure to know that, although they may not realize it, my neighbors down the road are able to enjoy a bountiful harvest from their garden due, in part, to the pollination services rendered by the honey bees in my care. I like the fact that my hives, working in conjunction with the local plants, help produce an abundance of nuts, seeds, fruits, berries, et cetera, so the birds and other wildlife living in the area around my bee yards have plenty to eat. Because many of the seeds and nuts that are not eaten will develop into a variety of beautiful flowers, herbs, bushes, and trees, my neck of the woods is more likely to be a beautiful and bountiful place to live for many years after my passing. By directly participating in the creation of numerous seeds, nuts, fruits, and vegetables through cross-pollination, the honey bee benefits plants, animals, and humans alike. Farmers know that as the number of bees in an area increases, so will the quality of the fruits and vegetables grown in the region.

All the while, the bee is going about the business of creating honey, a substance that not only tastes wonderful but can help heal burns, cuts, infections, and numerous other maladies of the body. So much has been written about the amazing healing aspects of honey that I will not go much further into this topic other than to encourage the reader to investigate the many benefits of this "first aid kit in a jar." Honey is something so precious and special; even with our highly developed technological sciences, we humans still have not been able to duplicate the efforts of the simple honey bee and create the same substance from what amounts to nothing more than sugar water.

The honey bee inspires me to work into my daily life this lesson: that we should take what we need to live in the world in such a way that we do little or no harm and at the same time give something back and improve upon things, thus making the world a better place. Many of the world's problems could potentially be solved in short order if every person took this single lesson from the honey bee to heart and worked to manifest it in daily life.

The indigenous peoples that inhabited the Americas prior to the arrival of Europeans shared a philosophy that regarded the natural world as a teacher. They saw that in the natural balance of life, extremely diverse organisms live in coexistence. Everything has its place in the order of things. Unfortunately, this tolerance seems to have been lost in the industrial model developed by our Western culture. Although the industrialization of agriculture offers much promise, it also has created many serious problems. This industrial model produces the zero-tolerance mentality that gives birth to the toxic chemical treatments typically used in agriculture and embraced by many in the beekeeping community.

A more holistic approach, one that views the honey bee and the hive environment in terms of a biological model, stems from a nature-based beekeeping philosophy. In this model, the concepts of coexistence and tolerance are included. This perspective precludes the use of toxic chemical solutions that seek to wipe out every pest in the hive. The holistic viewpoint recognizes the unique role

FIGURE 1-1. The propensity of the honey bee to focus on one type of flower during each foraging trip makes it ideally suited for crop pollination.

played by all of the creatures of creation. Everything has its place and reason for being, even if it is not immediately obvious. I must admit to not being privy to the reason for the existence of the mite known as varroa (*Varroa destructor*), which has become the bane of the beekeeping industry in most of the world. Nevertheless, I am certainly thankful for its presence, if for no other reason than because the trouble it has caused has inspired so many people to seek alternatives to the reigning beekeeping paradigm of our times: the industrial model. Once again, nature is teaching us to seek out a bigger picture.

• TOXIC CHEMICALS • INFILTRATE THE CLASSROOM

As of 1987, the keeping of honey bees was the only widespread agricultural endeavor in the United States that had not become reliant on toxic chemicals to secure a harvest. In that year the parasitic mite *Varroa destructor*, formerly *Varroa jacobsoni*, was first identified in hives located in the United States. This mite—known to infest honey bee colonies and suck blood from the bees, causing weight loss and deformities, spreading disease, suppressing immunity, and reducing life span—has since spread throughout America, assisted in part by the unwitting cooperation of migratory beekeepers. The overwhelming hive losses experienced by US beekeepers in subsequent years prompted an outcry from the beekeeping industry for assistance. The response followed the same path tread by other agricultural commodity groups that have had a need to control insect pests threatening their crops. Beekeepers turned to toxic chemical compounds to solve their mite problem. Unfortunately, hindsight shows that this approach is a short-term solution, at best.

Since the end of World War II, when pesticides first became widely utilized, insects have consistently developed resistance to the toxic chemicals that have been used to control them. The response from agribusiness has been to turn either to higher doses of chemicals or to more toxic compounds in an effort to keep pests in check. History repeated itself with the beekeeping industry's approach of inserting chemical-impregnated plastic strips into the hive to control varroa mites.

The first strip was released in 1992 and went by the brand name Miticur. The strip, when introduced into the hive, carried with it a dose of the chemical amitraz, which was designed to be toxic enough to kill the mites, but not so potent as to kill the bees on which the mites reside. This approach to pest control is similar to using chemotherapy drugs in the treatment of cancer. Chemotherapy drugs are toxic to healthy cells as well as cancer cells. However, the chemotherapy treatment protocol calls for a dosage that is toxic enough to kill the cancer before the drugs have the effect of killing the person hosting the disease. Unfortunately, many beekeepers continued to experience huge hive losses despite treating with Miticur, and it was removed from the US market after a short duration, although it recently has been re-released under the trade name Apivar. At the time of its discontinuation, some people believed there had been a problem with the strips containing the wrong dosage of pesticide, which failed to kill the mites outright and allowed the mite to develop resistance to the chemical. However, our lack of knowledge and experience with this new mite may have also played a role in Miticur's apparent failure. Unless the unlikely event of production problems during pesticide manufacturing occurred, the beekeepers who initially experienced major losses to the mite despite their use of Miticur were probably treating their hives too late in the season without realizing it. Remember, this was in the early 1990s, and beekeepers had not yet gained much experience dealing with the varroa threat.

The next pesticide strip to be utilized in the hive was Apistan, which relied on the active chemical component fluvalinate. Fluvalinate is related to the pyrethrin family of pesticides that are synthesized from plants and is unlike its more toxic and long-lasting cousins, which are derived synthetically from petroleum. In contrast to Miticur, this product initially gained success on the market. Unfortunately, after several years of Apistan use, resistance to fluvalinate became widespread among varroa mites, and the industry turned to a much more toxic synthetic chemical compound called coumaphos. Classified as an organophosphate, coumaphos is sold under the trade name CheckMite+.

Organophosphates are among the most toxic chemical compounds used in agriculture. In fact, organophosphates are so toxic and persistent in the environment that an effort has been made by the US Environmental Protection Agency (EPA) to wean agriculture off these compounds, with the goal of eventually removing them from the market altogether. CheckMite+ is similar to Apistan in that it relies on a plastic strip inserted into the hive to introduce the chemical control. Unfortunately, it took only a few years before varroa began showing resistance to coumaphos as well. A conventional approach evolved using a rotation between Apistan and CheckMite+, altering their use from season to season in the hope that mites resistant to one chemical will not be resistant to the other. Undoubtedly, additional synthetic toxic chemicals will be approved for use against varroa in the future in order to help further address the resistance factor that varroa has developed to existing treatments. Walking this chemical treadmill may prove effective in controlling varroa for a short period, until the mites become resistant to them all. Unfortunately, these toxic compounds have the ability to accumulate in beeswax and honey, and this increases the potential for chemical contamination, resulting in undesirable consequences for both beekeepers and consumers. Anecdotal stories of beekeepers using the active ingredients fluvalinate, amitraz, and coumaphos in unregistered and illegal ways further exacerbate this situation. The fact that varroa mites developed resistance to amitraz, despite the relatively short time it was originally available for legal use by beekeepers, supports such anecdotal evidence.[1] Increased costs associated with these "chemotherapy" treatments

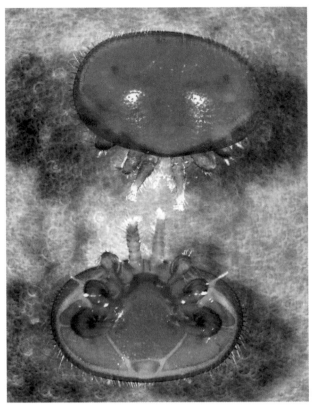

FIGURE 1-2. Australia is the only continent with bees that does not support a population of *Varroa destructor* . . . so far. PHOTO BY STEVE PARISE.

have reduced the profitability of the beekeeping industry. In addition, the backlash that may result if honey on the market is found to be contaminated with one or more of these pesticides, combined with their lack of effectiveness, greatly decreases the value of the chemical approach to mite control for beekeepers.

While the conscientious apiculturist will follow the directions on the pesticide label and wear gloves during application, exposure during use can still occur. Chemical contamination is most likely to result from secondary exposure, after the beekeeper touches objects with the chemically laden gloves worn while applying the pesticide. This is inevitable if the beekeeper works the bees without assistance, as is the case with most hobby and small beekeeping operations. I am reminded of the time I was assisting with the installation of some Check-Mite+ strips in a bee yard. One person's job was to smoke the bees and remove and then replace the hive covers, while the other person, wearing gloves, inserted the strips into the hive once the covers were removed. On this occasion, while the strips were being placed into a hive, a bee flew up and stung me on my eyelid. My initial reaction of shock and pain triggered my kindhearted coworker to want to help me. He instinctively reached up to remove the barbed weapon from my face as I was groping blindly at my pulsating eyelid and having difficulty removing it myself. The look of utter horror that crossed my face as I backed slowly away from the coumaphos-laden glove caused us to break out in hysterical laughter. This was undoubtedly the fastest recovery from being stung in a sensitive area that I had ever experienced. It also demonstrates how situations can arise when even a cautious individual can be exposed to the dangerous chemical pesticides used in conventional apiculture.

As a boy, I remember hearing it reported by the US Centers for Disease Control (CDC) that almost one out of every three American citizens could expect to get cancer at some point in their lifetime. Today, the CDC tells us that on average about 44 percent of the people in the United States can expect to get some form of cancer during their lifetime.[2] Meanwhile, the inventory of other degenerative metabolic diseases that plague our modern society continues to stack up like cordwood and is far too long to recite fully. They include not only cancer but attention deficit disorders, mental retardation, leukemia, male sterility, and birth defects—all of which can be linked to the poisons used in and around our food supply. It is said that beekeeping ranks high among the professions whose participants tend to live long and healthy lives when compared to people who make other career choices. It will be interesting to see if the potential for contamination due to the relatively recent advent of toxic chemical use within the beekeeping industry will adversely affect this favorable designation.

The effect of contaminated honey on consumers and their purchasing habits also becomes an issue when chemicals become an integral part of the beekeeper's management of pests. Research in Europe has shown that the chemicals used to control mites in hives are readily absorbed by beeswax. These pesticides and their chemical breakdown products have the potential to migrate into the honey that is stored in the wax combs. By approving the use of these miticides, the EPA has deemed the potential for such low-level contamination acceptable for both fluvalinate, coumaphos, and amitraz claiming that allowable residue levels in honey and wax are not exceeded when these materials are applied according to the label instructions. Nevertheless, as health food advocates have been pointing out for years, just because you can make it to the door after consuming minute quantities of these toxic compounds does not mean that they are benign and have no deleterious effects in the long run.

The effect of exposing honey bees to sublethal doses of these compounds is rarely discussed. Just as with people, the long-term health and vitality of the hive is likely to be compromised from such exposure, even if such detrimental effects are not readily and immediately evident. Because long-term tests have not yet been done, we may yet find that physiological and metabolic changes occur within honey bee populations over extended periods, and these detrimental effects may become obvious only after several decades of exposure. Research indicating reduced sperm counts in drones that have been exposed to coumaphos, and the negative effects on queens reared in cells constructed out of coumaphos-impregnated wax, validates these concerns.[3]

The effects of repeated exposure of bees and humans to small amounts of several toxic chemicals simultaneously is also largely unknown. The EPA requires safety studies be carried out only on individual pesticides in isolation, rather than in combination with one another. In actual practice, a beekeeper may utilize more than one chemical during the course of the season. Knowing that compounds may combine synergistically and have a greater effect when acting together than when used alone raises serious concerns about the effectiveness of the regulatory guidelines that are designed to safeguard our health and the health of our bees.

• THE MEANING OF • ORGANIC

Despite widespread belief to the contrary, the term *organic* does not mean that the final crop or product is free from toxic chemical contaminants. This mistaken perception has taken hold primarily due to the efforts of manufacturers and marketers, who have successfully promoted the notion that organic products are pure and chemical-free. Meanwhile, because of the success of this viewpoint, resentment has built up among many beekeepers, who feel that the organic label relegates their own commercially produced honey to second-class status. What was once considered a natural, healthy product is now deemed inferior, when, in fact, the final products of conventional and organic production may not be all that dissimilar in terms of their chemical composition.

At its inception, the organic approach traditionally referred to a management style and philosophy that is biological in nature. Rather than being a statement about product purity, *organic* was all about the big picture. It referred to approaches that care for the life in the soil and minimize the use of nonrenewable inputs and energy sources, such as those derived from petroleum. Organic principles embraced an attitude of fairness and care in regard to our common environment, as well as social concerns such as the welfare of farm workers. One of the original aims of organic agriculture was to establish a sense of stewardship for the Earth, embracing human-scale operations that fit harmoniously with the landscape and local community. Although it was certainly possible that organic management practices would result in a cleaner product, it was not the primary focus. Instead, organic management sought to mimic the natural world in its efforts to be sustainable, with the ability to be carried on indefinitely, as nature has proven herself to be. For example, organic

farmers have long relied on beneficial insects that feed on pests in order to reduce crop damage, such as the large-scale release of ladybugs to limit the population of aphids on a crop. Some farmers and growers use traps on a regular basis. These traps mimic the effect of natural predators by removing unwanted insects, often by luring the unsuspecting victim using a synthetic version of a natural pheromone attractant. These are just a couple of the approaches to pest control that copy the ways in which the natural world will spare some plants over others from the damaging effects of insect predation. Although these types of approaches may have little impact on the quality of the final crop, they are integral to the organic philosophy that stands behind the finished product.

In contrast, one of the guiding principles of the industrial model that has been developed by Western culture is the desire to maximize production. When applied to agriculture, this typically results in the drive to push biological organisms to the limits of their capacity. Unfortunately, the focus on increasing our harvest seems to distract our attention from the quality of the crop that is being produced and the health of the plants or livestock that are doing the producing. In the dairy industry, for example, the cow that historically produced 15 to 20 pounds of milk a day and lived for fourteen years or more in a healthy, relatively disease-free state has today been bred to pump out an average of 90 to 120 pounds of milk a day and has to be sent to the slaughterhouse within three or four years, simply because she becomes exceptionally prone to sickness and disease from the stress of the forced increase in milk production. The poor cows are literally worn out. The humble honey bee is similarly affected by our efforts to artificially boost the size of its honey crop. Activities such as the use of chemical mite controls or the feeding of sugar syrups and pollen substitutes, although beneficial to honey production in the short term, ultimately may weaken the vitality of the hive and increase its vulnerability to diseases and pests such as varroa. As a result, such management tools should be used sparingly, if at all.

Our industrial model encourages large-scale production under the "economy of scale" argument that has been the drumbeat of US schools of agriculture since the end of World War II. That is to say, "If you want to be profitable, you must grow larger." Although this makes sense in many industries, the fallacy of this approach when applied to farming—an inherently biological activity—is spelled out by Brian Halweil of the Worldwatch Institute in the book *Eat Here: Reclaiming Homegrown Pleasures in a Global Supermarket.*

In the post-war period, along with increasing mechanization, there was an increasing tendency to "outsource" pieces of the work that the farmers had previously done themselves—from producing their own fertilizer to cleaning and packaging their harvest. That outsourcing, which may have seemed like a welcome convenience at the time, eventually boomeranged: at first it enabled the farmer to increase output, and thus profits, but when all the other farmers were doing it too, crop prices began to fall.

Before long, the processing and packaging businesses were adding more economic value to the purchased product than the farmer, and it was those businesses that became the dominant players in the food industry. Instead of farmers outsourcing to contractors, it became a matter of large food processors buying raw materials from farmers, on the processors' terms. Today most of the money is in the work the farmer no longer does—or even controls. Tractor makers, agrochemical firms, seed companies, food processors, and supermarkets take most of what is spent on food, leaving the farmer less than 10 cents of the typical food dollar. (As noted earlier, an American who buys a loaf of bread is paying about as much for the wrapper as for the wheat.)

Ironically, then, as farms became more mechanized and more "productive," a self-destructive feedback loop was set in motion: over-supply and declining crop prices cut into farmers' profits, fueling a demand for more technology aimed at making up for shrinking margins by increasing volume still more. Output increased dramatically, but expenses (for tractors, combines, fertilizer, and seed) also ballooned—while the commodity prices stagnated or declined. Even as they were looking more and more modernized, the farmers were becoming less and less the masters of their own domain. On the typical Iowa farm, the farmer's profit margin has dropped from 35 percent in 1950 to 9 percent today. To generate the same income (assuming stable yields and prices), the farm would need to be roughly four times as large today as in 1950—or the farmer would need to get a night job. And that's precisely what we've seen in most industrialized nations: fewer farmers on bigger tracts of land producing a greater share of the food supply. The farmer with declining margins buys out his neighbor and expands or risks being cannibalized himself.

There is an alternative to this huge scaling up, which is to buck the trend and bring some of the input-supplying and post-harvest processing—and the related profits—back onto the farm. But more self-sufficient farming would be highly unpopular with the industries that now make lucrative profits from inputs and processing. And since these industries have much more political clout than the farmers do, there is little support for rescuing farmers from their increasingly servile condition—and the idea has been largely forgotten. Farmers continue to get the message that the only way to succeed is to get big.[4]

The same pressure for farmers to increase in size applies to the beekeeping industry. Aside from the inexperienced or inattentive hobbyist, it is the large commercial bee outfits that have had the hardest time preventing their hives from collapsing due to varroa, disease, or colony collapse disorder (CCD). Reports of winter losses exceeding 40 percent and attributed to either mites or CCD have been all too common among those with 600 hives or more. In truth, the number of colonies an experienced beekeeper can manage successfully in this new era is likely to be lower than those of pre-varroa/pre-CCD years, at least until new strains of bees have been developed that are either extremely resistant or outright immune to these challenges.

Part of the allure of chemical pesticide use is the economic benefit that can be reaped by a reduction in labor—a cost that reflects the investment in time, and the attention to detail, required by non-toxic organic approaches to pests and disease. With the chemical approach proving itself to be less than satisfactory in many ways, those in the beekeeping industry may find themselves having to decrease their hive-to-beekeeper ratio to match their colonies with the amount of labor available to keep their hives healthy. The number of hives that a single beekeeper can inspect and treat in a timely manner is limited. This is especially true when one considers all the unexpected issues, from bad weather to flat tires, that typically arise and cause delays, forcing one to fall behind schedule. As conventional chemical treatments become less effective, the shift to nontoxic, labor-intensive management techniques will require new approaches and technologies to make up for these increased labor demands.

Aside from chemical resistance in the case of varroa mites, another key factor that affects hive survival, even when such control methods work, is the timing of mite treatments. There appears to be a certain threshold level of mite infestation within a colony that must be reached before the hive will begin to show signs of stress. An even higher threshold must be reached before the collapse of the colony is imminent. The longer the mites have

to freely reproduce without hindrance, the more quickly these thresholds will be reached. By simply preventing the mite population from building up to critical levels, chances of colony survival are greatly increased, even when mites are present in the hive on a year-round basis. As a result, many beekeepers prefer to harvest honey and treat their hives earlier in the season than they used to, even though it may mean sacrificing a significant portion of the potential honey harvest.

In Vermont, this means we are usually finished taking honey off the hives by mid to late August, so that a highly effective treatment can be applied before the consistently cool weather rolls around. Traditionally the honey harvest would not begin until late August in our region, and often not until September, thus taking advantage of the late-blooming goldenrod. Goldenrod often provides a plentiful source of nectar-bearing flowers; the autumn honey flow they create results in rich yellowish-amber honey. Instead of seeing an early end to the season as a detriment, it means that the entire process of harvest and treatment can be completed prior to the start of the goldenrod honey flow. This in turn tends to result in a much lighter crop of primarily clover and alfalfa honey, which features the more delicate bouquets and flavors favored by many honey consumers. Alternatively, the late-season flows may be harvested simply by taking steps to remove a small percentage of the mites throughout the season. A key to successfully using this late-season approach is to have only so many colonies in your care that you are able to harvest and treat in time to prevent a deadly threshold of varroa mites from being reached. This can be accomplished even while honey supers—the boxes of frames specifically intended to be filled by the bees with honey and harvested by the beekeeper—are on the hive. Ongoing mite control is accomplished often by utilizing physical, mechanical, and cultural control measures, ones that will not expose the honey crop to the potential risk of chemical contamination. The consistent elimination of a percentage of mites early on will delay their population buildup and allow for the application of a traditional late-season mite treatment, without giving the mites the opportunity to overwhelm the colony.

It is possible that a strain of varroa might eventually evolve that learns to keep its reproductive rate low enough so its population within the hive does not overwhelm the host colony and cause its collapse, bringing about the mites' own demise in the process. We already know that some honey bees have adapted their behavior in order to coexist with varroa. The *Apis cerana* species of Eastern honey bee, for example, has learned to live in harmony with the varroa mite over the course of hundreds, or perhaps thousands, of years.

As we shall explore, beekeeping management techniques that help to develop similar behaviors in the Western honey bee (*A. mellifera*) may lead to a climate of coexistence where the constant presence of mites may be tolerated within the hive without leading to the collapse of the colony. The sustainably minded beekeeper wishing to follow an organic approach to beekeeping should embrace this type of model. By adopting beekeeping management techniques that encourage coexistence, we humans can play an important role in the eventual development of permanent coexistence between *Apis mellifera* and *Varroa destructor*.

As a novice beekeeper, I relied greatly on the advice and suggestions of other beekeepers I met. I also read numerous books and articles on beekeeping. As a result, I am indebted to the kind and generous modern-day practitioners of this ancient craft who helped me get started on the road to becoming successful in apiculture. Today, as an active and experienced apiculturist, I find that the questions I am most often asked concern how to keep and manage honey bees without the use of toxic chemicals and still have hives survive the winters here in the northern latitudes. Perhaps this is a direct result of the increased media exposure that organic agriculture has received following the implementation of the National Organic Program (NOP), which was passed into law in 2002 by the United States Congress. It may be that the increased interest in chemical-free beekeeping that I have noticed, from hobbyists and commercial beekeepers alike, is a response to the increasing reports of environmental decline that is a result of some of our conventional agricultural techniques used on other crops. Ultimately, though, the interest in organic beekeeping may simply be a result of the use of toxic miticides and the backlash resulting from the mixed results of conventional apiculture's reliance on these chemicals to control predatory insects within the hive. Increased media reports of the ineffectiveness of conventional mite control approaches and the resulting devastating losses of hives by beekeepers have inspired a new wave of people to take up beekeeping, many of them wanting to get involved in an effort to find a way to help the honey bee survive and thrive without relying on chemotherapy treatments.

· SOME SOCIAL · IMPLICATIONS

I believe it is important for us beekeepers to share information with the non-beekeeping community to help educate them about the many benefits of the honey bee. All too often, about the only thing the average person knows about bees is that they sting and they produce honey. The entire beekeeping industry stands to gain from the education of the general public regarding the benefits of the bee. On a professional level, the bond that can be forged from having direct contact with members of the honey-consuming public can prove invaluable in building a local market for hive products. This form of public relations becomes more and more important as communities continue to grow and agricultural land becomes more developed and densely populated. As humans and honey bees are forced to live in closer proximity to each other, more opportunities for unfortunate encounters are likely to develop. This has already resulted in the prohibition of beekeeping activities within the confines of many municipalities throughout the

United States. The importance of positive public relations as a preventive to public ignorance and prohibition of beekeeping activities will continue to expand. Meanwhile the growing demand for, and awareness of, organic foods by the general public may provide organically inclined beekeepers with a public relations edge over their conventional counterparts. The current state of the beekeeping industry certainly points to the need for new approaches and ways of thinking.

Following the development of the atomic bomb, Albert Einstein is said to have noted, "The release of atom power has changed everything except our way of thinking . . . the solution to this problem lies in the heart of mankind. If I had only known, I should have become a watchmaker." In many ways, the current dilemma facing the honey bee mirrors the challenge faced by the human race. In my opinion, some of the most difficult yet important work each of us can be involved in is our personal growth and evolution. If we no longer want to live in a world based on fear, lies, greed, and violence, and instead want to create a world where love, truth, peace, and compassion prevail, we must start with ourselves. Each one of us has the opportunity to self-evolve and play an important role in creating a more peaceful and harmonious world simply by living a life in which these values are expressed on as consistent a basis as we can possibly muster. By the same token, we cannot rely on those in positions of power and leadership to solve the myriad social and environmental issues with which we are currently faced. Just as the ultimate answer for solving the numerous

difficulties facing humankind lies in the raising of our society's collective consciousness, the most desirable and permanent solution to the difficult times the honey bee is now experiencing also lies in the bee's evolutionary process, through the development of resistance to disease and parasitic pests. The goal of raising humanity's collective consciousness requires the raising of each individual's consciousness to the point where enough of us evolve to affect society as a whole. This is not something that can be forced or imposed upon individuals. It is a responsibility each of us must choose to take upon ourselves in our own time, when we are ready. And so it is with the honey bee: the evolutionary process must take place one hive at a time. For, just as with us humans, the bee creates its own future with each seemingly insignificant daily decision and activity.

This realization has recently led me to start the process of examining and monitoring my own thoughts, feelings, and actions on a daily basis in an effort to identify those areas of my being that I wish to improve upon. Once they are identified, I can then begin the process of analyzing the source of my motivations and make the changes and shifts in my thoughts, beliefs, and actions necessary to bring about permanent change, whether it be altering unhealthy habits, changing ways of thinking that no longer serve me, letting go of false ideas and unproductive emotional responses, and so on. Albert Einstein is but one of many who have brought forth this message that I am only now just beginning to fully comprehend and embrace in my life . . . all with the help of the bees.

CHAPTER 2

Working with the Hive

· SOME BASICS · OF BEE BIOLOGY AND ANATOMY

A colony of bees is a single superorganism made up of thousands of individual bees. At the peak of summer, population numbers which average around thirty thousand can reach sixty thousand bees or more. Each honey bee that makes up this superorganism is like a single cell in your body. Thousands of your cells die every day, yet every day your body creates thousands of new cells to replace those that are lost. Just so, every day during the active season many bees in a colony will die, and many new bees will hatch out of birthing cells to replace them. It is the colony as a whole that is the individual.

While much depends on the time of the year and the state of the colony, a regular hive of bees is typically made up of male honey bees and two castes of female bees. The male bees are called drones. The females include fully developed female bees, called queens, and less-developed females, known as workers. The body of all three types of honey bee is made up of three primary parts: head, thorax, and abdomen.

The head includes the antennae, the eyes, and the mouthparts. The antennae are very sensitive detection organs through which the bee can feel and smell; it can also monitor the carbon dioxide levels in the surrounding environment. The honey bee has five eyes: two large compound eyes, which are the primary organs the bee uses to see with, and three small simple eyes (ocelli) located at the center of the head between the compound eyes. It is believed that bees use these simple eyes to detect diffused light and sense light intensity. The main components of the mouth of the honey bee are the jaws (mandibles) and the tongue and sucking tube (proboscis). The mandibles are primarily used for biting and chewing and for shaping wax and propolis (bee glue) during hive-building activities. The proboscis is used to transfer nectar and honey into or out of the mouth.

The thorax is the middle of the bee's body, between the head and the abdomen. It is composed primarily of muscle because it is the part of the body that enables locomotion. Both the wings and the legs are attached to the thorax. The honey bee has two forewings and two hindwings. The forewing and hindwing on each side of the bee's body hook together and act as one during flight. Like all insects, the honey bee has six legs.

The abdomen contains the digestive system and most of the breathing holes (spiracles). In the queen and drone, the abdomen is the part of the body that also houses the reproductive organs. In the worker bees, the reproductive organs tend to be undeveloped and the abdomen is primarily filled with the digestive system and circulatory system, along with the stinger and venom sac. The abdomen of the female worker bee also contains four pairs of wax glands, from which small flakes of wax are produced for building comb. Under certain circumstances, discussed in chapter 3 (pages 81–82), the worker can develop its ovaries to the point where it can lay eggs.

The queen bee is the largest member of the colony. You can spot a queen among the rest of the

RIDDLE:

What has four wings, five eyes, and six legs?

RIDDLE ANSWER: A honey bee.

drones, which tend to be lighter in color due to the large number of tiny hairs that grow there.

The queen is the reproductive organ of the hive: an egg-laying machine capable of producing around 1,500 to 2,000 eggs per day during the peak of summer. Although laying eggs is her primary activity, the queen also produces pheromones that communicate information throughout the hive about the health of the queen, among other things. The worker bees assist the queen by taking care of all other duties, such as making sure she is well groomed and fed. There is typically only one queen in a hive at a time, although in very rare instances, more than one queen has been found in a healthy hive.

hive's occupants because she is the only bee whose abdomen usually extends significantly past the end of her wings (it does so when she has recently been laying eggs and her abdomen is filled with them). Also, her back (the top of her thorax) is bald and dark compared to the backs of the workers and

FIGURE 2-1. Unlike other honey bees, the queen lacks hair on her thorax (back). Also, when she has recently been laying eggs, the tip of her swollen, egg-filled abdomen extends well past the ends of her wings.

The worker bee is the most populous member of a honey bee colony. Worker bees comprise about 97 to over 99 percent of the hive's population, depending on the state of the hive and the time of year. As its name implies, the worker bee takes care of all the work that needs to be done in the hive. The work that a bee will do is influenced by conditions both inside and outside the hive and by the age of the bee. The first jobs a worker bee carries out in its youth are house bee duties, after which the worker will "graduate" to more demanding and dangerous jobs, many of which will take place outside the hive. While individual worker bees don't take a turn at every job required by a hive, in a typical colony the sequence of job requirements as a worker develops would be something like this:

- a housekeeper bee cleaning cells
- a nurse bee rearing brood (eggs, larvae, and pupae) or tending to the queen
- a house bee receiving nectar and pollen and storing it in cells
- a comb-building bee
- a heater bee incubating the brood nest
- a bee supplying heater bees with honey for energy
- an undertaker bee removing dead bees, larvae, and pupae from the hive
- a ventilation bee

FIGURE 2-2. Larger than his sisters, the male honey bee (drone) makes up only a small percentage of a healthy hive's population. The drone is easy to identify because its eyes, which are larger than those of the queen or worker, meet at the top of its head. The tip of the drone's abdomen, rather than coming to a point like the worker and queen, is blunt and covered with fuzzy hair.

- a cementing bee working with propolis
- a guard bee and/or a emergency responder bee aiding guards in repelling unwanted intruders when necessary

Lastly, it is the oldest worker bees that typically become field bees that forage for nectar, pollen, and other substances or scout out nesting locations when swarming.

The male, or drone, honey bee is generally larger (both in width and length) than the worker, but is not as long as the queen. Usually a hive will contain only several hundred drones, and rarely more than a thousand, even at the peak of the season unless the colony is experiencing queen problems.

Drones are responsible for the vital task of mating with the queen to ensure fertilization and the propagation of the species. While this is the most obvious role that the drone plays within the colony, the male bee also has other important roles to play in helping to ensure the colony's survival. I discuss in depth his role as the sacrificial part of the hive in chapter 5, on pages 136–137.

All bees start out as an egg laid by the queen in a cell in the hive, and all eggs take about three days to hatch. If an egg is fertilized by the queen, it has the potential to become either a queen bee or a worker bee, depending on how it is raised within the hive. Unfertilized eggs are typically laid in cells that are a little larger than cells for worker bees, and these eggs

FIGURE 2-3. Patches of capped worker brood cells are usually a sign that there is a healthy fertile queen present in the colony.

give birth to drones (see figure 5-7 on page 130). Once the egg hatches, the larva begins to eat voraciously and grow rapidly. The larva, which resembles a pearly white grub, increases its body weight hundreds of times within a very short time. Once it is fully grown, the larva will begin to stretch out inside the cell, and the worker bees will cap the cell with a thin layer of wax, sealing the larva inside. The larva passes through a short pre-pupa stage during which it molts several times, shedding its larval skin before entering the pupa stage. During the pupa stage, the bee is still white, although many of the adult bee's features can be seen. Its color changes as it develops into an adult and chews its way out of the birthing cell. During this metamorphosis the developing bee spins a cocoon that adheres tightly to the cell wall and remains after the adult bee leaves. The remnants of the cocoon darken the beeswax that forms the cell walls. With repeated brood cycles, the yellow cell walls of the combs in the brood nest turn brown and eventually can become almost black in color.

In order to monitor the health of a colony of bees, a beekeeper must learn to identify the difference between capped worker brood and capped drone brood sealed within brood cells. The surface of capped worker brood will appear only slightly convex in shape; it is almost even with the surface of the comb. The drone brood, on the other hand, has a pronounced convex, bullet-shaped surface (see figure 3-34).

FIGURE 2-4. When building comb, the honey bee will always leave about ⁵⁄₁₆ inch of open space between combs; this is called the "bee space." This space serves as a passageway for the bees within the hive.

Another important concept that beekeepers need to be aware of is "bee space," which is the amount of open space that the bees instinctively leave between combs within the hive. This space measures about ⁵⁄₁₆ of an inch. Any space within the hive that is much larger than ⁵⁄₁₆ inch tends to be filled by the bees with comb. Any space much smaller than this "bee space" is often too small for the bees to access and clean efficiently, and the bees will fill such spaces with propolis in order to prevent mold or bacteria from contaminating the hive. A minister named Lorenzo Lorraine Langstroth is often credited with the discovery of bee space. He described bee space in a book, *A Practical Treatise on the Hive and the Honey-Bee* (modern reprints have renamed it *Langstroth's Hive and the Honey-Bee*), and he invented a removable-frame hive (patented in 1852) based upon the concept.

· SUGGESTIONS · FOR BEGINNERS

When addressing beginner beekeepers, I like to emphasize the importance of not getting started in beekeeping unless they are ready and willing to fully commit to taking the time to properly educate themselves about honey bee biology and proper care. I also ask people to carefully assess whether they are ready to devote adequate resources of time and money to ensure that the hive will be properly cared for year after year.

The initial learning curve in beekeeping is fairly steep, so it is wise to start learning well in advance about bees and the role that a beekeeper can play. Beginning to read beekeeping books and journals and take beekeeping classes and workshops a year ahead of setting up your first hives would not be overdoing it. Joining your local or state beekeeping association can be a huge help in this regard too. Busy modern lifestyles often get in the way, and if you allow too little time for education beforehand, you may find yourself unprepared when your bees arrive. At that point, your first year of beekeeping will end up unfolding haphazardly, and you won't

AN APICULTURAL ETHIC

In general, insects are not high on the list of life forms that our culture respects and expresses concern for. For example, if pet owners or farmers ignore the health and needs of their dogs, cats, horses, or cows, society intercedes on behalf of the animals. The negligent individual may be referred to the Humane Society for education on proper animal care, or the animals may be seized and placed in better conditions. In the most extreme cases of willful negligence, court proceedings may be in order. Yet a person can purchase a hive of bees and then ignore the bees' needs, allowing them to get sick or starve, and as a society we remain largely unaware of this poor stewardship of bees. Part of this is due to the prevailing attitude toward insects, and part is due to the fact that the colony is hidden inside a box and nobody can see what is happening to the bees. Nevertheless, just because one can treat bees harshly and get away with it does not mean it's right.

What's needed is an apicultural ethic that does not place the needs of the bees below human needs. I am not saying that beekeepers should put bees' needs above human needs. Rather, let's give them equal importance, striving for a give-and-take, a win-win situation where both parties benefit roughly to an equal extent. Following this approach, for example, one would not secure a hive of bees in the spring, rent them out for pollination, harvest all their honey, and then allow them to die in winter with the plan of purchasing new bees the following spring, simply because it is economically advantageous to do so.

have the kind of experience and good results you dreamed about.

The best way to learn about beekeeping is to work for a commercial beekeeper. I found out the hard way that is it much better to get paid to learn through an apprenticeship-type situation than to pay for the privilege of learning through a university or the school of hard knocks. If this is not an option, then it is very helpful if you can find a local experienced beekeeper willing to act as a mentor. A mentor can answer questions, guide you through the transfer of your bees into new equipment, and assist you with your first hive visits. When working with a mentor, it is good to acknowledge and appreciate the time and experience that is shared with you and not abuse the relationship. A mentor is there to act as a guide during your adventures with the bees, not to do everything for you. However, there is more than one way to keep bees and do so well. Therefore, don't be afraid to listen politely to your mentor's advice, smile, say thanks, and then make your own choice regarding how to care for your bees. For example, if your mentor tells you that you must use a certain antibiotic or toxic chemical in order to keep your bees alive, and this is not something you are comfortable doing, know that there are natural and organic alternatives that you can explore instead. Once you have a comfortable grasp of honey bee biology, as well as the biology of the various bee pathogens and pests, you will naturally reach the point where you can follow your intuition. Please, though, before you turn to intuitive beekeeping, be sure you have gained a clear understanding of what bees require in order to thrive. Then, follow your instincts: this book is designed to help you do just that. Another recommendation I offer the would-be beekeeper is to start with two hives instead of one. Two hives are only a little more work, and not a whole lot more money, than one hive, and yet the benefits are substantial. Two hives help greatly with the steep learning curve all beekeepers must go through: it doubles your experience level and allows you to make comparisons between the

hives. The importance of the number of hives in building experience was brought home to me during my second year working for one of the largest commercial beekeepers in Vermont. During that year I felt very much like the new kid on the block, unsure in my knowledge of the bees and my role as a beekeeper. Yet, at a meeting of the Vermont Beekeepers Association (VBA), I found myself discussing with a fellow beekeeper what I had seen and experienced with the bees I had worked with and the various issues, such as varroa mites, that I had dealt with. Imagine my surprise when a small cluster of beekeepers soon gathered to listen to what I had to say. I was especially perplexed because some of these beekeepers were old-timers who had been keeping bees for 30 years or longer. In reflecting upon this experience afterward, I realized that since the majority of the beekeepers in the VBA kept fewer than five hives, even after 40 years they would have gained only 200 hives' worth (5 hives × 40 years) of beekeeping experience. Due to their limited number of hives, they simply did not have the opportunity to encounter the range of experiences that I had in a single year helping to care for a thousand hives. By starting with two hives instead of one, novice beekeepers have the chance to increase their experience level 100 percent in the first year without the danger of becoming overwhelmed.

Another choice you'll need to make in advance is which type of hive to use. I recommend that novice beekeepers begin with a Langstroth-style hive. After you have a couple years of experience under your belt, you may wish to experiment with other styles such as the top-bar hive or Warré hive (the Warré hive, also known as the People's Hive, is a vertical top bar hive developed in France by Emile Warré and outlined in his book *Beekeeping for All*). There is much more information available on the use of the Langstroth hive than there is for top-bar hives or Warré hives, and there are far fewer beekeepers who have experience with the latter two options should you have questions or need the help of a mentor.

For reasons I will discuss later in chapter 4, whenever possible, it is advantageous to purchase local bees and even more important to buy types of bees that have some level of resistance to mites and diseases. And it is far easier to start with a nucleus colony or nuc (see glossary for a brief description of a nuc, page 269), if you can get one, than to start with packaged bees. Due to the high demand and short supply of bees, it is a good idea for beekeepers to place their orders for packaged bees or nucleus colonies early to ensure that the bees they want will be available at the desired time. Since the advent of colony collapse disorder, many beekeepers are finding that January is not too early to place orders for bees for spring delivery. On some occasions waiting until January may actually be too late, because some bee suppliers sell out well ahead of time, and the best they will be able to do is put you on a waiting list in case they receive order cancellations, or place you on the list for next year. Whenever you place an order, please be sure to assemble and prepare all your equipment *before* your bees arrive. The experience of long-distance shipping or being crowded in a small nuc box is stressful for bees. To keep stressed bees contained for several more days or weeks while you rush to order, assemble, or paint hive equipment is not a good situation.

The final piece of advice I like to offer first-year beekeepers is to open up and check on their hives

FIGURE 2-5. It is illegal to keep bees in a hive such as this one, which does not contain removable frames that allow for disease treatment and inspections.

on a regular basis. Some beekeepers might tell you to leave your hives alone and keep colony inspections to a minimum. Back in the old days before mites, small hive beetles, and a host of honey bee diseases surfaced on the North American continent, this advice would have been appropriate. Today, however, the environment that bees have to navigate and the interior of the hive cavity have been so manipulated and changed by humans that in most of the country, to leave hives alone is more likely to lead to colony death than to aid in its ability to thrive. As I describe in chapter 5, I normally check my bees every 7 to 14 days. While doing so, I try to keep the disturbance of the hives to a minimum. Nevertheless, I encourage new beekeepers to open hives, remove frames, inspect the brood area, look for eggs, try to find the queen, and observe the levels of pollen and honey within the hive every one to two weeks during the first year. (Refer to "Creating the Split or Nucleus Colony" in chapter 4 on page 103 for inspection tips.) That's right, every week or two if possible! Your goal is to get a sense of whether the hive is doing well and developing normally or not. There is a limit to how much you can learn about beekeeping from classes, workshops, articles, videos, and books. In the end, to become a successful beekeeper you have to actually open hives and handle frames of brood and bees. After the first year of handling the bees regularly, you should not have to go through your hives and disturb the bees so frequently. Instead, you can limit your weekly or biweekly hive checks to simply taking a quick look under the inner cover.

Your goal is to be a bee*keeper*, not a bee *haver*. This is accomplished by working in partnership with the colony and never abandoning the hive to its own devices, which might result in the hive starving or dying from disease. The ancient craft of beekeeping can be incredibly fulfilling and enjoyable as long as one is committed to following through, overcoming the various frustrations that may be encountered, and persevering even when a task occasionally seems overwhelming or intimidating.

OPENING AND INSPECTING THE HIVE

Here are the basic steps to take, in the proper sequence, whenever you open a hive.

1. Figure out why you are going to open the hive and what you want to accomplish.
2. Don clothing and headgear that will make you feel comfortable working with the bees; then light your smoker.
3. Approach the hive from the side or the back, not the front, where you may block the flight path of the foragers and make them defensive.
4. Smoke all entrances. Allow the smoke to always proceed you and announce your presence prior to entering any part of the hive.
5. Remove the inner cover, and always check it for the presence of the queen before you put the cover down. She could be anywhere!
6. If you decide to remove a frame, make sure to choose one of the outermost frames (or one next to the outermost) first in order to reduce the chance that you will injure the queen.

· OPTIONS · FOR OBTAINING BEES

When you are first starting out you may want to purchase bees. Honey bees, especially good bees, are in high demand these days. With beekeepers across the country experiencing some of their highest historical annual losses, the need for replacement bees has never been stronger. Orders often have to be placed months and sometimes years ahead of time in order to guarantee that the bees you want are available when you need them. This situation, combined with a surge in the number of new beekeepers who are getting involved in the craft, has created a huge demand for packages, nucs, and queens. Prices seem to go up every year, and still honey bee suppliers are running out of stock and are unable to fill the demand. In many ways, it has never been a better time to be a breeder or nuc supplier, while getting your hands on a great hive of bees becomes more and more challenging.

The easiest way to successfully get a hive of bees off to a good start is to purchase a complete hive that is already thriving. Many beekeepers are downsizing or giving up on the business, and complete hives have become more commonly available in many areas. The purchase of a complete hive eliminates risks and challenges associated with package and queen introduction, getting brood comb drawn out from foundation, and having to protect a weak and unorganized colony from being robbed. The downside is that a complete hive is heavy to lift and is one of the most expensive honey bee procurement options. There is also the risk of inheriting problems such as bee pathogens, antibiotic and pesticide contamination, varroa mites, tracheal mites, and small hive beetles.

When purchasing established colonies, it's wise to inspect the hive(s) prior to completing the transaction. "Buyer beware" should definitely be your guiding principle here. Be on the lookout for worn-out equipment, old combs, and disease and pest problems. How are varroa mites and foulbrood kept under control? How has the hive been managed during the past year or two? Answers to questions such as these can give you a hint as to how well the hive may fare in the coming year. If you don't feel qualified to assess hive condition yourself, ask your local bee inspector to give them a once over.

Compared to purchasing established colonies, buying nucleus colonies offers many of the same benefits, but nucs are easier on your back and are available at a much lower price. For many beginners, buying a nuc from a local bee supplier will be the best way to get started with bees. As with full hives, a nuc should not need feeding if purchased when a honey flow is on, nor does it require queen introduction. Simply transfer the frames into full-sized equipment, add additional frames for growth, and you're off and running. It is important to have your additional equipment ready to go before you pick up your nuc. Otherwise the bees may become overcrowded and decide to swarm.

Nucs are subject to the same potential problems as an established hive: bee pathogens, antibiotic and pesticide contamination, varroa mites, tracheal mites, and small hive beetles. In fact, a nucleus colony may be *more* likely to have such issues because it seems that many nuc providers choose to fill their nucleus colonies with older combs in order to reduce the buildup of pathogens and chemical residues in their breeding colonies. Honey bees are like flying dust mops, picking up all manner of materials along with the pollen and nectar they gather. Thus, they introduce chemical contaminants from the environment into the colony, even when their beekeeper refuses to use chemicals in the hive.

Admittedly, this is an age-old practice that has proven itself economically successful. Turn what would otherwise be waste and an economic loss or expense (in this case old, darkened, contaminated combs) into a moneymaker: components of a nucleus colony. This seems especially true of breeders who promote their bees as "survivors" that have not been treated and are chemical-free. Part of the reason such breeders are able to keep their bees alive without having to use any mite treatments is that they are regularly splitting up their hives to make nucs, and in so doing, removing the combs that have become chemical- and pathogen-ridden with age and replacing them with new foundation. The trap that many beginner beekeepers fall into is to purchase these survivors with the idea that they will not need to apply mite treatments to their bees either. Unfortunately, if the purchasers of such nucleus colonies do not also follow the same practice and regularly split up their hives and remove the old, darkened combs to make additional nucs, they should not be surprised when their hives die within a year or two.

To avoid falling into this trap, look for new equipment and frames of light-colored comb when making your nuc purchase. If your nuc contains dark combs that have no markings indicating how old they are, be sure to rotate them out of the hive by the following spring if not sooner.

The most common way to obtain bees is to purchase a package of bees. Packages are usually sold by the pound. A pound of bees is typically composed of 3,000 to 3,500 workers, and the standard package of bees available from bee supply companies weighs in at 3 pounds, for a total of about 10,000 bees. Reputable companies will err on the high side and provide extra to make up for bees that may die during transport, as well as to account for the honey and nectar in the stomachs of the bees when they are initially packaged and weighed. Bees are safely mailed in special boxes approved by United Parcel Service (UPS) and the United States Postal Service (USPS). Please note that a package of bees may not contain a queen and will contain only workers, unless a queen has been ordered at the same time as the package. Be sure to ask your supplier whether a queen is included. Either way, prepare your hive equipment and location before the package arrives in the mail.

Packaged bees are less expensive than nucs or complete hives and easier to ship. And while package bees may still carry diseases and pests, they are likely to be less infested or contaminated than nucs or established hives that are filled with comb simply because there are fewer places for pathogens, chemical residues, mites, or small hive beetles to hide in a package of bees.

Getting a package of bees started, however, is much more challenging than when starting with a nucleus colony. The package of bees must be transferred into your equipment; the queen and the bees are typically unrelated and must be "introduced" to each other before they will work together in unison; and then there is the need to provide feed to keep the nascent hive from starving, while at the same time stimulating bees to build comb if drawn-out frames of comb from a previous hive cannot be provided.

Choose a day when it is not raining and the temperature is above 60°F (15°C). To transfer the bees from a package into their hive, use a hive tool to pry open the top of the package and expose the feed can. Next, knock the bees to the bottom of the package by holding the package a few inches above the ground and then gently rapping the box on the ground. Use the hive tool to pry the feed can loose from the package, and place the open package full of bees without the feed can in an empty hive body placed on a bottom board. Be sure to hang on to the metal tab attached to the queen cage while removing the feed can. Once the bees are in place, put a hive body filled with frames on top of the hive body holding the package of bees. Then place the queen, still in her cage, between the frames in the hive body that will become the hive once the bees have moved in, as shown in figure 4-2 on page 102. You may have to remove one of the frames to make room for the queen cage to fit. Then put the hive covers in place and let the bees crawl out of the package up to the queen and release her. The following day, lift the hive body now full of bees off the empty hive body and remove the mostly empty package from inside (any bees still lingering inside can be shaken out). Be sure the queen cage does not fall from its position between the frames. Set aside the empty hive body too, and place the full hive body on the bottom board. Warning: do not leave the package in the hive too long or the bees may begin building comb inside of it.

If the queen has not been freed from her cage within two or three days, it is usually safe to manually release her, as the bees will normally have had

FIGURE 2-6. With the cork removed from the candy end, the bees will typically eat away the candy and release the queen from her cage. One colony had not released their queen within four days and the screen was removed from one cage in order to release the queen manually. Undamaged queen cages can be reused by simply replacing the screen and the sugar candy and cork in the opening.

FIGURE 2-7. A metal push-in queen cage like this is easy to make out of ⅛-inch hardware cloth. Just cut the edges of the cloth and fold them down to form sides that are about ½ inch deep and that can be pushed into the surface of the comb. The queen (with a white dot painted on her thorax) can be seen in this cage along with a number of worker attendants.

plenty of time to acclimate themselves to her. Observe how the bees react to the queen once she is released. If you see worker bees acting aggressively toward her, you have several choices. You can place her back in the cage for a few more days and try again. Alternatively, you can place the queen in a cage made of ⅛-inch hardware cloth for a few days. This type of cage will allow the queen to start laying eggs, helping to increase the likelihood that the bees will accept her. If they don't, the bees will at least have the opportunity to raise an emergency queen from the eggs recently laid by the rejected queen. It can also be helpful to include a few young workers in the cage with the queen and place the queen cage over an area of comb that contains cells of worker brood that will hatch within a day or two.

A third approach that has worked for me when workers are continuing to act aggressively toward a queen is to cover all the combs of bees in the hive with a light spray of sugar syrup and a mixture of lemongrass and spearmint essential oils. There are two products commercially available that contain these oils in emulsified form (so that they will mix with sugar solution and not simply float on top). They are Pro Health, distributed by Mann Lake Ltd. in Minnesota (see resources on page 273), and a product called Honey-B-Healthy, which is sold by independent distributors throughout North America. Between the distraction of the sugar being sprayed on the bees and combs and the strong scent of the essential oils, the workers seem to ignore the queen for a time until the mess is cleaned up, and by that time they will usually have gotten used to her presence and accept her.

If you don't have an extra hive body to encase the package of bees, then simply turn the package upside down and shake the bees out of the open package onto the hive body of frames. When using this approach it is good to make sure that at least some of the workers fall onto the queen cage suspended between the frames. Unfortunately, this process may tend to get the bees riled up, so be sure to wear a veil while doing this.

Caution should be observed when shaking several packages into hives all in the same location. Since the workers are not familiar with the area and have not established themselves in a hive yet, there will be a tendency for shaken bees to fly around wondering where to go. When more than one package is being shaken out at a time, the airborne workers are likely to drift to a neighboring hive. In some cases the drifting has been so bad that a hive was left without enough bees to keep the queen alive and she died, while the neighboring hives benefited from the extra bees being added to their hive population. To help reduce the tendency of the bees to fly during this process, you can turn the package on its side, with the screen side up, and spray sugar syrup onto the screen before opening up the package. Then spray a light mist of syrup onto the bees (enough to dampen their wings).

Once all the bees have been shaken out of the package, cover the hive and don't disturb the bees for two to three days to give them a chance to free the queen from her cage. It is a good idea to check the hive about three days later to remove the empty queen cage, or release the queen if she is still in the cage, and add the tenth frame that was removed to make room for the cage back into the hive body. If the hive is being created without any frames of sealed honey, additional feeding is necessary to help the colony get started and prevent starvation in the event that the weather turns cold and rainy and the bees cannot forage for nectar.

Making your own nucleus colonies from bees you have successfully overwintered is less expensive than buying packaged bees. Rather than spend money on obtaining bees, your money need only be spent on equipment to accept the nuc. By timing your nuc making with the natural swarming season in your area, you help to increase the chances of success and reduce the need to feed the nascent colony.

If you don't already have access to bees and don't want to shell out a bunch of money in order to obtain bees, swarm catching is the way to go. There are no worries about queen introductions here; all you have to do is get the swarm into your

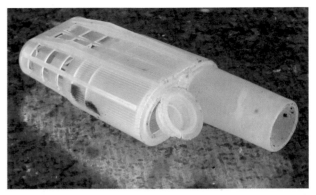

FIGURE 2-8. Queen cages can be made of plastic, wood, or metal.

hive body, but that can be a challenge. Swarms are primed to build new comb fast, and you should always take advantage of the opportunity to have a newly hived swarm draw out new frames of foundation, foundation strips, or naturally built combs. If captured early enough in the season they may be able to store enough honey on their own so that autumn and winter feeding will not be necessary.

With swarms, however, you have little control over the quality of the bees. Some swarms are composed of wonderful bees; some swarms are lousy. While you will save cash, you must invest the time to chase them down—and hope that they don't fly off before you arrive.

One way to avoid being led on a wild bee chase is to make use of baited hives to lure in a swarm. Drawn-out beeswax combs make any empty box with a cover more noticeable and attractive to scouts seeking a new home for the colony. If no comb is available, a few drops of lemongrass oil on a cotton ball can take the place of a drawn frame as bait since this essential oil is very attractive to bees. For those with an added sense of adventure, you may want to try your hand at some old-fashioned bee lining and track down a feral hive in a bee tree. Just be sure to bring along someone with logging skills if you are not handy with a chain saw, an ax, and splitting wedges.

Another way to obtain bees is to provide a honey bee removal service. Basic carpentry skills and tools will come in handy should you find yourself removing bees from inside the walls of a building or attic space. Like the errant swarm or colony of bees taken from a tree, the quality of removed bees can vary. At least with bee removal, if you charge for your services there is the consolation of monetary gain should the bees you take home prove to be less than hoped for.

• DEALING WITH • THE BUSINESS END OF THE WORKER BEE

Among the myriad things the honey bee teaches us is the value of strength in numbers. Whereas a single honey bee may prove merely irritating and bothersome, dozens of bees present a formidable and intimidating front that can be enough to drive off predators many times their size. Thus, the honey bee earns respect. Conventionally this respect expresses itself by way of a full-body bee suit that includes a veil and covers every part of the flesh except the hands and feet, which are protected by thick gloves and sturdy shoes. While such an outfit may provide a certain level of emotional security, its physical protection all too often proves tenuous, as at least one bee always seems to eventually find an opening in which to deliver a sting, either around the cuffs, at an ankle, through a ripped seam, or through pinched material that is pressed tightly against the skin—usually in the crook of a bent elbow or knee. As a result, the beekeeper has numerous opportunities to learn to both accept and conquer the fear that shadows every move within the apiary. Although unseen, this fear is like a constant summons to the beekeeper, calling upon the inner strength and resolve that each of us possesses, bringing it forward with the confidence and calming knowledge that we can handle whatever crisis might take place, including the occasional shattering pain that stands in our path, like a sheet of glass that we must pass through as its sound echoes among the hives.

The beekeeping supply companies are not going to appreciate me for this, but from my perspective a pair of old trousers and a long-sleeved shirt combined with a veil offer as much protection as most bee suits—with a lot less expense. Indeed, given the right approach, the veil, the shirt, and even the pants are not strictly necessary. It has been well established that stinging insects are much less likely to sting light-colored, smooth-textured clothing than clothing that is dark and rough, such as wool, which may too closely resemble the fur of a bear or other predator, thus making the bees unruly. As a result, in choosing what to wear, light-colored cotton clothing is the preferred attire when you are dressing for a day at the apiary.

The one exception to my bias against bee suits is found in the unique design of the coveralls sold by Golden Bee Products of Metairie, Louisiana (see the resources section). This suit comes complete with a built-in veil and is made of two loosely woven layers of fabric that sandwich a third layer. This center layer is a nylon mesh that acts as a spacer between the inner and outer layers of fabric. The three loose layers result in a suit that is too thick for a bee to sting through due to the space within the nylon mesh. Nor does the suit even offer many places that a bee is capable of stinging. The large number of little holes in the outer fabric creates a suit with a minimal amount of fibers that are woven tightly enough to allow a honey bee to embed a stinger; the stinger has nothing to catch on. The only areas that are vulnerable to being stung are the pocket areas, which, out of necessity, are made of fabric composed of tightly woven fibers. This airy design also provides plenty of ventilation, which makes the suit a lot more comfortable to wear in hot, humid weather than conventional suits made of cotton.

No matter what you wear, if you work regularly with honey bees, sooner or later you are going to get stung. Accepting this fact of life and learning how to deal with it are essential prerequisites for any apiculturist-to-be. The anatomy of a bee's stinger consists of a venom sac and muscles attached to a shaft, which is composed of two barbed lancets lying side by side, with a hollow opening down the middle that acts as the poison canal. When the stinger is embedded in flesh, the attached muscles

FIGURE 2-9. Beekeeper in full bee suit. Due to its unique construction, this bee suit made by Golden Bee Products outperforms the competition when it comes to reducing the possibility of the honey bee stinging through the suit. In addition, its increased ventilation makes this suit preferable for use in hot weather. As you can see, suits age with use, and duct tape works well to seal patches where holes start to wear through. Note that the legs of the suit are tucked inside the boots to prevent the persistent bee from crawling up the pant legs.

cause the lancets to work their way deeper into the sting site, while the venom sac pumps venom down the poison canal and into the victim. The act of pulling the stinger out with your fingers can cause more venom contained within the sac to be squeezed out into the poison canal and delivered into the skin, causing greater inflammation, swelling, and irritation. Therefore it is recommended that you remove bee stingers by scraping them off the skin with a fingernail (or the flat edge of a hive tool), rather than by grasping and pulling. By not squeezing the venom sac, you minimize the amount of venom that is deposited in the site of the sting, which will help limit the associated unpleasant side effects.

When a bee stings, it gives off an alarm pheromone that marks the sting site like a bull's-eye, attracting additional bees intent on defending their hive. So if you get stung, once you've removed the stinger, immediately use a smoker to blow smoke on the area of the sting. This will mask the warning pheromone and reduce the defensiveness of the rest of the bees. (I will discuss the use of smokers in depth in a moment.)

For the roughly 1 percent of the population who suffer from hypersensitive allergic reactions, wherein the whole body responds negatively to bee stings, taking up the craft of beekeeping is not advisable. Luckily for the other 99 percent of us, this is not the case. Unfortunately, some beekeepers, and even some doctors, will overreact to the body's reaction to a bee sting. One thing I learned from Charles Mraz and his use of bee venom in apitherapy is that although having one's leg swell up after being stung in the foot is certainly a severe reaction to bee venom, it is not life-threatening. Nor is it the type of reaction that would warrant carrying around a dose of antihistamines and an EpiPen. A truly dangerous reaction is one where the entire body reacts, rather than just areas near the site of a sting. Symptoms such as reduced blood pressure and breaking out all over in hives are the types of hypersensitivity reactions that can lead to a fatal outcome and legitimately call for a medical antidote to be carried as a precaution.

Besides setting up the wearer for the unreasonable expectation of never getting stung, the full regalia of the beekeeper comes with additional drawbacks. Thick bee-proof gloves make it difficult to perform actions that require much dexterity. A screened veil can be a challenge to see through, especially when you are trying to view small objects, like bee eggs, nestled deep inside a wax cell. Plus all that extra gear can be very uncomfortable on those hazy, hot, and humid days of summer. Inevitably, you will be tempted to shed part, if not all, of your bee armor at some point in your beekeeping career. Luckily there are ways of doing so that can ensure adequate protection, while at the same time creating the impression to the casual observer that you are much braver than you actually may be.

Smoke

There are two primary tools that beekeepers can use in place of some, or all, of their defensive clothing when working with honey bees. The first is the time-honored practice of using smoke to confuse and distract the residents of the hive. According to some of our most ancient historical records, smoke has always played an integral

part in working with honey bees. Although the types of materials used to create smoke and the various forms of applying smoke to hives have varied throughout the centuries and from culture to culture, the basic concept has remained the same: to distract the bees and disrupt their communication. Because smoke by definition contains pollutants and toxic gases, the type of fuel used in the smoker is extremely important, and in fact it is regulated by most organic standards. Recognizing the necessity of using smoke when working with bees, modern-day regulations allow the use of only those smoker fuels that are derived from natural sources. In this context, the definition of *natural* means "in its original form, without added chemicals or ingredients." Hence, materials such as newspaper, cardboard, and treated burlap—as well as any other material that contains petroleum-based or synthetic substances—are prohibited, whereas wood shavings, pine needles, leaves, dried grasses, and sawdust are all considered acceptable.

Of all creatures in the wild, the moth is the only one famous for not having a healthy fear of fire. Most animals and insects—the honey bee included—will head the other way when they smell smoke. They do not understand fire in the same way people do. For humans, fire has become a highly relied-upon tool. We have come to appreciate the many benefits and uses of fire, from cooking our food to heating our homes, powering engines of internal combustion, and helping us to clean up and dispose of unwanted materials. Fire can be a wonderful resource; however, fire also has its wild side and must be treated with respect, for when uncontrolled it can cause massive amounts of destruction and harm. Animals are familiar primarily with fire's destructive powers, and as a result their natural instinct is to run from anything that even remotely resembles a blaze. The beekeeper capitalizes on the honey bee's natural fear of fire by sending smoke on a mission to spread alarm deep within the dark recesses of the colony. After sensing smoke within their chambers, the worker bees can often be seen gorging on honey stored within the cells of the honeycomb in preparation for a hasty retreat, because they have been misled to believe that their hive may soon be consumed by fire. When the bees are distracted by the message of imminent hive destruction, the colony's inhabitants are much less likely to worry about any intrusion into their midst. At the same time, the warning pheromones that are given off by the agitated bees guarding the entrances to the hive—and that serve to warn the rest of the colony in an effort to rally the troops to their common defense—are masked by the scent of the smoke. Thus, the hive's primary form of communication and organization is temporarily shut down.

Corporate America many years ago learned the benefit of limiting the ability of workers to come together and unite in defense of their mutual interest. The inability of workers to unite and unionize can lead to exploitation of the workforce by management. Fortunately the realization by many of today's employers that happy and unexploited workers are more productive and profitable in the long run has led to fairer treatment of employees by most companies. By the same token, the use of smoke to distract the bees and prevent them from organizing in force and chasing us away with their venom-tipped barbed stingers needs to be tempered by fair treatment. Unfortunately there are beekeepers who share the exploitative philosophy and who will remove too much honey, ignoring the bees' needs in favor of their own desires. These shortsighted individuals remove all the honey and simply allow the colony to die, calculating that the cost of a replacement package of bees purchased in the spring is less than the price that can be gained from the sale of the additional honey harvested. Other beekeepers will allow the bees to replace the lost honey by feeding them sugar syrup so that they are more likely to survive the winter months. These approaches show a lack of respect for the honey bee and are prohibited by many of the organic standards that are now in place.

When it comes to such an integral part of beekeeping as the use of smoke, the proper and effective use of the gaseous products of combustion cannot be overestimated. Now, some may feel that the following information is so basic that it does not require explanation. Or perhaps they think that because the smoker, the hive tool, and the veil are such basic necessities to beekeeping—incredibly common and seemingly simple in their design and use—any instruction as to their proper use is unnecessary. It is certainly true that through trial and error even a novice is able to learn many of the basics of what works and what doesn't when using these tools of the trade. Teaching yourself about beekeeping through the school of hard knocks, while not necessarily easy, can provide an excellent education—though the tuition is high, because the bees are quick to let you know when you're not making the grade.

I have found the following approach works best for me. Keeping in mind that heat rises, clean out all the ashes and partially burned fuel in your smoker and place a little fresh, well-dried fuel in the bottom of the smoker cavity. Light the fuel and work the bellows with short, quick squeezes until the fuel is burning well. I look for flames shooting up out of the top of the smoker as a sure sign that there is "fire in the hole." It is important to be conscious of the direction of the wind and how close you, your clothing, and your equipment come to the exposed flame. You don't want to melt the screening on your veil or singe any hair, if it can be helped.

Once you have a good fire going, add another handful of fuel to the smoker, and again work the bellows with rapid, short blasts of air until this additional fuel is well lit. Now that you have a good hot fire burning in the bottom of the smoker, fill the rest of the fire chamber with more fuel, close the top, and you're ready for action. Care must be taken not to pack this final addition of fuel too tightly into the fire chamber or you run the risk of smothering the fire. Should you find that, after setting the smoker down for a period of time, the amount of smoke produced is reduced to a mere wisp, then making short, rapid squeezes of the bellows for 30 to 60 seconds should be

FIGURE 2-10. The well-made smoker is constructed from stainless steel, will hold plenty of fuel, features a shield to prevent accidental burns, and has a stout nose that can act as a wedge to hold supers apart once they have been separated.

enough to rekindle the burning embers within the smoker's fire chamber, unless the fire has gone out completely.

Once you have the smoker lit and the fire chamber full, making the most effective use of the smoker requires some practice. A smoker can be used not only to distract and calm the bees but also to push the bees in a certain direction, either toward or away from an area. A typical mistake that many beginners make is to use too much smoke, perhaps in the mistaken notion that if a little smoke is good, then a lot must be better. It takes just the right amount of smoke, at just the right time and from just the right direction, to prod the bees into going where you want them to go. Use too much

smoke and the bees will just run around in circles, confused as to where to find a breath of fresh air. Use too little smoke and the bees will act as if there were no smoke at all and place their focus on things that we would prefer they would ignore, like the tip of our nose. For the bees to move, they need to be able to identify where the smoky areas are (places they will move away from) and where the areas clear of smoke exist (toward which they will naturally gravitate). The beekeeper can use this behavior to "herd" bees in the desired direction. Again, noting the direction and speed of the wind can play a critical role in either helping or hindering your efforts, and care should be taken to position oneself advantageously.

FIGURE 2-11. An army surplus ammunition box (minus the ammunition, of course!) provides a safe, fireproof container for storing your smoker between visits to the apiary.

It is also important to pay attention to the quality of the smoke that issues from the smoker. The smoke should exit in thick white clouds and feel cool to the touch. A sure sign that the fire within the smoker has made its way to the top of the fuel pile and that your fuel needs replenishing is the appearance of hot, thin, gray smoke. Miss this telltale sign and you will likely end up with sparks, or perhaps flames shooting out of the smoker and singeing your bees. Trust me, blowing hot gray smoke, flames, and red-hot cinders onto your bees is a great way to make them upset with you really fast. If the fire is approaching the top of the fire chamber and you don't have enough fresh fuel to refill the smoker, one trick is to place a wad of green grass on top of the burning fuel. This will serve to cool the smoke and will help prevent flames and sparks from exiting the smoker and overexciting the bees. When all the work is completed and you're ready to put away the smoker, it is a good idea to plug the smoke hole with a wad of green grass or a cork to limit the amount of fresh air available to the burning material inside and help smother any hot embers that remain. To be extra safe, it is wise to store your smoker in a metal can that has a lid. It can be very disconcerting to be driving down the road, look in your rearview mirror, and see flames shooting up from your bee equipment, all because a smoker was not fully extinguished before being stowed in the back of your vehicle. Keep a fully charged fire extinguisher handy in any vehicle or building that is used for beekeeping purposes on a regular basis. When properly used, the smoker becomes one of the best pieces of protective beekeeping gear you can own. Just make sure you don't run out of good-quality fuel that is dry and ready to burn!

An alternative to the use of smoke as a means to make the bees manageable is the use of a one-to-one dilution of sugar in water. This solution, applied in the form of a fine spray, not only distracts the bees as they lick the sweet syrup off one another's bodies, but also serves to wet down their wings—in effect grounding them so they can't fly after you. Although not a viable option in cold weather (it is unwise to allow the bees to become cold and wet, because they too can die from a form of hypothermia), the use of a sugar spray can come in handy when no matches or dry fuel are to be found, so long as you are a beekeeper who doesn't mind providing support to the sugar industry.

There are beekeepers who promote the idea of not using smoke at all. Their reasoning is that smoke is toxic and that the bees don't like having smoke blown in their faces, and thus smoking the hive is too disruptive and not good for the bees. However, any beekeeper who has smoked a hive and then continuted to work there for 10 to 15 minutes can attest that the bees ventilate the smoke out of the hive and reorganize themselves relatively quickly. Nevertheless, some folks continue to claim that not using smoke is not only better for the bees, but also better for the beekeeper, who won't be exposed to secondhand smoke. Some beginner beekeepers even assume that the increase in pitch and volume of the sound emanating from a hive when smoke is applied is evidence that smoke makes the bees angry. This is a misconception; the sound is simply the result of increased wing fanning by the bees to vent the smoke out of the hive.

I concede that smoke is toxic, bees don't really like it, and it would be better if beekeepers didn't have to use it. However, in most cases, the potential for harm from *not* using smoke outweighs the harm caused by *using* smoke.

When a beekeeper wearing protective clothing does not use smoke or a smoke substitute (or does not use the smoker properly), the bees may become defensive and start to sting. In the process of stinging, many bees will die, because their stingers will lodge in the protective clothing and tear loose from the bees' bodies. When smoke is used properly to distract the bees and disrupt their communication, far fewer bees are incited to sting, and thus to die, in defense of their hive. I believe that whatever harm may be done by exposing a colony to smoke for a brief period during a hive inspection is far less damaging than having bees die during hive manipulations where no smoke is used.

Attitude

The second tool you can use to help keep the bees calm is your head. Keep in mind that honey bees are kind of like you and me. When it is warm and sunny, most of the foraging-age honey bees are out and about running errands and foraging for goodies. The majority of the bees left behind in the hive are younger individuals, many of whom have yet to develop their stinger and venom reserves. These young workers are busy storing the incoming nectar and pollen, feeding the brood, cleaning the hive, and attending to the myriad other tasks that need to be accomplished to keep everything running smoothly within the colony. However, on cool, damp, and rainy days, when it is too cold for the bees to fly and there is no nectar or pollen coming in, the hive's occupants may seem grumpy, all cooped up inside waiting around for something to do. It is not the best time to pay them a visit if you don't want to incur their wrath. This is one of the reasons why living in Vermont I prefer to position my apiaries in locations that receive lots of sunshine. The extra warmth provided by the sun's golden rays could be enough, when the temperature is on the cool side, to make the difference between working with a calm hive and one with a quick temper.

Once favorable weather has arrived and you have scheduled a call on a bee yard, one of the most important things you can bring along is your peace of mind. Early on in my work with bees, I noticed that I would get stung a lot more often on days when I was irritated and not in a good mood than when I was feeling good, present, and focused on my work within the apiary. As a result, my experience has taught me that honey bees are like dogs, farm animals, and even people: they can sense when we are calm, cool, and collected, or when we're irritable, fearful, or upset, and they respond accordingly. With every visit to the apiary, the bees reinforce within me the importance of my presence of mind. The bees will waste no time in providing me with immediate feedback when I allow myself to be grumpy, distracted, or off-balance. In their own way, they provide me with the motivation to do the work I need to do on myself in order to stay present, calm, and focused, even in situations where I could potentially be stung to death. (The lethal dose for a healthy adult is typically 10 stings per pound of body weight, and ranges anywhere from about 200 to 2,000 bee stings, depending on age, state of health, et cetera.) It is all too typical that when two or more people enter a bee yard, the person who is afraid of bees is the first one stung. Once you are able to maintain a calm, peaceful, focused, and present state of mind in the bee yard, the real trick is to be able to sustain that same harmonious presence when you return to the world of humanity.

Protective Gear

Armed with a beautiful sunny day, a well-lit smoker, and a calm, focused demeanor, I rarely find it necessary to use a veil, let alone a bee suit, when working with bees. Now, if this is the case for me, I am sure there are others for whom these same protections work as well. I must confess, however, that I did not start out keeping bees this way. The first day that I worked with honey bees I wore a veil and gloves, along with a long-sleeved shirt and jeans with the cuffs tucked into my socks. After all,

although it is good not to be afraid of being stung, it is not something you typically want to encourage. Thus, a veil and gloves are prudent for those just starting out. As I slowly got used to working with the bees and became more comfortable around them, I became braver and more willing to try working without protective clothing. First to go were the gloves. They were so thick and bulky that it was hard to feel anything through them, and I found that I accidentally crushed many more bees while wearing gloves than I did when I didn't have them on. I suggest that beekeepers who wish to use gloves when working with bees consider purchasing a tool called a frame grip. This scissorlike device will allow gloved hands to grasp and lift frames of comb covered with bees without needlessly crushing large numbers of bees. You may also discover, as I did, that hands are a lot less sensitive to the pain of a bee sting than many other areas of the body.

It was during a particularly hot and sweltering week working in the bee yards—with sweat stinging my eyes as I viewed my more experienced coworkers from under my veil, looking much cooler and more comfortable than I as they worked without similar headgear—that I first decided to try removing my veil to see what would happen. Thus began my lessons on the importance of one's attitude and mood when working in the apiary.

Readers new to beekeeping may start to think that I have been stung in the head one too many times when I state that it is incredibly beneficial to have developed enough comfort around the bees to be able to work with them on most occasions without any special protective clothing. I make this statement as a result of both personal experience and my observation of others. Now don't get me wrong. I fully recognize that a good-quality veil is handy to have around when faced with the realities of our fast-paced modern lifestyles, which may require us to perform hive manipulations under less than ideal circumstances due to our time constraints. Nevertheless, the benefit of not requiring protective equipment when working with bees goes well beyond the conveniences of saving

FIGURE 2-12. A frame grip, available from most beekeeping supply companies, will allow the beekeeper to handle bee-covered frames without crushing and injuring the bees, even when wearing gloves.

money, having unimpaired vision, and flirting with heat exhaustion inside a bee suit during the dog days of summer. When I put on a veil, I am putting up a barrier between the bees and myself. Once I have separated myself behind this protective wall, it is easy to get lazy and forget to show respect for the little ladies with whom I am working. Human history has shown over and over that, with separation, all too often one side will eventually start to dominate the other. Without such a division, there is greater opportunity for prolonged and peaceful cooperation.

On too many occasions, I have observed beekeepers decked out like astronauts in full protective gear carelessly crushing bees, knocking into hives,

and generally handling the colonies and equipment in a rough manner. Often these are the same beekeepers who will complain that their bees are "mean" or overly aggressive, whereas I suspect that they are simply not treating their bees with the proper respect. It is much easier to be disrespectful when the consequences of such contempt have been largely removed. I recall that one day I was asked to help a local beekeeper with his "angry" bees. Despite that description of the bees, I approached them as usual, with just a hive tool, lit smoker, and cap on my head to prevent the occasional curiosity seeker from getting tangled in my hair. On this occasion the bees seemed not to be aggressive at all, and I went through all the hives without getting stung once. Unless the bees have recently been disturbed, the primary variables that account for differences in European honey bee behavior seem to be the individual working with the bees and how he or she goes about the work. I suspect another factor in this so-called angry-hive syndrome is that beekeepers making such complaints are often not washing their bee suits on a regular basis. As a result, their veil and coveralls are impregnated with the smell of venom from previous visits to their bee yard, and this is all that is necessary to rile the bees when the suited beekeeper decides to pay them a visit.

A beehive is not built by a multitude of single bees toiling away in isolation, and thus the bees have much to teach us about the value of cooperation. The honey bee acts selflessly in relation to the rest of its hive family. The vast majority of the bees' activities are done for the good of the colony as a whole, rather than to benefit any individual alone. As a result, we see that the strength of the honey bee—an insect that has survived since long before the time of the ancient Egyptians—can be traced to its ability to cooperate. This line of thought leads me to believe that Darwin's "survival of the fittest" theory, which refers to the strongest individuals as being the most likely to survive, is flawed. Closer observation indicates that those individuals who have the greatest ability to cooperate and adapt are the ones that also have the strongest likelihood of succeeding in the long run. In his later years, even Darwin himself seems to have become convinced that love (or altruism) was a stronger force than the survival instinct. Numerous examples of this phenomenon occurring in nature support his later theory, such as the behavior of packs of wolves, schools of fish, flocks of birds, and herds of buffalo. Groups of animals learned long ago that working together in cooperation greatly increases their long-term chances for survival, compared with acting alone. These are some of the thoughts that make me passionate about the craft of beekeeping—a passion that inspires me to carry on keeping bees despite the hard work, difficulties, and challenges involved. It's a passion that results in my wanting

to have honey bees around for as long as possible. I would hope that everyone in the world might find something they can feel so passionate about during their lifetime.

Separating oneself from the bees with protective equipment also prevents the beekeeper from making subtle observations that may provide vital clues to problems among the hives. Without a veil, beekeepers are much more sensitive to the temperament of the colonies they work with. Not only do the bees provide instant feedback if, from their perspective, they are not being treated properly, but uncalled-for aggressiveness can point out problems that need to be addressed. To walk into an apiary and immediately have several bees angrily buzzing about your head provides an important clue that something is not right, and more investigation is needed to get to the bottom of the issue. Because of the angry bees, your heightened state of alert may call your attention to signs that may indicate, for example, a recent dearth of nectar and thus no honey coming into the hive; or that skunks have been feeding regularly at your apiary; or that there is an infertile queen or a worker that has started laying infertile eggs that mature into drones (known as a drone layer) amongst your hives. In all these cases, the bees will be extra defensive or demoralized and let you know that they are not pleased about the situation.

On one occasion, while entering a bee yard located near the intersection of two dirt roads one sunny spring day my coworker and I were immediately set upon by a number of angry bees and were forced to retreat into the cab of the truck and put on veils before we could proceed with our work. This unusual reception prompted us to theorize as to the cause of such unusual activity by our typically friendly bees. While surveying the apiary, we noticed a fair number of good-sized rocks randomly scattered among the hives. The rocks matched, in both color and size, those alongside the dirt road that ran past this particular location. A little while later, a school bus pulled up to the corner and dropped off several children making their way home at the end of their day. The sheep-ish looks on the children's faces told us the story of why there were so many rocks scattered around the bee yard and why the honey bees were so jumpy.

I am pleased to report that, even though I rarely wear a veil these days, I do not get stung on a regular basis. When I do suffer a sting, I find that it is typically because I am not paying proper attention to my attitude or to what I am doing and have grabbed a bee while picking up a piece of equipment or something similar. I view these occasions as reminders to pay closer attention to the task at hand, and I take solace in the idea that, typically, the more stings one receives, the more immunity one builds up to the effects of bee venom. Nevertheless, despite the development of immunity to venom's effects, there seems to be no similar increase in one's tolerance for the pain involved—bee stings always hurt. If I am not care-ful and focused, it is likely that I may get stung several times during my visit to the bee yard. The good news is that it is usually only the last bee sting that hurts the most.

By leaving myself unprotected by a veil, I find that I am forced to become a better beekeeper. In addition, the knowledge that bee stings can reduce or eliminate symptoms of diseases such as arthritis and multiple sclerosis helps me to accept the occa-sional bee sting I receive as beneficial to my health and a form of preventive health care. It's all part of the employee benefits package I receive from my job working with honey bees.

If I lived in the southern regions of the United States, it is likely that my ability to work with bees without a veil would be impossible, or at the very least ill-advised, because of the Africanized honey bee (AHB). Africanized bees arose from the 1956 importation of 50 African queen bees into Brazil for research purposes. Due to the large number of natural honey bee predators that exist in Africa and the short, sporadic honey flows, the African bee developed a very aggressive defensive nature in order to protect its limited honey stores and deliver a robust response whenever its colony felt threatened. This has led to the sensationalistic moniker "killer bee" being attached by the media to this strain of honey bee, which has migrated northward over the past 50 years to the point where it can now be found in almost all southern states in the United States, in a band that stretches across the continent from California to Florida. How much farther north Africanized honey bees will travel is anybody's guess. AHBs have a dif-ficult time surviving the winters of the Northern Hemisphere. They have a tendency to nest out in the open, leaving them little insulation and protection from the elements other than the bees themselves as they cluster. Because it is in their nature to keep the size of their colonies small and to swarm frequently, AHBs are unlikely to build up a population large enough to form an adequate cluster for the winter, even if they were sufficiently sheltered; nor are they likely to store away enough honey to last throughout the cold season. That said, their maximum northward expansion will likely be augmented by the general global climate change pattern of recent years.

And yet the northern beekeeper cannot rely on the cold to keep the AHB at bay. The ease with which packages and queens can be shipped from breeders located in areas known to be colonized by Africanized bees, and the large number of migratory beekeepers that move between southern areas known to harbor Africanized bees and the northern latitudes, pres-ents the potential risk that bees with AHB genetic traits will be shipped into northern areas. As a result, many cool-climate beekeepers are already focusing on ways to become more self-reliant in the rearing of replacement bees and queens in order to avoid the damage that a stinging "incident" is likely to cause to the image and industry of northern beekeeping. Never mind that the vast majority of AHB stinging episodes are nonfatal. The relatively few fatal inci-dents that have occurred were typically a result of an individual's hypersensitivity to bee venom, or they happened when animals were tied up or when people panicked or were unable to move quickly away from the area once the bees started to become defensive.

Another issue to consider when deciding how much protective clothing to wear, or not to wear, is the susceptibility of some beekeepers to suddenly developing a life-threatening allergic reaction to bee venom. This is in direct contradiction to the experience of others, who have observed that repeatedly getting stung by bees allows for the development of resistance to the bee venom, so that the side effects of bee stings become less severe and eventually virtually unnoticeable as the number of stings received increases over time. Judging from some of the accounts I have heard, the use of drugs and medications may play a role in the sudden allergic disposition to bee venom that some bee-keepers develop. Certain medications can change body chemistry and affect immune response, and I suspect that such changes may make a person more prone to developing a life-threatening aller-gic reaction to bee venom, even if that person has never had any allergy issues in the past.

One final but related word of caution: there are reports that members of beekeepers' families who don't work directly with honey bees themselves may nevertheless develop severe life-threatening allergies to bee stings. Some people believe that these responses come from repeated exposure to the minute amounts of bee venom the family member is exposed to from the protective clothing worn by the beekeeping member of the family. As such, it may be advisable for beekeepers to do their own laundry and clean their beekeeping clothing in loads that are separate from the rest of the wash. (By the way, the trick for removing propolis stains is to soak the stained area with rubbing alcohol, which will dissolve the propolis.)

While standing amid a cloud of bees can be quite intimidating, especially in the beginning, over time you will become comfortable working among the bees and may reach the point where you can hold a bee in your ungloved hand. When you are standing by your hive with a bee cradled on your palm, you will find that the world appears more beautiful than usual. This is because beauty is in the eye of the bee holder.

CHAPTER 3

Hive Management

A swarm in May is worth a load of hay, a swarm in June is worth a silver spoon, but a swarm in July ain't worth a fly.

—OLD NEW ENGLAND BEEKEEPER'S SAYING

We like to believe that advancements in our society take place as old beliefs and customs slowly give way to the changes brought about by science, experimentation, education, and practical experience. Nevertheless, this approach works only when the new approaches improve on old methods. In our history, agricultural advances that have boosted production all too often have overlooked long-term consequences in the rush for immediate profits and have resulted in the creation of larger, more complex problems. The world of beekeeping is no exception.

Many of the modern organic hive-management techniques that are emerging are not all that different from those of conventional apiculture. This is simply because the basics of honey bee biology and behavior must still be observed. In fact, to be successful at organic beekeeping, it is even more important to learn to work with the natural biological processes and instinctive behaviors of the bees. Whereas conventional apiculture tends to force the beekeeper's will upon the hive organism, the organic beekeeper is more inclined to work with the colony in partnership rather than in domination. As such, many of the ways of keeping bees that were common 20 to 30 years ago are being edged out in favor of new methods and techniques, which continue to change as new research and information becomes available.

One key distinction of the organic approach is that it does not hold the maximization of the honey crop as its top priority. This one seemingly small shift in perspective makes a world of difference. All of a sudden, the *quality* of the honey harvested and the health of the hive become just as important as, if not more important than, the *quantity* of honey that the beekeeper can coax out of the hive. This emphasis on quality over quantity is perhaps the defining notion of the organic agricultural movement. The attribute of quality is not limited to simple cosmetic factors such as size and looks, nor is it applied only to the resulting crop that is harvested. In organic agriculture quality is applied to the effects of the overall operation and takes into account the health and well-being of the farm, the land, the animals, and the people involved, from farmer to consumer. Of course, quality also has to do with the taste, color, and smell of the final product. Over time, we discover that when proper attention is paid to quality, the quantity aspect takes care of itself. In the world of beekeeping this translates into the fact that, by keeping your bees alive and healthy through organic practices from year to year in the face of chemical-resistant mites, in the long term you will realize larger harvests and increased profits over those of beekeepers who rely on toxic approaches that ultimately fail the bees. This can occur even though your harvest of organic honey may be less, pound for pound, than what might be obtained through conventional means, simply because the organic beekeeper may not have to spend as much money and time replacing hives every year or applying mite and disease treatments, and also because of the premium price that organic honey commands.

• APIARY LOCATION •

Traditional considerations that a beekeeper needs to take into account when choosing an apiary site include not only the availability of pollen and nectar sources but other factors as well. Among these other concerns are nearby sources of clean water, yard accessibility by vehicle, good drainage, exposure to the south or southeast to provide adequate sunlight, windbreaks in the direction of the prevailing winds, local laws and zoning regulations, and a location that will discourage vandalism and theft. Another factor that is sometimes overlooked in locating hives in an apiary is the importance of leaving enough working space around the hive so that there is room for beekeeping activities to take place. Placing hives right up against a fence or a row of bushes can make it difficult to carry out required inspections and hive manipulations. As if all this were not enough, the organic beekeeper has additional concerns to consider when locating a bee yard, and this has become one of the most controversial aspects of organic apiculture.

Honey bees are semidomesticated wild creatures. They will choose to forage anywhere and at any time they want, and we humans have little to say about it. As a result, bees may forage for nectar and pollen on flowers that have been raised organically; or, they may visit plants that have been sprayed with chemicals or have been genetically modified and therefore do not fit within the framework of organic agriculture. As a result, current standards for organic honey often require that hives be located on a certified organic farm. Standards also require that all areas near an apiary where the bees are likely to forage must be free of activities that in any way could compromise organic integrity. Some folks have a real problem with organic regulations that address hive and foraging location and think that, as long as the hive itself is managed using natural methods that utilize no synthetic chemicals or antibiotics, where the bees forage should be of no concern. They believe that organic standards for honey and honey bees should address only hive management and manipulations—factors that are clearly within the beekeeper's control and are thus verifiable. The reason is that, even when a hive is placed on land that is managed organically, the bees are likely to fly right past the organic blossoms and forage on non-organic flowers that are within reach of the colony. A hive's foraging radius can extend significantly during times when there is a dearth of nectar. It has already been shown that about 10 percent of the honey bees in a colony may fly well over 5½ miles from the hive, and some can travel as far as 8½ miles or more in search of food during periods when forage is scarce.[1] Additionally, if a plant in bloom has been doused with severely toxic chemicals, the foraging bee is not likely to make it back to the hive alive.

Folks who take this position are espousing what may be called a liberal viewpoint of the term *organic*. They believe in a loose reading of the requirements that does not follow a strict "by the book" interpretation of what organic has traditionally come to mean when applied to livestock. Because the biology and culture of the honey bee are significantly different from those of other farm animals, it is argued that bees warrant separate regulations regarding feeding, foraging, and so forth.

In the other camp are those people who stress that the integrity of organics requires that not only must hives be managed without synthetic chemicals and antibiotics, but as much effort as possible should be made to ensure that the bees do not forage on plants that have been exposed to prohibited materials. Such areas that are considered to adversely affect organic integrity include city and town centers; industrial sites; conventional farms that use chemical fertilizers and pesticides or genetically modified seed or plants and grow crops that could potentially be used as a foraging source; garbage dumps and sanitary landfills; non-organically managed golf courses; nuclear power plants; incinerators; major highway, power-line, and railroad right-of-ways; contaminated water sources; and conventionally managed apiaries. These folks take a conservative approach to the organic standards, in that they believe in a traditional interpretation of the organic philosophy when applying it to honey bees. They argue that organic standards should dictate constraints similar to those that affect other livestock, and they believe that beekeepers should follow this traditional approach to the letter. Not surprisingly, organic certification agencies have tended to take a position that is more in line with this conservative approach.

Those who share the liberal view argue that they are being more realistic in their perspective, given that the vast majority of apiary locations within the United States are not able to comply with organic regulations as currently written, and most of the locations that do conform tend to be in areas where bee forage is extremely limited, and resulting honey crops are likely to be small or nonexistent. As a result, they point out, little certified organic honey is currently being produced within the United States due to the overly restrictive nature of the standards. Conservatives counter that it is important to meet the expectations of the organic consumer and not dilute the meaning of the term organic just because, unlike other livestock, bees cannot be penned in and their foraging activities controlled as easily as cows or farmed fish, for example.

As with most arguments, each side's position has a piece that is true, but the truth lies somewhere in the middle. Most standards, as currently written, incorporate a distance limitation with respect to areas that may affect organic integrity, in an effort to find a compromise position. Requirements range anywhere from 1 mile to about 3¼ miles in all directions from the apiary location, even though honey bees have been known to fly much farther under severe circumstances when searching for food. The thinking is that honey bees are like people: they are not going to work harder than they have to. If adequate nectar and pollen sources are available within a mile or so of the hive, the vast majority of foraging bees are unlikely to fly much farther afield in search of additional food. This solution thoroughly pleases neither the liberal nor the conservative points of view. If honey bees have the ability to access

prohibited foraging areas just outside the regulated distances specified in organic standards, then from the conservative perspective this jeopardizes organic integrity. From the liberal viewpoint, if the bees are likely to have access to prohibited areas anyway, why bother placing any limits on apiary location at all? Doing so simply creates a false impression in the marketplace when, in these people's opinion, organic honey is not going to be all that different from conventionally produced honey.

These two conflicting points of view have yet to be fully resolved, and, as a result, a wide range of criteria governing apiary and foraging locations has been written into the organic standards enforced by the numerous organic certification agencies authorized by the United States Department of Agriculture (USDA). Until the USDA National Organic Program (NOP) establishes a clear set of national regulations for apiculture that will supersede all other standards, the mishmash of rules that now exist is likely to continue well into the foreseeable future, and no resolution to this controversy is likely anytime soon. In fact, I suspect that it is precisely the issue of apiary location and forage availability that is responsible for much of the delay in a set of national organic regulations for apiculture being codified by the NOP.

From the honey bee's perspective, the best location is one that provides access to a wide variety of flowering plants that yield nectar and pollen throughout the growing season. Due to the fluctuations of the protein and mineral content of plants, there is no single source of pollen that will supply a colony with everything it needs, and a diversity of pollen and nectar sources helps ensure that the hive will be able to adequately provide for its nutritional needs.[2] Access to many different types of plants will become even more important as Earth's climate changes accelerate. As various areas of Earth become warmer or cooler, the bloom cycles of plants will fluctuate. This will create more and more situations in which bees are unable to rely on the pollen- or nectar-bearing vegetation that they have relied on in the past.

An example of this is reports from parts of New York State where, during an unusually mild winter, the pussy willows flowered in March instead of April. The temperatures then dropped back to seasonal norms and were too cold for the bees to fly consistently, so they were unable to collect pollen from the pussy willow blooms. This lack of access to an important source of early spring pollen adds stress to colonies. By locating hives in areas where a wide variety of plants grow, we increase the chances that colonies will be able to find adequate forage to meet their needs in spite of changing weather patterns.

• EQUIPMENT •

The long, cold, and dark winter months are the perfect time for puttering around in a woodshop making, assembling, and repairing the hive equipment that will see you through the upcoming honey season. There are several ways to obtain bee equipment. It can be purchased from beekeeping supply companies unassembled, or conveniently preassembled for a little extra money. Taking the time to assemble the equipment yourself gives you more control over the quality of the workmanship. You can also purchase used equipment from someone else, or, if you have the tools, you can build your own from scratch. When buying used equipment one should be extremely wary of purchasing frames of drawn comb. It is within the wax of the combs that diseases and chemical residues can lie hidden. Both of these can adversely affect the quality of your honey crop and the health of your bees.

The types of materials used for constructing honey bee equipment are addressed by most organic standards, because certain materials can adversely affect the overall health and vitality of the colony or the quality of the products harvested from the hive. In the organic hive, all painting and weatherproofing are confined to the exterior of the hive, while the interior is limited to bare wood, metal, and beeswax. Due to the presence of chemicals, conventional pressure-treated lumber is not approved for use in constructing beehives. Commercially available hive bodies and supers are typically made of pine or cypress. Some folks like to use cedar when making hives. The smell of the cedar will be an irritant to the colony and will result in their coating the interior with propolis in order to cover up the wood and reduce the scent. Not only are the types of materials used to make bee equipment important, but also the quality of the craftsmanship that goes into the construction plays a huge role in equipment usefulness and longevity of service. It is this extra attention to detail, or lack thereof, that separates the well-crafted hive from the poorly assembled one.

Choose products carefully when preserving wooden hive components. When the hive bodies that make up the hives in an apiary are painted different colors, the ability of the individual bees to identify their colony among others is greatly improved. Paint should, however, be used only on the parts of a hive that are exposed to the elements, because the honey bees' interior decorating preference is bare wood and beeswax, both coated with a very thin layer of propolis. Because bees give off heat and moisture, the exterior of hives should not be painted with oil-based paint; it will tend to crack and peel over time. Latex paints are not only less toxic and better for the environment than oil-based paints but will breathe and allow for the passage of moisture, while at the same time repelling water and its damaging effects. As a result, latex paint holds up much better than any other wood protection I have tried. It is interesting that the bee equipment I painted more than ten years ago still looks great—aside from some dirt and scuff marks—even though I used no primer and only a single heavy coat of exterior-grade latex. This is in contrast to most wooden buildings, which are typically primed before being painted yet end up needing to be repainted every four to five years due to peeling, cracking paint. It makes me wonder: if bucking conventional wisdom and forgoing the use of a primer coat on hives works so well, how well would it work on my home? One of these days I hope to get adventurous enough to try it and find out.

PAINTING SUPERS AND HIVE BODIES

If you don't have a ventilated room available in which to use spray paint and you have more than a few boxes to paint, here is a way to quickly paint a bunch of hive bodies or supers: Stack the boxes on a stand of some kind so they are off the ground and square them up. Place a board on top of the supers and top the whole thing off with a heavy object that will prevent the stack from shifting while being painted. Now you can use a roller to quickly paint the four sides of the stack from top to bottom. Use a regular paintbrush to touch up the handle areas.

FIGURE 3-1. Speed up the job of painting by stacking the supers and hive bodies and using a paint roller.

Rather than using paint, many commercial beekeepers dip their hive equipment into vats of beeswax mixed with various wood preservatives. Unfortunately, most of the preservatives on the market are petroleum-based and quite toxic, having potentially sublethal effects on bees. Linseed oil works well as an alternative to paint or petroleum-based preservatives and can be mixed with a little natural turpentine, which will help the oil to dry faster and soak deeper into the wood for better protection. Be sure to read the label, because turpentines derived from petroleum distillates have replaced much of the naturally produced wood turpentine on the market.

Some biodynamic beekeepers make a lacquer by dissolving propolis in rubbing or grain alcohol and use it to varnish the outside of the hive, much like the bees do on the inside of the hive. Dr. Marla Spivak of Minnesota State University has conducted studies that indicate that varnishing the interior of the hive with a propolis tincture can help improve bee health and boost the colony's disease resistance. This is something you may want to consider, especially if you decide to use aromatic cedar wood for your hive. However, use only food-grade alcohol when making a varnish for use on the interior of the hive, because rubbing alcohols contain a toxin. Of course, making varnish is a wonderful idea only if you happen to have lots of propolis to spare. In my view, propolis's healing properties make it far too valuable for me to use it as a varnish. I prefer to leave the propolis in the hive to improve bee health and tend to collect only modest amounts of propolis from my bees for personal use (usually only when I come across a large chunk while extracting honey supers).

Plastic or Beeswax?

There is a trend toward increased use of plastic hive parts because they do not rot and are incredibly durable. Plastic has become a favorite material, especially as a replacement for some or all of the beeswax normally used for foundation and combs. This is an extension of a broader trend toward the ubiquitous use of plastic in modern society following the rapid development of petrochemicals in the early part of the twentieth century.

Plastic foundation and combs have the benefit of being a lot more rugged than wax, resulting in fewer lost frames from the comb breaking and collapsing while the frames are spinning around at high speed in a honey extractor (centrifuge). Also, plastic is not as temperature-sensitive as wax, which reduces shipping and storage damage. For example, plastic frames and foundation are less likely than wax to become brittle and crack in cold temperatures, and they are not easily destroyed by mice, wax moths, and other pests. Frames composed of plastic foundation and wood, or all-plastic comb, take less labor to assemble than traditional wax foundation in wooden frames. Such savings are especially important when running a large beekeeping operation. In addition, for those with fading eyesight, frames made with black plastic foundation make spotting bee eggs a lot easier.

Research cited by Roger Morse and William Coggshall in their seminal book, *Beeswax*, indicates that honey bees must consume somewhere between 6.6 and 8.8 pounds of honey in order to produce 1 pound of beeswax. Thus, frames of beeswax comb represent a substantial investment of time and energy on the part of the bees as well as the beekeeper, who now has less honey to harvest. The use of fully drawn plastic frames of comb significantly reduces the consumption of honey to support comb-building.

Despite beekeeper enthusiasm, no plastic frames or foundation have been manufactured to date that the bees prefer over beeswax. Bees will typically refrain from drawing out plastic foundation unless it is coated with beeswax or sprayed with sugar water or both. Why the bees don't take to plastic foundation as well as beeswax is a bit of a mystery. It may be that the bees sense minute amounts of chemicals (due to off-gassing from the plastic), or perhaps they simply don't like the texture of the smooth plastic surface. Or it may be more difficult for bees to transmit messages through their dance language if the vibrations that the bees send out into the comb while they dance are not transmitted by plastic in the same way as by beeswax.

Even for beekeepers who choose not to follow the bees' lead and listen to what the bees tell them, there are plenty of other reasons to avoid using plastic as part of the hive. Increased reliance upon plastic parts means an increased reliance upon petroleum that is used in their manufacture—and increased harm resulting from the plastics manufacturing process, which releases numerous pollutants and toxins into the environment. This issue is so pervasive that there are stretches along the Mississippi River, where many of the nation's plastic factories are located, that are referred to by some as "cancer alleys."

Another ongoing problem with plastic is disposal. Most plastics are extremely stable and do not break down or biodegrade readily. Thus, one of plastic's most valued benefits is also one of its most difficult challenges. While plastic frames and foundation stand up well to handling, they can become badly warped if left out in the sun too long. Despite this drawback, plastic frames and foundation have the potential to be used and reused for decades, if not centuries. What happens, however, when a hive containing plastic frames becomes infected with American foulbrood disease? According to the law in most states, the hive must be burned, but the burning of plastic releases even deadlier toxins into the environment than are emitted during the manufacturing process. Such toxic pollution is not good for bees, people, or any other living thing.

Styrofoam, foamed or expanded polystyrene (another form of plastic), is also becoming more common for use in constructing hive bodies, covers, bottom boards, and hive-top feeders. Polystyrene is promoted especially in northern climates as providing greater insulation for bees compared to standard wooden hives.

Unfortunately, highly insulated hives can create more harm than benefit to the colony they contain. When a midwinter thaw occurs, it will take the hive longer to warm up inside, because insulation is just as effective at keeping cold in as it is at keeping cold out. This delays the ability of the bees to warm up during a thaw and make cleansing flights, thus decreasing their chances of surviving and coming out of winter strong and healthy.

Unlike plastic or wooden hive components, polystyrene does not hold up to rough use and is easily dented, cracked, and broken. While this also means that polystyrene readily breaks down into small pieces in the environment, it unfortunately does not biodegrade. Polystyrene presents a similar problem to the "green plastics" that incorporate cornstarch into their structure and break down from a single large piece of plastic into numerous smaller pieces. This problem, combined with the enormous amount of non-biodegradable plastic that is disposed of daily, has led to the formation of a Texas-sized island of plastic flotsam that is currently floating around out in the ocean.

The increased use of Styrofoam in beekeeping is occurring simultaneously with an effort to find alternatives to polystyrene foam, especially in restaurant settings, as society becomes more environmentally conscious. Restricting the use of foamed polystyrene take-out food packaging has become a priority of some environmental organizations. Given that beekeepers tend to be among the most environmentally conscious individuals in society, this paradox is puzzling.

One possible explanation is that plastic and polystyrene hive components are heavily promoted and sold to beginner beekeepers by some beekeeping supply companies because it is easier to deal with than wax foundation. It is also noteworthy that plastic foundation and fully drawn plastic frames, along with polystyrene hive bodies, supers, and other components, tend to be more expensive than their natural wood and beeswax counterparts. Beekeeping supply companies tempt beekeepers by offering "beginner kits" made up of plastic or Styrofoam components at a discounted price. Once new beekeepers have started using plastic, they will often continue to purchase more of the same as the need for new and replacement equipment arises, despite the greater cost.

Why would a beekeeper decide to forgo the use of plastic hive components and spend extra time assembling hive parts, accept the need to replace some of the parts more often, and give up a significant portion of the honey crop by requiring bees to convert honey into wax? Putting aside the larger social and environmental consequences that might inspire such a decision, the bees themselves may provide the reason: they like wax and wood better. Their behavior tells us this. While there has yet to be any scientific research to support this theory, I believe that the production of beeswax by worker bees that are approximately 12 to 18 days old is an important part of the honey bee's biological development and health. Just as people eliminate toxins from their systems when they perspire, it may well be that the honey bee is able to excrete toxic chemicals and toxins from its body during the process of wax production. If this is true, then encouraging bees to build out new combs on a regular basis may well benefit the colonies in more ways than the simple removal of old, diseased, and chemically contaminated combs from the hive.

Responsible disposal of removed combs is simplified when they are entirely composed of wax, wood, and a small amount of metal, all of which biodegrade. Beeswax is a valuable commodity that has many uses—it is more valuable pound for pound than most honey—and it can provide a good alternative income stream for a beekeeping operation. Another reason you may want to use beeswax foundation rather than plastic is to simply support your fellow beekeepers rather than the petroleum/plastics industry.

Additional Equipment Details

One key hive construction detail that some beekeeping supply companies often overlook is the direction of the heartwood in relation to the exterior of wooden supers and hive bodies. When a piece of lumber warps, it will bend away from the heartwood. As a result, when the heartwood is facing the inside of the hive, any warping that occurs will cause the nails at the the corners of the board to pull out, eventually resulting in gaps between the boards at the corners of the hive body or super. By positioning the heartwood to face toward the handhold on the exterior of the equipment instead, the force of the board's natural tendency to bend is countered, and openings between boards at the corner joints are much less likely to occur.

Another beekeeping equipment detail that could stand to be changed is the frame rest protector designed to hang down into the hive. This design seems to have become the standard industry practice. Regular cleaning is required in order to remove the wax and propolis buildup that occurs within the frame rest area. It would make more sense to design the frame rest protection so that it protects both the

horizontal and the vertical sides of the ledge rather than just the bottom edge. If the nails that fasten the metal frame rest protector in place are driven down a little below the surface, then a hive tool will not catch on them when the area is being scraped clean. Frame rest protection that covers both sides of the frame rest can add many years to the life of your equipment and is well worth the investment.

One of the greatest equipment challenges faced by today's organic beekeeper is obtaining organic beeswax foundation. Beeswax acts like a sponge in that it easily absorbs and accumulates not just honey bee pathogens, but also toxic chemicals and their residues. As a result, many organic standards require the use of residue-free wax foundation, and

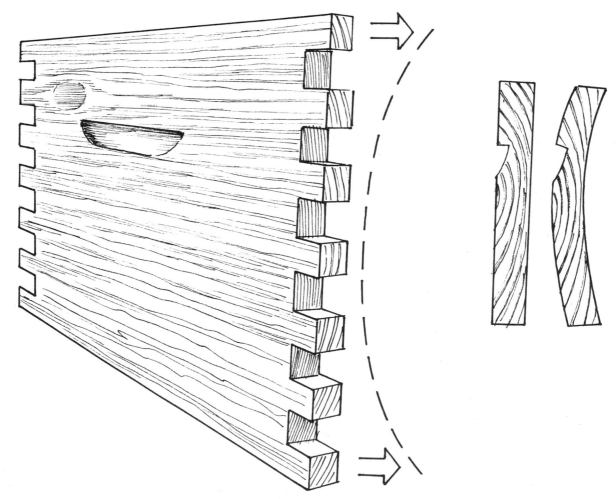

FIGURE 3-2. The well-made super features handholds on the heartwood side of the lumber.

FIGURE 3-3. The typical rabbet protector sold by most bee-keeping supply companies provides protection to only the bottom edge of the rabbet cut that makes up the frame rest area.

FIGURE 3-4. Reinforcing both faces of the frame rest area results in a significant increase in the working life of the equipment by reducing wear and tear when this area is regularly scraped clean of wax and propolis.

until the infrastructure is in place to supply organic beeswax, meeting this standard will be difficult. Initially, any organic wax foundation that is available is likely to be both scarce and expensive. Sheets of foundation made from certified organic beeswax can be cut into 1- to 2-inch strips and used to guide the bees' comb-building process. The bees will draw out the strips as they build their comb, forming whatever cell size and type (worker or drone comb) that the bees feel like building at the time. The beekeeper must be willing to give up a level of control over the type of comb the bees will build in order to capitalize on this method of avoiding chemically contaminated commercial beeswax within the hive.

If clean beeswax or foundation is unavailable, beekeepers drip a bead of wax along the underside of the top bars of wooden frames in the hopes that the bees will use it as a starting point to build natural combs without the aid of foundation. Because the bees always build comb parallel to the force of gravity, special care must be taken to level hives from side to side when a starter strip of foundation or a bead of wax is being used to entice the bees to build comb within the frame. A 2-foot level placed on top of the hive with the inner cover off (or resting on the sides of a top bar hive) works well for checking whether a hive is level. If the colony is positioned on a slant, the combs built within the frames will stick out in relation to the frame's edges,

creating problems when combs are moved around and during honey extraction and processing.

Starter strips of foundation inserted horizontally along the top bar (see figure 3-5) tend to work best when mounted in shallow or medium frames. When horizontal strips are used in deep frames (about 9¼ inches deep), the bees can sometimes have a hard time making contact with the foundation strip, which is some 8 inches above their heads, unless the colony is a package, nuc, or swarm being placed into its initial hive body. This is especially true when the hive is weak, but is not a concern with top bar hives (more on top bar hives next). When bees that are established in their initial hive body are unable to easily make contact with the starter strip in a hive body placed above them, the bees will actually build their comb up from the bottom bar rather than down from the top bar (as is usual). To avoid this problem, foundation strips can instead be inserted vertically between the top bar and the bottom bar. The temptation will be to use a single 1- to 2-inch strip that runs from the middle of the top bar to the middle of the bottom bar. Unfortunately, this will often result in the bees attaching comb to an adjacent frame of comb as they build. To avoid this situation, use two vertical strips, one located at either end of each frame, butting up against the end bars. The bees will start building comb at either end of the frame and connect it in the middle without the tendency to attach the comb to the adjacent

FIGURE 3-5. Horizontal strips of foundation like this work best in shallow frames, top bar frames, and deep frames when they are in the first box that bees are placed into. Be sure the hive is level from side to side or the bees will not build straight, flat comb. The thrifty beekeeper will appreciate this method of stretching the use of foundation and thus reducing the expense of purchasing it.

frame. Because the vertical starter strips are installed flush against the end bars and the bottom bar, the resulting comb will end up being secured fairly well to both the end bars and the bottom bar, and the final comb will be stronger than when a horizontal starter strip of foundation is used.

While we're on the subject of frames, a mention should be made to let the uninitiated in on a way to help keep the top bars and end bars of a frame from being pulled apart while trying to pry it loose from a hive. The trick is part of frame assembly: after the standard two nails are hammered through the top bar down into the end bar, a third nail should be put through the end bar and into the top bar from below. This third nail is key to preventing the top and end bars from becoming easily separated from each other. Another method for nailing frames,

FIGURE 3-6. When using starter strips in deep frames for supers and hive bodies above the first hive body, better results can be achieved by using two vertical strips rather than a single horizontal strip.

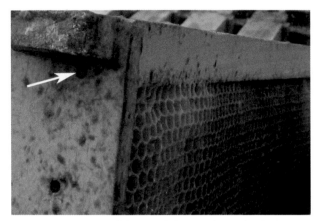

FIGURE 3-7. Frames will last much longer if we take the time to add this one strategically placed nail.

FIGURE 3-8. As long as the grain in the wood is clear and straight, time and nails can be saved by nailing through the end bars into the top bar as shown. Be sure the joints are fitted tightly together before you set your nails.

promoted by New York beekeeper Chris Harp, is to nail the end bars to the top bar with ¾- or ⅞-inch nails (see figure 3-8). The two nails placed through the sides of the end bar replace both nails that normally go down through the top bar, as well as the third nail that goes through the side of the end bar into the end of the top bar. This method works well and saves time and nails as long as the grain in the wood is straight and clear; otherwise the wood has a tendency to split when nailed.

Due to its simplicity, the inner cover is perhaps the easiest piece of equipment for the do-it-yourselfer to build. Unlike the majority of commercially available inner covers, which feature a raised lip around the edge on both sides and a single, oval-shaped hole in the center of the cover, I prefer covers that have a lip on one side, are flat on the other side, and feature two holes for ventilation and feeding. By positioning the cover on top of the hive with the flat side down, the colony's bee space is maintained, and this typically results in less buildup of burr comb in the space between the top of the frames and the bottom of the inner cover. Because I prefer to use an escape board rather than a bee escape when removing the harvestable honey from the hive (as explained in chapter 10 on page 222), a single oval-shaped opening in the inner cover is unnecessary for me. Instead, I like to drill two 1½- to 1¾-inch holes in the inner cover. Not only does this provide increased ventilation through the cover, but it allows me the option of being able to feed a hive two buckets of feed at once on those occasions when feeding may be necessary.

Many manufacturers of beekeeping equipment have taken to fashioning their inner covers with a notch cut out of the rim on one end. If the outer

FIGURE 3-9. Whether it is a hole drilled into the side of your hive body or a notch cut out of the rim of the inner cover, an upper entrance hole can improve hive ventilation.

FIGURE 3-10. Homemade hive equipment can be modified to more closely fit in with your particular beekeeping management style. This inner cover can accommodate two feed buckets at once.

cover does not fit too tightly when flat and there is room to slide it forward on the hive so there is sufficient space for the bees to come and go, the notch can be used by the bees as an upper entrance, as well as for ventilation.

It's ideal, though not critical, to keep the notched side of the inner cover in the up position during the summer and in the down position for the winter. Paying attention to this detail allows the airflow to pass through the middle of the hive and the brood nest and then through the hole in the middle of the inner cover during the summer, maximizing cooling while venting the hive. In the winter, with the notched side of the inner cover facing down and the outer cover tightly covering the hive, air cannot flow through the hole in the middle of the inner cover and instead is drawn along the inside wall of the hive as it travels between the bottom board and the inner cover. This avoids creating a draft through the center of the cluster during the cooler months, which helps the colony conserve heat and maintain proper brood-nest temperatures during winter.

• TOP BAR HIVES •

Since the release of the first edition of this book, the popularity of the top bar hive (TBH) has grown significantly in America, along with general interest in beekeeping. While most top bar beekeepers are part-time backyarders, there are a few human-scale commercial top bar operations (typically consisting of about 100 to 300 hives) scattered across the United States.

Many claims are being made, both for and against the TBH. Some top bar beekeepers tout the benefits of the TBH while denigrating the Langstroth hive. To listen to such folks, you would think that beekeepers who use Langstroth hives are damaging the bees' ability to be healthy and are behind the current colony collapse disorder (CCD) epidemic—whereas TBH beekeepers are the world's answer to the declining health of the honey bee. Then there are keepers of Langstroth hives who deride the TBH as a fad that may increase the likelihood of colony starvation over winter and that such hives are a waste of time. Can both these extreme views be accurate? Not totally: let's look at some of the benefits and challenges of the TBH in more detail and see what's what.

Unlike a Langstroth hive, the TBH has a horizontal rather than vertical orientation—it looks like a long, rectangular box with a lid. The earliest documentation of top bar hives, dated 1678, describes Greek hives that were round.[3] The shape of today's top bar hive is most often rectangular or trapezoidal with the sides angled, in the style of the Kenyan top bar hive (KTBH). The KTBH was developed by a research team from Guelph, Ontario, in the 1970s to provide movable frames for the traditional Kenyan hollow log hives (often

FIGURE 3-11. The top bar hive comes in various shapes and sizes, from inexpensive models you can build yourself with scrap lumber to top-of-the-line models like the one pictured here, which sells for $300.

tied to a horizontal tree branch) in an effort to allow for a transition to Langstroth-style hives. The other primary feature of the TBH is that the combs are not contained within a frame. Instead, the beekeeper places a series of top bars into the hive (up to 40 or more in a full hive), and the bees are allowed to build their comb freely by simply attaching it to the top bars without the aid of end bars or bottom bars (see figure 3-14 on page 51). The comb is suspended freely within the open space of the hive. The angled sides of the hive body decrease the tendency for the colony to attach the comb to the side of the hive.

Despite claims to the contrary, I am not aware of any evidence that the elongated horizontal orientation of the TBH is a more natural hive shape and preferred by the bees. My own observation is that European honey bees seem to prefer to work vertically rather than horizontally. Whenever I have seen a swarm settle in an open area where comb building was not limited in any direction, the bees have tended to build 6 to 12 long combs rather than the 30 to 40 short combs typically

found in a top bar hive. The TBH shape may be perfect for the African or Africanized honey bee given its habits and the climate it lives in, but I am not convinced that it is the best cavity shape for the European honey bee in North America or Europe. The vertical orientation of the Langstroth hive mimics the shape of a standing hollow tree, which is the natural cavity that feral European honey bees colonies most often call home, better than a TBH does. That said, honey bees are incredibly resilient and can adapt to whatever cavity shape they decide to use as a residence.

The TBH needs to be level in order for the bees to build straight free-form combs, just as when allowing bees to build natural comb without foundation in a Langstroth hive. Honey production will be limited when combs are regularly removed from the hive and the bees are forced to rebuild them, whether using a Langstroth or top bar hive.

The top bar hive certainly offers benefits that are not available with the Langstroth design. A TBH is very simple to build, requiring a minimum of materials and tools. This is part of what makes this hive ideal for rural Africa, where it is widely used. The KTBH design requires a minimum of lumber, a scarce resource in Africa, and can be made from many naturally occurring materials—sticks and the like are common. The tools typically available to a village carpenter in Africa (if they are lucky) are a handsaw and hammer, and these two tools are sufficient for constructing a TBH. The simple design also reduces breakage and makes repairs much easier should they be required.

The ease of construction and simplicity of design also help make the TBH inexpensive to build compared to the conventional Langstroth design. While one can certainly spend several hundred dollars on a fancy, well-built top bar hive that comes complete with a viewing window and stand, it's also possible to build a basic, no-frills TBH with scrap lumber for well under $20. This is a big savings over conventional Langstroth hive equipment, which typically costs $100 to $200 depending on the supplier and the number of honey supers purchased.

The traditional TBH design utilizes top bars that fit snugly against one another, unlike the Langstroth hive that features a bee space between each frame's top bar. The top bars not only provide an anchor for the bees to build their comb, but it also provides a kind of roof underneath the outer cover that is usually made of metal or wood. Since there are no spaces between the bars, the only part of the top bar hive that is exposed and disturbed during hive inspections after the outer cover is removed is the comb being inspected and the combs adjacent to it. The reduction in colony disturbance can reduce stress on the hive that may result during an inspection, as well as the stress on the beekeeper. By limiting the disturbance of the hive to the comb bars that are being inspected and the ones next to it, the number of bees that are likely to become defensive is greatly reduced, which is a huge advantage when working with aggressive Africanized genetic stock! While this advantage isn't as great in TBHs that house less aggressive European honey bees, it is still beneficial and can

allow some top bar beekeepers to work their hives without using a veil, smoker, or sugar syrup spray.

Without the need for numerous honey supers, the TBH reduces the amount of equipment that must be stored, making this hive design ideal for urban beekeepers, who often have limited storage space available in their apartment buildings. The unusual shape of the TBH also makes the hive less recognizable as a bee hive, which can be a huge benefit in highly populated areas where the need to maintain good neighbor relations is greatly multiplied and there are likely people living nearby who fear bees or insects. The low profile of the TBH can also help decrease the incidence of vandalism and theft. The top bar hive's design relieves the beekeeper from having to worry about damage from mice, because these hives tend to be mouse-proof by design.

Being relatively new to common use, TBH design has not yet been standardized. Top bar hives are being constructed with varying depths and lengths. Some have entrances at one end of the hive, while others feature an entrance located in the middle of the hive. Some entrances are positioned near the top of the hive, and some are placed at the bottom of the hive. Top bars vary in width as well, and some designs include space between frames as in the Langstroth hive, or even spacers placed between the bars to accommodate combs of varying widths. Sometimes a top bar hive design will include an observation window that allows the bees inside to be seen without having to open up the hive. All these options make building the TBH a lot more fun and much more forgiving when compared to the exactness of dimensions required when constructing a Langstroth hive based upon bee space. It is also easier to position the TBH at waist height, which makes the hive easier to inspect and manipulate, simply by designing an appropriate hive stand.

Some other benefits often promoted by top bar enthusiasts are not exclusive to the TBH, because they can also be accomplished with standard Langstroth hives. Such benefits include allowing the bees to build their combs naturally while

FIGURE 3-12. Follower boards like these are used to adjust the size of the hive's cavity to match the population of bees. The follower board on the left, which has a hole drilled through it, is used when a feeder is placed inside the hive, so that the bees can gain access to the feed.

FIGURE 3-13. An empty jar feeder has been placed in the end of this brand new and unoccupied top bar hive with a follower board positioned between the feeder and the rest of the hive. Note the hole in the follower board that allows bees to access the feeder.

limiting the use of foundation, which can be done in a Langstroth hive by using narrow strips or going without foundation (as described in the preceding section, "Additional Equipment Details").

Another benefit cited is the removal and replacement of comb on a regular schedule to help reduce pathogen and chemical buildup in the beeswax. Traditionally top bar beekeepers cut honeycomb out of the hive at harvest time and either use it unprocessed as comb honey or crush the comb in order to separate the honey from the wax. Some beekeepers, however, are working on developing an extractor in which combs from a TBH can be placed in a horizontal position for spinning in order to preserve the comb for reuse. This would increase the honey yield from the TBH, because the bees would not be forced to consume additional honey in order to build new comb, but it would also eliminate the benefit of regular removal and replacement of old comb as a built-in feature of top bar hive management.

The process of producing free-form comb in a TBH allows the bees to build cells of any size and shape they like. (This can occur in a Langstroth hive as well if you construct it with thin strips of wax or foundation to help guide the bees' efforts, as previously noted.) Rather than using strips of foundation, many top bar beekeepers utilize the shape of the underside of the top bar to help guide the hive's comb-building activities. This may be done by building the top bars so that the underside of each bar is wedge-shaped, forming a narrow edge on which the bees can attach comb. Other beekeepers may insert a popsicle stick or tongue depressor in the top bar groove rather than using foundation. Sometimes a beekeeper will coat the bottom edge of the top bar (or popsicle stick) with a thin layer of beeswax to further encourage the bees to build comb in the desired place and in the desired shape. These techniques can also be used in a Langstroth-style hive. Unfortunately, relying solely on the shape of the top bar or replacing the foundation with a tongue depressor to guide comb building is not as consistently reliable as using full sheets of beeswax foundation or even 1- to 2-inch starter strips of foundation. It is especially tricky in a top bar hive, because the bees will want to build their comb across the underside of the bars, connecting one bar to another, if given too much room to build. To reduce this problem in TBHs, one can use a follower board which takes the place of a top bar and divides the hive, restricting the colony's access to the part of the hive on only one side of the follower board. Use the follower board to limit the number of top bars the bees can work to no more than eight or ten initially. Once the hive has

FIGURE 3-14. The comb from a top bar hive, naturally built without the aid of foundation, will take on the shape of the hive cavity.

established comb, adding additional space by moving the follower board back four to six bars at a time will go far in limiting cross-comb production.

Whenever a colony is able to build comb free-form without the aid of foundation, the bees will tend to build significantly more drone comb than is typical in a Langstroth hive filled with worker foundation. The instinct to raise drones is so strong in hives that when bees are allowed to build comb naturally as much as 30 percent of the comb may end up as larger drone-sized cells. On average, though, a healthy colony devotes about 15 percent of its brood-rearing area to the production of drone comb. For more about drones, see chapter 5.

Naturally built comb in a TBH will start out in a round pattern similar to the shape of combs built by bees in the wild. Once the comb is fully drawn out, it ends up taking on the shape of the cavity within the hive. In a KTBH, this will result in a comb with a flat bottom and two slanted sides. This shape is closer to the natural oval shape (known as a catenary curve or "hanging chain") that bees would normally construct when building comb that is not contained by the sides of a hollow cavity. It is not, however, all that different from the rectangular shape of a Langstroth frame.

The TBH is said to help save the beekeeper's back because the single frames of comb are manipulated individually—no need to lift boxes full of heavy frames. Nevertheless, a number of older beekeepers who are not able to lift full supers and hive bodies manipulate Langstroth hives by removing one frame at a time from the hives, often placing them into empty boxes. A slower process, but very doable.

FIGURE 3-15. An example of a Kenyan top bar hive with sloping sides and permanently mounted legs. This hive was in the process of being inspected—that is why some of the frames are separated and there are bees on top of the top bars. An inner cover from a Langstroth hive is sitting on top of the rear of the hive.

Since manipulation and inspection of the TBH must take place one frame at a time and is a slower process compared to working with a Langstroth hive, many believe that this translates into better care of the bees by the beekeeper, who is likely to be more observant and less likely to crush and injure bees during forays into the hive. My own observations, however, are that there is nothing to stop a beekeeper from inspecting a Langstroth hive in a slow and methodical manner, and that a slower process does *not* necessarily translate into better care of the bees.

TBH enthusiasts also contend that the TBH design forces regular hive inspections (which is a good management practice for any hive) because of the need to regularly harvest honey to provide space for expansion and help prevent swarming. However, it is possible to remove frames of honey without bothering to check the brood area in a TBH hive. And, similar to the discussion above about pacing of inspections, a well-organized self-motivated individual can certainly make regular hive inspections of a Langstroth hive a part of their schedule.

Perhaps the biggest attraction of the top bar hive is that it is an alternative to the industrial model of beekeeping. The TBH represents a clear break from this industrial model. Nevertheless, there are many beekeepers, myself included, who choose to keep bees in Langstroth hives but are decidedly not following the industrial model of beekeeping.

Managing bees in top bar hives presents some special challenges. For example, due to the lack of end bars and a bottom bar support, handling top bar comb without breaking it is tricky. The key to working with top bar combs is to *always* keep them vertical. This is not necessarily a drawback unless you are used to handling Langstroth frames and have to unlearn old habits. However, I do not recommend that beginners start their beekeeping career with a top bar hive. Generally it is best to have some experience under your belt before trying to keep bees in a TBH. Hive management activities such as feeding or treating for mites can be more challenging for the TBH beekeeper because there is typically no bee space above the combs in a top bar hive. I discuss some of the feeding options for

FIGURE 3-16. The honey bee is incredibly versatile and resilient. A hive of bees can live in almost any shape of cavity as long as it is big enough to hold enough food for winter, protected from the weather, and easy to defend from predators and robbers.

TBHs later in this chapter; in chapter 5, I review some of the mite control techniques used by top bar beekeepers. It is, however, a good idea to design the cover on a top bar hive so that it sits an inch or two above the top bars. This insulating air space will help keep the hive cooler in summer, a serious concern in southern and desert regions.

Due to the limited amount of room for honey storage, bees that are conservative in their honey use while overwintering are more desirable as TBH residents; otherwise additional feeding may be needed. As a result, the often favored Italian honey bee (*Apis mellifera ligustica*) is not a good candidate for the TBH unless you already have some experience managing colonies in this type of hive. In addition, the limited size of the cavity within a TBH means that they require closer management if the swarming instinct is to be slowed down by adding top bars that provide additional room for colony expansion, both between the entrance and the brood nest as well as in the honey storage area.

As winter approaches in northern climates like Vermont, it is important for the brood area to be located at one end of the TBH with honey stores laid out toward the opposite end, just as significant stores of honey should be above the brood nest in a Langstroth hive.

One should also be aware that some of the potential benefits of the TBH may also become liabilities. A major difference in managing a TBH is the need to visit the hive regularly and continually harvest honey to provide more space for the bees. Because this is not always convenient or possible, the colony housed in a TBH may be more prone to neglect than those kept in Langstroth hives. Since it takes more time to inspect a TBH compared to a Langstroth hive, it may be more of a challenge for commercial and sideline beekeepers who utilize this style of hive. There is also the issue of having to clean harvesting and extracting equipment a lot more often when colonies are located in areas with abundant nectar flows, unless combs filled with honey that are removed from the hive are used to create nucleus colonies or splits (see "Creating the Split or Nucleus Colony" in chapter 4 on page 99).

In addition, the lack of standardization, while allowing wonderful creativity in design, can result in problems. I have encountered TBHs built so that the top bars rested on wooden cleats attached to the sides of the hive about an inch below the top of the hive sides. This made it difficult to slide a tool down the inside of the hive to cut combs free from the sides of the hive and remove them without damaging the comb. As mentioned above, TBHs with slanted sides and smooth inside walls are less prone to having combs attached to the side walls. Since side-wall comb attachment typically occurs in the upper corners of the comb, using starter strips that are a couple inches short of the ends of the top bars can help alleviate this problem as well. In Langstroth hives, bees also attach their combs to the sides of the hive, but since there are only the two outer frames that are adjacent to the sides of the hive, this problem occurs much less frequently.

Besides having the top bars rest on top of the hive's sides, other standardizations that make sense to me and should be considered during construction include:

- Making all top bar hives the same width. My suggestion would be to use the same top bar dimension as the standard Langstroth frame (19 inches) so that frames could not only be moved easily between all TBHs but also between TBHs and Langstroth hives and vice versa. This would be a big help when introducing a commercially purchased nucleus colony into a TBH.
- Making all TBHs the same depth. The ability to move frames easily between top bar hives would allow for nuc making, equalizing hives, and boosting weak or queenless hives with brood and eggs. While the deeper the frame the better, I would think that a minimum of 10 to 12 inches would be adequate for most climates.
- Entrances should be located at one end or the other of the hive and closer to the bottom of the hive, which is where bees prefer to enter a hive, rather than near the top.[4]
- Top bars should fit tightly together, with no spaces between bars.
- Unless they are being used to make nucleus colonies, hives should be at least 3 feet long in southern climes and 5 feet long in northern states.
- It would be handy to have Kenyan-style rails protruding from the ends of the hive to provide convenient handles for moving the hive and a place to hang frames that have been removed during inspections.
- A screened bottom to facilitate ventilation and help reduce mite loads would be helpful.
- A removable stand could be incorporated to help make the hive easy to relocate.
- A spacer between the outer cover and the top bars to provide an insulating air space during summer when temperatures are high.

I would love to see a group of committed TBH suppliers and users work together to come up with a top bar hive design that everyone could agree on as the best all-around design for everyone to use. But the nature of beekeepers makes this unlikely: trying to get two or more beekeepers to agree on the solution to any beekeeping question tends to be problematic at best.

To sum things up, from my perspective the top bar hive is not a whole lot better or worse than the Langstroth hive. Under certain conditions, various races of honey bees and types of beekeepers may do better with one or the other. The TBH is simply another way to keep bees. It may work better for you given your situation, or it may not. In the end folks who vilify one style of hive or the other are not doing anyone a favor. What is important is the way in which the bees are being cared for, not the type of hive they happen to be housed in. I recommend that you choose the style of hive that makes you feel most comfortable so you will most fully enjoy your relationship with your bees. With a few exceptions, most beekeepers are interested

FIGURE 3-17. Using top bars that are the same size as the standard Langstroth frame allows the beekeeper to more easily move frames of bees, brood, or food between the different types of hives when needed.

in doing everything they can to support the health of their bees, whether they use top bar hives or Langstroth hives. It is this common ground that we need to nurture while supporting one another in the process, because no one has all the answers.

• FEEDING •

All living organisms, including honey bees, need high-quality nourishment to reach their maximum potential. As with all living things, insects that are well fed are less susceptible to disease than malnourished insects. Pure honey provides the bee with calories for energy, while pollen provides the protein and minerals required for optimum health.

If the bees do not have access to sufficient quantities of honey of their own making, honey is the preferred nourishment for feeding them, with sealed combs of honey taking precedence.

Extracted honey *from a disease-free* source comes in a close second, though it can be hard to sacrifice all the hard work that went into harvesting, extracting, and packing the honey only to give it back to the bees. It is far better to leave the honey in the comb and not take honey bees may need than to have to feed harvested honey back to a hive (see "Honey Inventories" in chapter 10 on page 228). The best foods to feed a honey bee hive are unheated honey and bee bread. Organic beekeepers should note that organic honey standards dictate that only certified organic honey and/or certified organic sugar be used as supplemental feed.

When the source of honey for feeding bees is in doubt, I used to consider it wise to mix the honey with a little water and boil it for about two hours in order to kill any disease spores that might be present. This is how we fed bees during the six years I worked at Champlain Valley Apiaries. The

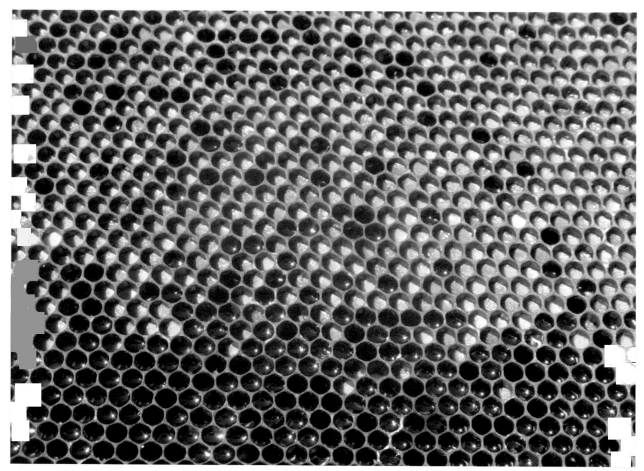

FIGURE 3-18. To create bee bread, bees fill cells three-quarters full with pollen and allow it to ferment. The bees sometimes store this highly nutritious bee food for later use by filling the final quarter of each cell with honey and then capping the cells with wax.

scrapings from the top of honey barrels (which contained bits of wax, pollen, and propolis) would be collected and liquefied by pumping steam directly into the honey to form a syrup. The bees seemed to do fine on this feed; however, I have since learned more about honey and its reaction to heat.

Hydroxymethylfurfural (HMF) is a compound formed when fructose degrades after being exposed to heat while in the presence of an acid. In general, the warmer the temperature, the greater the production of HMF. The presence of HMF in honey has been shown to cause ulceration of the honey bee gut, leading to dysentery issues and premature death.[5]

Fructose and acids are naturally present in honey, so the production of HMF is always taking place in honey and accelerates when honey is heated. The level of hydroxymethylfurfural in honey is used as a gauge to determine how old a sample of honey is and whether it has been exposed to heat either during processing or while in storage. The international tolerance standard for HMF in honey is 40 mg/kg (or 4 mg/100 g) which can be reached after 230 days at 68°F (20°C).[6] This standard is enforced to help ensure honey's quality and potentially to address concerns about possible human health risks from HMF.[7] The ease of HMF formation in honey depends upon the botanical origins of the honey, with locust, fir, and chestnut honey being among those most resistant to HMF buildup.[8] In general, honey heated to approximately 122°F (50°C) experiences a relatively slow increase in HMF. Honey has a high increase of HMF when heated up to about 144°F (62°C), and honey becomes seriously contaminated with excess HMF when exposed to temperatures of 180°F (82°C) and above.[9]

Since the heating of honey is standard practice during honey harvesting and processing, the formation of HMF is something that the beekeeping industry should take a long hard look at. When hot, honey thins out and flows more easily through pumps and filters during processing. In addition, heating and filtering delays honey's natural crystallization process. Unfortunately, heat also tends to change the color of honey. The flavor of honey is affected by heating, and as explained above, heat degrades the quality of honey through the increased formation of HMF. Considering the growing evidence that HMF is harmful to bees and beekeepers, folks who are concerned with maximizing the quality of their honey will modify their operations in order to use as little heat as possible.

This also means it is more desirable to use honey mainly in recipes that call for little or no heating, such as salad dressings, dips, spreads, and toppings. When cooking with honey at temperatures over 120°F (49°C), it is best to use recipes that involve diluting the honey (for example, coffee or tea), because significant dilution of the fructose and acids in honey will prevent the formation of HMF.

When honey is not available for feeding bees, many beekeepers unfortunately opt to use high-fructose corn syrup (HFCS). Sweeter and less expensive than sugar, high-fructose corn syrup is responsible for one of the largest changes in the diets of both the average American human and the average US-dwelling European honey bee over the last 40 years. HFCS has the potential to harm bees because it contains two types of sugar that are mildly toxic to honey bees: stachyose and farinose.[10] And when HFCS is heated, HMF is also produced, with concentrations jumping dramatically at temperatures of 120°F (49°C) and higher.[11]

To make matters worse, most of the corn grown in the United States today is genetically modified to produce a Bt (*Bacillus thuringiensis*) toxin to protect the corn from corn borers and other insects. This pesticide is produced in every cell in every part of the corn plant. As a result this poison also ends up in the final corn-based products that are consumed, including HFCS. As if all this weren't enough, studies have found toxic levels of mercury in almost half the samples of HFCS tested, and in about a third of the food products studied which contained corn syrup as an ingredient.[12] The most likely sources of the mercury contamination are mercury-containing hydrochloric acid and caustic soda, both of which may be used in the production of HFCS. All of this does not reflect well on the use of high-fructose corn syrup, for bees or humans.

The US Corn Refiners Association (CRA) compares high-fructose corn syrup to honey with the statement "the saccharide composition (glucose to fructose ratio) of HFCS is approximately the same as that of honey, inverted sugar, and the disaccharide sucrose (table sugar)."[13] This statement may be true with regard to honey, depending on the type of corn syrup you are referring to, since HFCS is available in three different formulations. HFCS containing 42 percent fructose is used primarily in processed, packaged, and baked goods. HFCS containing 55 percent fructose is used by soft drink manufacturers. Finally, an extremely sweet HFCS containing 90 percent fructose is used in low-calorie "diet" products. Honey, on the other hand, tends to be composed of a mixture of primarily fructose and glucose. Given that the National Honey Board lists the fructose range of honey as between 30.91 and 44.26 percent, the comparison of HFCS to honey may be valid, but only between certain types of honey and corn syrup containing 42 percent fructose.[14]

The Corn Refiners Association also attempts to refute the negative studies and reports on HFCS and hydroxymethylfurfural. They question the quality and accuracy of the studies that point to potential human or honey bee health issues and cite other studies that seem to reach conflicting conclusions with regard to the effects of HFCS. They respond to the mercury contamination issue with misleading statements such as, "Our industry has used mercury-free versions of the two reagents mentioned . . . , hydrochloric acid and caustic soda, for several years," without referring to the fact that not all members of the industry have made the switch to using the mercury-free processing agents. The CRA will also point to Food and Drug Administration (FDA) and Environmental Protection Agency (EPA) approval for genetically modified corn as proof that products made from GM corn is safe for human and animal consumption.

Corn processors would like the public to believe that the fructose in HFCS is the same as the fructose found in natural foods like fruit and honey. However, most of the fructose found in fruit and honey is in the form of L-fructose or levulose; the fructose in HFCS is D-fructose, which has a slightly different chemical structure. Fresh fruits can contain small amounts of D-fructose, but "the D-fructose in HFCS has the reversed isomerization and polarity of a refined fructose molecule."[15]

All in all, the industry response to the growing concerns over high-fructose corn syrup is eerily similar to the tobacco industry's efforts that deceived consumers into believing that cigarettes were safe, and in some cases even healthful, to smoke. The evidence, however, seems to indicate that a prudent approach would be to avoid honey bee or human consumption of HFCS in all its forms. As individuals with free will, we can make such choices for ourselves. Unfortunately, the honey bees in our care do not get to make an informed choice. When we feed HFCS to them, they will doubtless consume it.

If you do decide to feed HFCS to your bees, be sure to purchase syrup that is produced using the enzyme hydrolysis process, which tends to result in less HMF ending up in the syrup and avoids the opportunity for mercury contamination— as opposed to acid-hydrolyzed inverted sugars. HFCS purchased as bee feed should be used up as soon as possible and stored at temperatures well below 120°F (49°C) in order to limit the buildup of HMF that occurs with time and temperature. And

be aware that numerous cases of "bad batches" have been reported by beekeepers who purchased HFCS that had been sitting in a tanker truck out in the sun too long.

A much better alternative to HFCS if unheated honey (preferably in the comb) is unavailable for feeding bees is sugar syrup. It is the next-best thing to unheated honey. Only white sugar should be used and brown sugar must be avoided because it contains material that bees cannot digest. If you must use raw or brown sugar, and I don't recommend it, do so only in the spring or summer. This allows the bees to make cleansing flights in order to void the indigestible waste—and they are less likely to store the feed for winter use.

In addition, the sugar used should be cane sugar. Estimates are that about 90 percent of non-cane sugar is manufactured from genetically modified sugar beets. This sugar has hit the market illegally, because courts have ruled that a proper environmental review of this new life form has not been completed.[16] Recent studies have shown that genetically modified organisms (GMOs) have sublethal effects on honey bees, from both the toxins that they are designed to produce and because genetically altered material can transfer directly from the GMO into the bacteria that line the honey bee's stomach (see chapter 8, "Biological Pollution" on page 188). Therefore, it is best to avoid sugar made from beets in order to help reduce any unintended stress that may be produced as a side effect from feeding. Unlike crops developed through traditional plant breeding and hybridization, genetically modified (GM) crops incorporate genes from completely different species (e.g., viruses, bacteria, animals, etc.) into the plant's cells. Due to the fact that these types of changes would never occur naturally and there are a growing number of studies indicating potential harm to organisms (and people) who consume GM foods, organic standards ban these products from use. Only sugar made from sugar cane and labeled as "cane sugar" should be used by folks who want to avoid GM sugar. Certified organic beekeepers must use organic cane sugar.

Honey bees are meant to consume honey, and sugar water is a poor substitute. The white sugar used to make sugar syrup to feed bees is 100 percent carbohydrate. Sugar syrup made from table sugar and water does not contain the trace minerals, vitamins, and enzymes found in honey. The water used to make syrup may have some minerals in it, but the sugar itself is composed of "empty calories" that nutritionists warn us about. It can be compared to junk food. Many people who eat at fast-food restaurants enjoy a meal that tastes great, and they can still make it to the door to live another day. Nevertheless, as the documentary *Super Size Me* makes abundantly clear, if a person eats such food for breakfast, lunch, and dinner, seven days a week, over a period of three or four weeks, the body becomes severely stressed, and the person's health is compromised to the point that they may die if they don't change their diet fast. This is similar to what can happen to bees that are constantly fed an artificial diet and deprived of real honey and pollen from a wide variety of plants.

One way to help reduce the negative impact of artificial feeding on the health of the hive is to follow suggestions made by the man widely viewed as the father of biodynamic agriculture, Rudolf Steiner (1861–1925), and feed the colony bee tea.[17] I only like to use this tea mixture when the bees are in danger of starvation, which can occur when a hive's honey stores have been robbed out, during times of prolonged dearth such as during winter (if the hive does not have enough honey stored away), or when a colony is shaken onto foundation (such as when hiving a package of bees) and there is no honey stored in the hive.

Unlike feeding plain sugar syrup or HFCS, feeding this tea to the bees will not only help prevent starvation, but I believe it can help boost the colony's immune system when compared to plain sugar syrup or HFCS. Commercial beekeeping outfits are unlikely to spend the extra money and time to feed their bees a tea, but for small-scale and backyard beekeepers it may be a viable option. Conventional feeding recipes call for a 1:1 sugar-to-water mix in spring to stimulate brood rearing, and a 2:1 ratio in autumn when the bees have only a short period of time to convert the feed into honey stores before cold weather sets in. I always use a 2:1 ratio when I feed bees because I feed only when they are in danger of starving, and I want to get as much food into the hive as fast as possible. I do not use the tea to try to stimulate brood rearing in order to pump up the bees artificially and maximize honey production. When possible it is best to plan ahead and feed in late summer and autumn and avoid having to feed the bees during the dead of winter because when temperatures are low, the bees are not always able to get to the feed. Also, in wintertime, the moisture from the tea is difficult to evaporate and may add significant stress to the hive.

Numerous contraptions are available to feed syrup or bee tea to hives. They are all variations on entrance feeders, top hive feeders, division board feeders, and feeder pails. As their name implies, entrance feeders (aka Boardman feeders) are placed at the hive entrance. Entrance feeders are easy to use, but they require the workers to travel far from the brood nest, and therefore they lose their effectiveness when the temperature drops into the 50s or below, because at that temperature worker mobility is reduced. Such feeders may also encourage robbing, where other bees or insects sneak into the hive intent on taking food from the colony. Entrance feeders do, however, allow for easy feed-level checks and refilling compared to other feeder designs, which is a good thing, given that this style of feeder doesn't usually hold a very large amount of syrup. This is an important consideration that makes the entrance feeder the least desirable option in my opinion, because one should think in gallons when feeding bees.

Top hive feeders are shaped like a super and placed above the brood nest underneath the inner cover. Although the bees still have to travel a significant distance to reach the food, heat rising from the brood area facilitates collection activities when outdoor air temperatures dip to levels where an entrance

BEE TEA

All sugar and tea should be from a certified organic source whenever possible. If not organic, use pure *cane* sugar only to avoid putting genetically modified materials into your hive.

Ingredients
(makes about a gallon):

16 cups white cane sugar

6 cups hot tap water

fresh or dried chamomile and/or
   thyme—enough for 2 cups of tea

1 teaspoon natural sea salt with minerals

Directions:

1. Combine sugar and salt.
2. Add 6 cups hot tap water to the mixture.
3. Stir thoroughly (don't boil sugar; it may caramelize).
4. Boil 2 cups water and add herbs. Let steep, covered, for at least 10 minutes.
5. Strain and mix herbal tea with sugar-salt solution.
6. Mix thoroughly until all sugar is dissolved, and fill feeders.
7. Allow the tea to cool before feeding it to bees.
8. Refrigerate unused bee tea.

Observe how fast the bees are consuming the tea. If they are not using it up fairly rapidly (at least 1 pint every two to three days), then they probably do not need it (or it may be too cold for the bees to feed, or the bees may have a bad case of nosema and thus aren't eating). Only feed the bees when they need it and will take it. The bees should not need to be fed if there is capped honey stored in the hive and daytime temperatures are warm enough for foraging and there is an abundance of blossoming plants available to forage on. (The exceptions being when the blossoms are not producing nectar or cold weather is approaching and the hive has not drawn out their foundation and filled up the combs with enough honey to get them through the winter.)

Do not feed bees stored tea that has started to ferment. Adding the essential oils mentioned below will help prevent fermentation and mold growth. After each use, wash the containers you use for feeding your bees.

Suggested optional formula: Use emulsified lemongrass/spearmint essential oil mix, such as Pro Health manufactured by Mann Lake Ltd. or Honey-B-Healthy available through independent distributors (see resources on page 273), as a feeding stimulant/immune booster and to help prevent fermentation and mold growth in bee tea. Add 1 teaspoon per quart (4 teaspoons when following the recipe above). Both Mann Lake's Pro Health and Honey-B-Healthy are now available formulated without sodium lauryl sulfate (SLS) or genetically modified ingredients, but you may have to specifically ask for this formula.

Note: Like all recipes, this one can be played with. Some beekeepers like to replace a larger percentage of the water with tea, for example. If you decide to add honey to your bee tea, do so after the temperature has dropped to lukewarm and stir the honey in well. Be absolutely sure that your honey source is free from American foulbrood disease (see chapter 9).

feeder would not be adequate. Top hive feeders have the potential to hold the greatest amount of syrup of all the feeder types, reducing return visits by the beekeeper to refill the feeder. In the past the design of top hive feeders allowed many bees to drown in the feed; however, improvements have been made to most models that eliminate this problem.

The division board feeder replaces a frame, or two, in or next to the brood nest. This feeder offers the benefit of locating feed right by the brood area so bees don't have to travel far to reach it. Small pieces of wood floating in the syrup provide a life-saving platform that will allow the overeager bee to drag herself out of the sea of sugar. The biggest drawback to using a division board feeder is the extra work involved with checking the feed level and refilling the feeder. It also reduces the amount of room within the hive that can be used for food storage and egg laying, because normal frames of comb are removed to make room for the feeding apparatus.

Pails, cans, or jars that have a lid perforated with small holes can be filled and placed upside-down over the hole in the inner cover to allow the bees to suck down the syrup. An empty hive body placed on the hive with the outer cover on top will protect the feeder from the elements and prevent robbing. Feeder pails typically come in gallon and half-gallon sizes. This approach offers ease of access for both the beekeeper and the bees, because the position of the hole in the inner cover is typically right above the brood nest. The biggest problem with feeder pails is the potential for sugar syrup to drip down and drench the bees at a time when cool weather prevails. Turning the pail upside down and letting the liquid run out until a vacuum is formed, thus stopping the syrup from flowing, before placing it on the hive will help avoid soaking the bees.

Another critical feeding detail is to be sure the colony cluster is centered directly under the inner cover hole where the pail, can, or jar is inverted. I have seen weak hives that required spring feeding starve with a half-gallon of feed on the hive, all because I failed to notice that the cluster was located off to one side of the hive. When temperatures

FIGURE 3-19. Ensuring that hives are full of honey as they head into the winter months is one of the beekeeper's primary responsibilities.

dropped and the bees had to tighten up their cluster to keep warm, they were unable to reach the feed bucket and its life-giving calories because the hole in the inner cover was a couple inches to the left of, instead of directly above, the bees.

Some folks like to fill an empty frame of drawn comb with sugar syrup and place it in the hive. This is best accomplished in the bee yard to reduce spillage during transportation. A paintbrush dragged across the face of the comb will help to break the surface tension and allow the syrup to collect in the cells.

Plain dry table sugar can also be used, but only as an emergency measure, such as during extremely cold weather, when it is not desirable to open the hive to install or refill a feeder. When sugar is sprinkled around the hole in the inner cover, the bees will be able to use moisture to dissolve the crystals and consume the sugar. The use of sugar candy (fondant) or queen candy for emergency feeding has also reportedly met with some success. Especially during winter, when the addition of significant amounts of moisture to the hive in the form of syrup may increase stress on the bees, feeding is typically done by mixing syrup made of sugar and water with a small amount of tartaric acid, which inverts the sugar. This inverted sugar is then combined with powdered sugar until it is transformed into a stiff dough. Queen candy is even easier to make. Simply combine honey from a disease-free source with powdered confectioners' sugar until it has a stiff, doughlike consistency, then flatten it out into a patty about ¼-inch thick and place it directly on the top bars over the brood nest.

Leaving a bulk container of bee feed sitting open in the bee yard with a cover over it to keep out the rain is the simplest method of feeding bees. Bulk feeding in this manner takes the least amount of work, but it is also the least efficient. The strongest hives will typically benefit the most from such general feedings, while the weaker hives that need the food most are least able to fatten up on such an offering. Straw, pine needles, or bits of wood must be added to the container to provide a life raft for the feeding bees. This method can also end up providing food for a host of other insects in the area, and it may also encourage robbing behavior among honey bee colonies.

Bees should be fed only when starvation is imminent. This can occur when, due to weather conditions or some other factor, the colony has not stored sufficient honey to last the bees through the winter, or because their keeper has taken more than his or her share of the harvest. Colonies that need to be fed in the spring are typically a sign that the beekeeper harvested too much honey and did not leave enough on the hives in the fall, that the late-fall honey flow that was anticipated to fill out the empty space within the hive did not materialize and adequate supplemental feeding did not take place, or both. Another instance when a hive typically requires feeding is when a nuc, swarm, or package of bees is starting out with nothing but foundation or empty combs.

Whenever the feeding of hives is taking place, no honey supers that will be harvested should be allowed to remain on the hive. This is simply a matter of integrity. Pure, natural honey is defined as plant nectar that has been collected and transformed by honey bees. The beekeeper who offers "honey" produced from sugar syrup, has compromised the integrity of the final product, as well as his or her moral character. It is questionable whether the apiculturist who removes the real

honey the bees need for winter and replaces it with "honey" from an inferior source such as processed white sugar, corn syrup, and so forth is deserving of the name "beekeeper." The term "bee exploiter" may be more fitting.

The feeding of pollen substitutes falls into the same category. Bees consume pollen for its high protein and mineral content, and the colony requires an abundant supply when raising brood. In the Northeast, the supply of pollen is most critical in late winter/early spring when the hive's pollen stores are typically running low and the first pollen-producing plants—the soft maples, various willows, and wildflowers—have yet to bloom. As a wild-grown, nutrient-dense food source, pollen is difficult to replicate, though many have tried, typically by using soy powder or nutritional yeast (brewer's yeast) as substitutes. Patties containing soy flour, brewer's yeast, or both are fed to hives early in spring to jump-start brood-rearing in an effort to ensure a maximum worker population that will take full advantage of the first major honey flow of the season. Nevertheless, an evaluation of the nutrient and protein differences between pollen and soy or brewer's yeast reveals that imitation pollen patties are a poor substitute for the real thing. Whether one is replacing honey with sugar syrup or pollen with soy flour or nutritional yeast, the overall health and immunity of the colony will be weakened. This will occur even if no noticeable effects are immediately evident. Research has shown that the beneficial bacteria and microorganisms that live in the hive and form a symbiotic relationship with the bees cannot thrive on artificial protein patties. They do, however, thrive on bee bread made from pollen. Thus, protein patties are never stored in the hive like pollen is, but will only be consumed or removed from the hive by the bees. These patties are also extremely attractive to small hive beetles (see chapter 6, "Small Hive Beetles" on page 162). I believe that non-organic and commercial operations that utilize such feeding techniques tend to experience more problems among their hives than those that don't, especially if sugar and supplements are used over prolonged periods. Common sense dictates that, like all creatures, honey bees thrive on wholesome natural foods, and they suffer when forced to consume the equivalent of junk food. If protein supplementation is required, the natural approach, in keeping with the honey bee's biology, is to mix powdered pollen with just enough honey so that it forms a dough. Such a patty can be left inside the hive near the brood area for utilization by the bees.

• REVERSING •

Reversing a hive is a technique used primarily to reduce, but not eliminate, a colony's swarming impulse. Swarming is the act of reproduction wherein a single superorganism, in this case the colony of bees, breaks into two groups and creates a replica of itself. It is a natural survival instinct of the species *Apis mellifera*.

A number of factors are believed to influence a colony's decision to swarm. Among them are: a crowded hive lacking space for additional brood rearing; an aging queen (two years old or older); and an abundance of honey and pollen available to foragers combined with insufficient storage space.

Preventing an early swarm helps to ensure that the bees will have time to stow away enough honey to last the winter before they divide themselves up and thus reduce their nectar-gathering abilities. This can be crucial during poor honey years when the total amount of honey produced by the hive is low. Reversing also provides an opportunity for the beekeeper to conduct a thorough inspection of each colony, so that potential issues may be resolved before they become acute.

Just like it sounds, the process of reversing a colony simply involves restacking the boxes that make up the hive so that the bottom hive body ends up on the top, and vice versa. The intention is to end up with the brood and honey positioned as close to the bottom of the hive as possible, with extra space above. The act of reversing a hive may also break up the brood nest, creating more space

for the queen to lay eggs. When using a system where the main hive consists of two shallow hive bodies with a deep hive body sandwiched between them, the positions of the shallows are reversed. If the hive proper consists of two deep hive bodies, then the position of each deep is switched. In the management system I use, this is how a shallow "overflow" hive body ends up beneath the deep hive body. The nuc or package of bees begins the year in the deep hive body. By the time winter is approaching, the hive has a shallow (or two) of honey above the deep box containing the brood nest. In the spring, most of the bees are up in the shallow and the deep is mostly empty, so the position of the boxes is reversed and the shallow on top is now placed on the bottom. Given that the process of reversing a colony involves taking the hive apart, this is a good time to clean up the equipment a bit. The following is one example of how to go about reversing, cleaning, and inspecting a hive in early spring.

First of all, when you walk up to a hive, it is best to approach from either the side or the rear of the colony. It's not a good idea to position yourself directly in front of a hive, because the foragers that are coming and going from the front entrance may interpret your actions as trying to block their path and may feel threatened enough to defend their turf with their stingers. After blowing some smoke into *all* the entrances of the hive, lift the outer cover and send a puff or two of smoke under the cover before removing it. The idea is to let the smoke precede you as you go through the colony, in effect announcing your presence before you arrive. Place the outer cover on the ground a couple feet behind the hive with the bottom side up. After applying a puff of smoke to the bees gathered around the opening of the inner cover, use your hive tool to pry up a corner of the top super or hive body. As your hive tool lifts up the corner of the super, send some smoke into the crack between the supers and continue to pry up until you can stick the entire nose of your smoker into the space between the boxes. This will allow you to remove the hive tool without

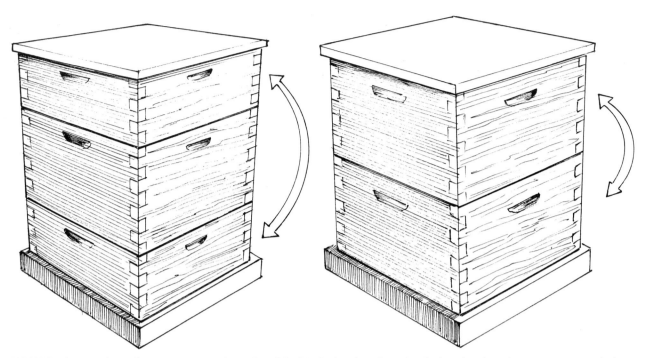

FIGURE 3-20. In northern climates, reversing the order of the hive's chambers in spring, before they have become overcrowded, can delay the bees' swarming instinct, helping to ensure that the colony will have the workforce available to store enough honey to see them safely through the winter.

the two boxes coming back together and possibly crushing some bees in the process. With the nose of the smoker inserted between the supers, squeeze the bellows a couple times while you reach over with your free hand to the handhold on the raised end of the upper chamber and lift it up. If the back end of the top box is too close to the edge of the box underneath it, you may have to slide the hive body being lifted forward a bit to keep it from slipping off the back end of the hive. After allowing some smoke to drift over the exposed bees, the top super with the inner cover still attached can now be lifted off the hive and placed on the telescoping edges of the outer cover that is on the ground, so that there is little chance that the bees on the exposed underside will be crushed (see figure 3-24). If more than two boxes make up the hive, repeat the procedure as outlined for the top hive body and place the second hive body off to the side, standing on end.

If a colony is ignored for too long and allowed to become crowded, the bees will often build burr comb between the boxes on the hive, even though

the distance between the top bars of the frames in the bottom box and the bottom bars of the frames in the box above respect the bee space. This can result in the frames below sticking to the frames above when the boxes are being pried apart (see Figure 3-22). When this occurs, bees get crushed and the top box cannot be lifted off the hive. This usually causes the beekeeper to drop the box back down on the hive, squashing even more bees, which results in the hive taking on an ugly mood.

Whenever you handle supers or hive bodies containing bees, it is best to keep them as level as possible. When it is not possible or convenient to keep a box full of bees in a horizontal position, it is important to always tilt the hive section so that the short end faces down toward the ground, thus preventing the frames from collapsing upon one another. If a nuc or hive body is carried or set down with one of the long sides facing the ground, gravity will force the frames together and in the process kill or injure the bees that are unfortunate enough to get pressed between the leaves of honey, like flowers in a

FIGURE 3-21. A well-built smoker can aid the beekeeper by acting as a wedge that prevents supers from collapsing together once separated and crushing bees. PHOTO BY ALICE ECKLES.

FIGURE 3-22. When frames stick to inner covers or supers above, blow smoke between the supers (or super and inner cover) and pry the frames down away from the inner cover or super without dropping the box (or inner cover) and crushing more bees.

FIGURE 3-23. Allow a super to slip off the hive during an inspection, and the fun really begins.

FIGURE 3-24. The bees appreciate it when the beekeeper makes use of an inverted outer cover to prevent bees from being crushed and grass, dirt, and debris from getting all over the hive's boxes and frames when separated.

book. If the queen is one of the victims, the hive may well suffer a lethal blow at the beekeeper's hand.

Once the hive body sitting on the bottom board has been removed, scrape the bottom board clean of dirt and debris before returning it to its place as the floor of the colony. Now you are ready to rebuild the hive by replacing the hive bodies in reverse order as compared to their previous position. Unless the colony is exceptionally strong, or you're late in getting around to reversing them, the former bottom portion of the hive will likely contain few bees, little honey, and no brood. Next to the bottom board is precisely where you *don't* want additional space to be within the hive. Because the bee's natural inclination is to move upward, an empty hive body or super should be located on top of the hive, and the boxes filled with the most honey, pollen, bees, and brood should be positioned as low in the hive as

possible. This way, the underutilized space within the hive is kept to a minimum, and your equipment is used most efficiently, because the lion's share of the honey gathered will be stored above the brood nest in preparation for winter. Sometimes you will find that a colony that goes into winter with three hive bodies will, by spring, have all their bees, brood, and honey confined to the uppermost story of the hive. In this case the bottom hive body can be taken away and the middle section placed on top of the hive to provide the needed room for growth.

As a general rule, it is a good idea to match the size of the cavity within which the colony is housed (the hive) with the number of bees in the hive. Additional space should not be added, in the form of a new hive body or super, until the bees fill up most of the space they already have. For example, if a hive consists of four frames of bees and brood and six empty frames in a ten-frame hive body, then adding another ten-frame hive body on top would be tantamount to moving a four-person family into a mansion with 20 rooms. The additional unused space is not only unnecessary given that they still have six frames to expand into, but the extra work of patrolling and maintaining the space in the additional hive body can actually be detrimental to the colony.

Positioning the free space within the hive on top simplifies the process of inspecting the colony in the future, since all it takes is a peek under the

FIGURE 3-25. Allowing the frames of a hive body to collapse together is a good way to squash a queen.

inner cover to tell whether additional room is needed. Once a colony has filled all but the outermost frames in the top super or hive body, an empty super should be placed on top to provide additional space. Back in the old days, beekeepers placed empty supers just above the brood nest and beneath the honey supers that the hive had already filled. This is because the bees will tend to draw out foundation and fill empty combs located in the middle of the hive a little faster than when they are placed at the very top of the colony. Nevertheless, the disruption to the colony, not to mention the labor involved with checking the supers and removing and replacing the full honey supers every time an empty super needed to be added, eventually caused this approach to be discontinued in favor of supering from the top. The one exception is during the production of comb honey, when

new supers of extra-thin foundation, or frames that allow for naturally drawn comb, are added below the full supers to reduce the incidence of "travel stain" that occurs when bees constantly walk across the filled and sealed combs to get to the new foundation above. This occurs in much the same way that dirt builds up on carpeting in high-traffic areas of your home. By positioning the new supers below the previously filled ones, the appearance of the final comb honey harvest is greatly improved.

Incidentally, queens generally prefer to lay their eggs in the larger spans of comb found in the deep hive bodies rather than in the smaller shallow or medium-sized supers. Nevertheless, this preference will not stop the queen from laying in shallow supers, especially if the brood nest becomes congested and additional space for egg laying becomes difficult to find. Although it is inconvenient to find brood in your honey supers, it is a healthy thing to allow the queen all the room she requires to lay her eggs. A gentle way of encouraging the mother queen to keep her brood confined to the lower part of the hive, without forcing her by using a queen excluder, is to start off the honey supering process with shallow supers. Because the bees will be more likely to fill a shallow super with nectar, rather than eggs, two or three shallows should be built up upon the colony before any deep honey supers are used. The relatively large span of sealed honey provided by several filled and capped shallow honey supers acts as a natural queen excluder and discourages the queen from crossing all that honey to lay eggs in a distant part of the hive.

Getting back to our hive reversal process: Prior to placing the top hive body, which is chock-full of bees and honey and still has the inner cover attached to it, on the bottom board, the bottom of the frames of the hive body should be scraped clean of all burr comb. Care must be taken to scrape off burr comb and brace comb clear down to the wood, since any wax remaining will act as foundation and provide the bees with guidance and motivation for rebuilding in the same location. If you find your hive tool is gouging and peeling off layers of wood while scraping your equipment clean, your tool may be too sharp. Dragging the sharply honed end of the hive tool along a rock or brick a dozen times or so will dull the edge enough so that, when held at the proper angle, it will remove wax and debris without stripping off the wood underneath like a planing tool. By taking the time to thoroughly clean burr comb off the frames and other areas of the hive early in the season, a lot of trouble and inconvenience can be avoided later on. As with the process of cleaning up dead hives, a significant amount of beeswax can be collected for rendering by scraping burr comb from the tops and bottoms of the frames into a box before setting the hive bodies and supers back into place. This also helps to prevent the crushing of many bees when repositioning the hive bodies on their stand. You would be surprised at how much beeswax can be collected from a hive during such a process.

· WRAPPING UP· THE SEASON

A major concern with keeping bees in northern locations is winter survival, but I think far too much attention is given to concerns about the cold temperatures that the hive will have to endure. As long as the hive is strong and healthy with plenty of adult bees and brood, has an abundance of sealed honey stored in the combs, and is able to stay dry, it should have little trouble withstanding the frosty breath of Old Man Winter. Honey bees are very intelligent. Rather than trying to keep the entire interior of the hive warm, the honey bee cluster serves to focus and conserve the colony's energy by keeping warm the minimum space necessary for survival. They do this by warming only the area of the hive that they are occupying. The temperature of the rest of the hive interior is not that different from the ambient temperature outside the hive. As such, I find most efforts to insulate and wrap hives to be a waste of time. If I kept my bees in a climate colder than USDA plant hardiness zone 4a, where it typically gets down to about -25°F (-32°C) for a maximum of a week or two each winter, then I might view the practice of wrapping or packing hives with insulation differently. As it is, I have never wrapped or packed my hives for winter, and yet, even since the arrival of varroa mites, I can achieve good overwintering results by simply focusing on the three key ingredients of winter survival: strong, healthy colonies; plenty of honey stores; and dry bees.

Because I have already covered ways to ensure a strong, populous colony with plenty of honey stores for winter (and I will go into these topics in even more detail later on), I will focus here on the issue of moisture within the hive. A secure, waterproof outer cover and hive bodies that are solid and tight-fitting are the hive's primary defenses against moisture. Many beekeepers will use a rock or brick to weigh down the outer cover so that strong winter winds will not pry the cover loose, flinging it aside and allowing precipitation to enter. This usually works fine unless the apiary is located in an area where farm animals like cows or horses may use the hive as a scratching post and inadvertently push a colony over while rubbing up against it. Wet areas that experience severe frost-heave activity can potentially cause our little bee condos to topple over as well. During major storms, floods or powerful winds can pick up a hive and transport it a considerable distance. To help defend against these types of occurrences, or when you have so many hives that the number of rocks or bricks needed to weigh down the covers is unwieldy, outer covers may be either strapped down or tied down securely with thin rope or twine. The twine should wrap around at least the two uppermost hive bodies and hold them together, to increase the chances of colony survival in the event that the hive becomes dislodged from its base during the winter. If the twine or string used to tie down the covers in the fall is made from a natural material such as jute or hemp that is not treated with chemicals to prevent rot or repel rodents, it can be saved after being removed in the spring and reused as smoker fuel. (This is explained more fully in "Tying Down Hives for the Winter" on page 72.)

Since the first edition of *Natural Beekeeping* was published, the bear population in Vermont has doubled and bears have started visiting some of my bee yards that had been free of bears for over a decade. Because of this, instead of tying up my hives, I now strap my hives, which helps discourage bears, as well as ensuring that the covers on my hives do not come off in winter (see the section on bears in chapter 7, page 175).

FIGURE 3-26. If you remove the outer cover during winter when the bees are too cold to fly, they will arch their abdomens and aim their stingers up into the air as a warning to leave them alone.

FIGURE 3-27. Provide your bees with a hat for winter: a sheet of insulation cut to the size of the inner cover and placed under the outer cover will help to prevent moisture from freezing underneath the inner cover and then dripping down on the bees during a thaw.

Another source of moisture that may find its way into the hive during the winter comes in the form of water that runs in the bottom entrance and collects on the bottom board. Placing a piece of wood or some other material under the rear of the hive will tilt the colony forward so that water from rain or melting snow will not run into the hive. This is especially important for colonies with bottom boards that extend out in front of the hive to provide the bees with a landing area, because these extensions can catch a significant amount of snow and rainfall. Colonies outfitted with screened bottoms made of wire mesh or hardware cloth open to the ground don't need to be concerned with this possibility, since any water that runs into the hive will drain through the screen rather than collecting on the bottom board.

A large amount of moisture is created within the hive during wintry weather as a by-product of honey bee respiration. The importance of adequately ventilating the hive to remove this vapor during cold weather cannot be overstated. This is why the reduction in ventilation created by most hive wraps, blankets, or packing materials often causes more harm to the hive than any insulating benefits they provide. Packing insulation can also serve to lock in the cold and prevent the colony from responding to temporary warm periods that would normally allow the cluster to shift its position within the hive and keep in contact with crucial honey stores. The colony's ability to take advantage of short winter thaws is also important because they provide the bees with the opportunity to go on cleansing flights and relieve themselves of the indigestible products of feeding activity. By the same token, the well-packed hive will take longer to warm up in the spring, delaying the ability of foragers to take advantage of the earliest pollen sources that are so crucial to the colony's spring buildup. (For these reasons, it is also preferable to place apiaries where hives will receive direct sunlight during the day.)

FIGURE 3-28. This hive has two upper entrances, one drilled into the side of the hive body and one in the rim of the inner cover.

To help guarantee adequate airflow during the winter, I do not recommend reducing the opening in the bottom entrance of the hive. This advice will go against much of what you are likely to read in other books, and it certainly runs counter to how we maintain our own living spaces during the winter. In my opinion, though, entrance reducers should be used only during the spring and early summer when a colony, captured swarm, or nuc is small and weak and needs to be protected from robbers. If a weak colony is discovered in autumn, many beekeepers will combine it with another hive rather than try to carry it through the winter by itself, since such efforts typically end in failure. A ¾-inch hole drilled into each hive body, below or to the side of the handhold, will provide an upper entrance that can greatly improve ventilation. Some beekeeping suppliers manufacture inner covers with a notch cut out of the rim of the cover. This notch can provide a good upper entrance as long as the outer cover is not positioned tightly up against the opening.

Should the bottom entrance become clogged with snow, ice, or dead bees, an upper entrance becomes especially important, not just for ventilating, but also to allow egress for the colony. The use of screened bottom boards open to the ground also improves hive ventilation. Should the hive become buried after a heavy snowfall, it is a good practice to dig the colony out so that the moist air in and around the hive can be ventilated away, and so the bees will have the opportunity to make cleansing flights on the next warm day that comes along. Adequate ventilation that will help the bees to remove excess moisture from the hive ranks right up there—along with plenty of honey, a large population of healthy bees, a secure hive cover, protection from the wind, and exposure to direct sunlight—as one of the most important wintering requirements for a hive located in a northern climate.

Aside from tying down the outer covers on hives, the only other winter preparation I make is to insert a 1-inch-thick sheet of foam insulation between the

TYING DOWN HIVES FOR THE WINTER

Wrap the cord around the hive so that it catches on the second or third box down from the top. Cross the loose end under and then back over the opposite piece about four to five inches from the end (A). Now, pass the working end down behind the standing part (B) and feed it through the loop created by it and the opposite piece. This ties the loose end around itself in a figure-eight pattern as shown (C). Tightening this knot will create a slip knot that will allow you to pull the twine tight, like a lasso around the hive. Run the loose end over the top of the hive and down the other side, and cut the string so it hangs down a few inches below the horizontal strand. Pass the cut end over and then under the horizontal cord that is strung across the opposite side of the colony (D). Push down on the outer cover while simultaneously pulling the twine tight. Then pull the free end forward and place a thumb or finger on the junction of the cords to prevent them from loosening up. With your free hand, thread the working end across itself and up and under the horizontal piece of twine opposing it (E). Loop the cord and pull it through the hole created by the horizontal strand, the vertical strand, and the working end so that it tightens on itself (F). The cover of your hive is now securely tied down with a knot that quickly and easily comes apart by simply pulling on the loose end like a shoelace (G).

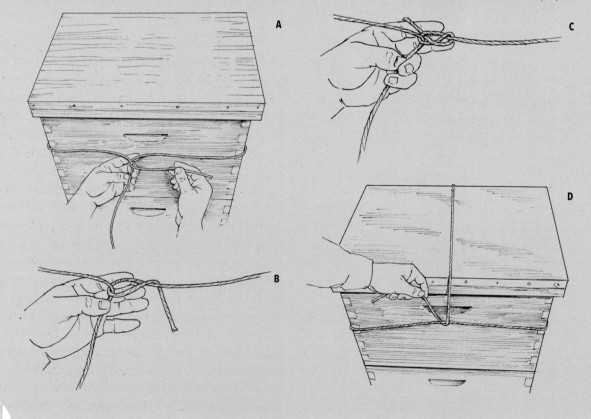

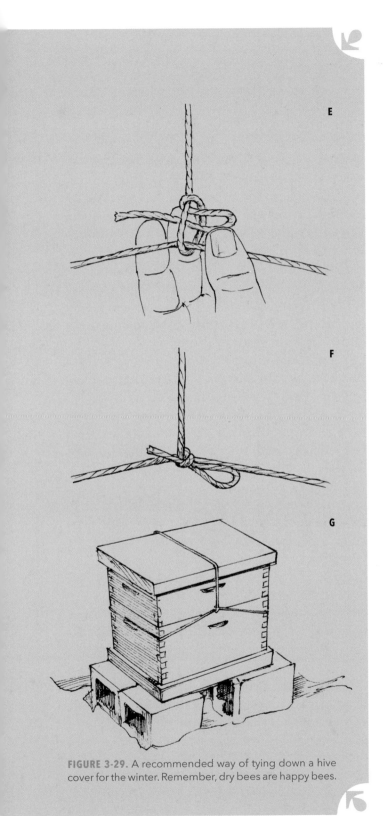

FIGURE 3-29. A recommended way of tying down a hive cover for the winter. Remember, dry bees are happy bees.

inner and outer covers. An empty grain bag with a piece of 3-inch fiberglass insulation inside of it that has been cut to the size of an inner cover also works well. I have even heard of folks using other materials such as loose straw or dried leaves as insulation. Such material placed between the outer and inner covers acts like a hat, keeping much of the heat created within the hive from leaving as it rises from the winter cluster. More importantly, it helps prevent the moisture riding on the breath of the colony from collecting on the underside of the inner cover, freezing during a cold spell, and then thawing out later on, only to drip back down on the bees. The bee cluster can handle being wet when it is warm, and it can deal with very cold temperatures as long as the cluster can stay dry and maintain its access to honey. However, if honey bees both become wet and are exposed to cold temperatures, it is a recipe for disaster.

· UNPACKING THE HIVE · IN SPRING

The first warm, sunny day in early spring when the temperature reaches 45° to 50°F (7° to 10°C) is a great time to pay a visit to your apiary. This first visit of the year involves primarily a quick check of the hives, simply to make sure they are still alive. A trip to see the bees can provide great peace of mind as one plans for the upcoming season. If it is early in the day and the hive has not had time to warm up, no bees may be visible around the entrance or the opening in the inner cover. In such instances, a sharp knock on the side of the hive should produce a reply that will tell you if somebody is home. If after knocking you don't hear an answer, a quick peek underneath the inner cover is usually enough to confirm whether the hive is still occupied. This is also a good time to make sure that the colony has at least several full frames of sealed honey to tide the bees over until the first honey of the new season begins to flow. The "heft test" is an expeditious way to judge whether the bees will need to be fed at this time of year. Although colonies will obviously be lighter in the spring compared to the fall, they

FIGURE 3-30. When it comes to providing increased hive ventilation for the winter and an emergency exit for the bees, an upper entrance is tough to beat.

should still have considerable weight to them. By getting a good feel for the weight of hives within which you have confirmed that there are at least four to five capped frames of honey, you can learn to accurately estimate a hive's stores with a quick lift.

One of the most sorrowful parts of beekeeping is the discovery that a colony has died during the winter. I find it helpful, however, to view a dead hive as a gift from the bees—the opportunity to become a better beekeeper. As with all dead hives, the first thing to do after coming upon a dead hive in the spring is conduct an autopsy. If we can figure out why the hive has died, we can review how the hive was managed during the past year and figure out what we need to do differently so we can prevent the same thing from happening again. It is an unfortunate fact that the only way one gets to be a really good beekeeper is by killing a bunch of bees, because we tend to learn more from our mistakes than from our accidental successes. Knowing why a colony has expired can help prevent the same fate from afflicting other hives, especially if the brood nest shows signs of American foulbrood. If the equipment is free of disease, however, any supers or frames filled with honey that are found can be used to feed other hives that are low on food, thus saving you the work of cooking up a batch of sugar syrup or bee tea.

The discovery of a dead hive also provides an opportunity to take care of the important housekeeping detail of cleaning the equipment. This cleaning will prevent the frames from getting so glued up that they become next to impossible to remove. Take out every frame in each of the hive bodies and supers that made up the former colony, scraping away all of the burr comb and propolis attached to the sides and frame rests of each empty box. Then clean the excess wax and propolis off each frame before placing it back in position. The most efficient way to remove frames from a hive body in which the frames have been securely propolized into place is to drop the empty hive body upside down on an upturned outer cover so that the box comes to rest on the telescoping edges of the cover, and the frames are able to drop down into the space below.

While cleaning up a dead colony, it's a good idea to collect the scrapings from the equipment as you work. With the price of wax and propolis significantly higher than the price of honey these days, the value of these precious substances should not be overlooked. An easy way to gather these items is to scrape the equipment above a collection box that will catch the debris as it falls. A simple version can be made by attaching a piece of window screen to the bottom of a hive body. The screen will allow honey and water to drain, while retaining the propolis and wax pieces within the box.

I like to wait until daytime temperatures have reached 50°F (10°C) or higher, on a fairly consistent basis, before I untie the colonies and remove the insulation underneath the outer cover. If a colony is coming out of the winter in a weak condition with only a handful of bees, leaving the insulation in place a little longer will help the colony weather the cool nights until it's time to reverse the hives. The transition from winter to spring is the time of year when honey bees in the North Country are at their most vulnerable. Hives coming out of winter have been weakened from a significant loss in numbers as individual bees have died of old age or other causes after being cooped up inside for most of the past six months. The temperature swings that occur during this season can also wreak havoc on the bees as they invest precious energy and limited resources in expanding the brood nest during warm spells, only to have the cold return with a vengeance. Due to their decreased population, the bees are often unable to adequately cover the expanding nest area and keep it warm during cold snaps, resulting in the brood getting chilled and dying, a condition known as chill brood.

This is also the time of year that can be lethal should the bees lose contact with their stored honey. On too many occasions, I have inspected the brood nest of a dead colony that still had plenty of honey stored within its combs, but the honey was located several inches away from where the bees were clustered. The telltale sign that the bees have starved from want of nourishment is that many of them are found with their tails sticking out of the cells, as they scraped the back of the cells that form the comb with their tongues, desperately looking for a drop of golden ambrosia. This is why the first visit of the year to the apiary is so important. Should you find, upon initial inspection, that the colony is still alive but has "eaten itself into a corner," and that the brood area is surrounded by empty comb, move a frame full of sealed honey so it sits next to the cluster. It would probably also be a good idea to feed a hive in this situation in order to be sure they have enough food to last them until the first honey flow of the year gets underway. By making sure that the bees are free from disease and have plenty of honey within reach, many unnecessary deaths can be avoided during the critical transition from winter to spring.

FIGURE 3-31. Heavily propolized frames can be difficult to remove when cleaning out a dead hive. By turning the hive body upside down and dropping it on an inverted outer cover, gravity will do the work for you.

BEEHIVE AUTOPSY RESULTS

Characteristics/Symptoms	Probable Cause
A dramatic reduction in hive population within a short period of time that results in few or no adult worker bees left in the hive either on combs or dead on the bottom board or out in front of the hive. A larger brood area than the workers can adequately cover and maintain. A noticeable delay in scavenging activity by other bees, small hive beetles, or wax moths.	Colony collapse disorder
Pupal mass under cappings is brownish in color and has a ropey or elastic viscosity. Sunken brood cell cappings that are dark brown or black in color and have a greasy appearance. Some cappings may also contain small pinholes. Combs have a distinct foul smell. In later stages the pupal mass will have dehydrated and formed black scales on the bottom side of the cells.	American foulbrood (*Paenibacillus larvae*)
Small pinholes in brood cell cappings. Numerous dead bees with deformed wings and/or short abdomens. Numerous dead varroa mites found on bees, in sealed brood cells, or on the bottom board.	Varroa mites (*Varroa destructor*) and associated parasitic mite syndrome (PMS)
Remains of dead cluster contain bees that are positioned headfirst in cells. Any honey left in the hive is usually located two or more inches away from the remains of the cluster.	Starvation
Remains of numerous drone brood cells sometimes scattered within worker brood on the same comb.	Old or failed queen/drone layer
Combs, brood, or dead bees covered with mold or mildew.	Indicates that the hive died a while ago or was too weak to maintain combs.
No honey left in hive. Wax cappings that covered areas where honey was stored have been ripped open—jagged wax capping pieces litter the bottom board.	Hive died out or was too weak to defend honey stores from robbing by other bees, wasps, and/or hornets.

Characteristics/Symptoms	Probable Cause
Significant brown spotting or large patches of brown staining on frames, comb, or in front of the hive.	Nosema disease or dysentery
Numerous dead bees lying in front of the hive, maybe combined with brown spotting on inside or outside of hive entrance. Bees that have disconnected their two pairs of wings and rotated them into an orientation that resembles the letter K.	Tracheal mites (*Acarapis woodi*)
Buildup of webbing on combs containing small black pieces of debris. Remains of old cocoons and rounded elongated indentations in wooden hive parts. Damaged/disintegrated combs. Grayish moths either dead or alive.	Greater or lesser wax moths (*Galleria mellonella, Achroia grisella*) moved in once the hive became too weak to defend itself or died out.
Small hard larval remains that are white, gray, or black within the brood comb or on the bottom board.	Chalkbrood (*Ascosphaera apis*)
Combs are riddled with holes (but no webbing is evident). Inside of hive is covered with slime and any honey left in the hive is fermented and runny. Some beetle larvae or small dark beetles may still be present.	Small hive beetle (*Aethina tumida*) moved in once the hive died out or became too weak to protect all areas of the comb.
Sudden collapse of hive. Numerous dead bees lying around in front of the hive with their tongues sticking out.	Pesticide/chemical poisoning
Dead brood (may be in any stage of development) that do not show signs of distress other than perhaps being dead for an extended period of time (mold, etc.)	Chill brood, caused by the temperature in the brood area dropping too low for too long (usually as a result of something weakening the hive and decreasing the population of bees so they are unable to fully cover the brood area and keep it warm).

· IDENTIFYING · AND WORKING WITH QUEEN ISSUES

With the arrival of spring, known as "mud season" here in Vermont, comes the time for inspecting hives and evaluating their condition. A critical part of the colony evaluation process is judging the condition of the queen. The state of the queen has a major impact on both the short- and long-term health of the colony.

Direct observations of the queen herself can give us some information. Clearly, if you see a queen, then there is a queen present. But what is her condition? Are the queen's wings frayed and tattered at the ends, indicating that she is old and perhaps close to the end of her life? Is she the same queen that was installed in the hive during a past season? Unless you have a photographic memory (and only a hive or two to keep track of), the easiest way to be sure she is the same queen is to mark the back of her thorax with paint or a tag of some kind when she is first introduced to the hive. When I want to mark a queen (typically for use in an observation hive that will be displayed in a school), I like to use water-based White Out correction fluid. It does not last more than about a year, but it is nontoxic to humans and bees.

Unfortunately, finding a single queen amongst tens of thousands of workers and drones can be a difficult task. Thus it is often much easier and quicker to evaluate the condition of the queen by observing what is going on in the brood nest, where much more can be determined about the condition of the queen and the state of the hive in general. In the ideal hive with a fertile laying queen, the brood area will be filled with mostly worker brood. The worker brood will typically appear in large patches on the comb, and all the young workers being raised on each comb will appear similar in age. This will normally be seen as either a large patch of unhatched eggs or eggs intermingled with recently hatched eggs; a wide section of comb filled with pearly white larvae all of a similar size; large

areas of sealed worker brood; or comb filled with sealed worker brood intermixed with empty cells and perhaps the occasional worker in the process of chewing her way out of her birthing cell.

Traditionally, beekeepers and queen breeders have always prized queens that lay solid brood patterns with few or, preferably, no empty cells within the patches of brood. However, in his book, *The Buzz About Bees: Biology of a Superorganism*, author Jürgen Tautz points out that empty cells scattered among the brood are critical for warming the brood nest most efficiently. When a heater bee sits on the surface of the brood comb and increases its body temperature to warm the brood beneath it, she can only cover and therefore transfer heat directly to about two brood cells. However, when the same heater bee is able to crawl into an empty cell surrounded by brood on all sides and then heats up its body, all six cells of brood adjacent to the empty cell are warmed, as well as the two cells directly opposite the empty cell on the other side of the comb. We used to think that the bees sitting in the empty cells scattered throughout the brood area were just resting . . . now we know better.

An ideal colony that has a fertile laying queen is what we call "queen-right." Honey bee eggs, when laid properly by a fertile queen, are laid one per cell, standing on end in approximately the center of the back end of the cell. This is true for both the fertilized worker eggs and unfertilized drone eggs. When it comes to drone brood, its layout within a queen-right colony will be similar to the layout of worker brood. There will be patches of similarly aged brood all clumped together, but the areas of comb containing drone brood will be smaller and less numerous. There will also be a clear delineation between the patches of drone brood and the more common, and usually larger, patches of worker brood. The patches of drone brood and worker brood, though they may appear next to each other, will not be mingled together.

Recognising what the brood and brood nest of a healthy hive looks like, as described above, is critical to being able to identify when a problem

FIGURE 3-32. A healthy fertile queen can lay her weight in eggs every day (1,500 eggs or more) during the peak of the summer season. This means that, around 21 days later, about 1,500 new bees will be joining the colony every single day!

exists with a queen. The biggest challenge with a queen-right hive is to provide space for expansion during the busy seasons of spring, summer, and autumn when foragers are bringing in nectar and pollen. It is during spring and summer especially that the colony can expand its population and fill up space within the hive very quickly.

Beekeepers who do not realize how rapidly a hive can build up can easily fall behind and allow hives to become so crowded that worker bees begin to store nectar in the birthing cells recently vacated by young adult bees. This condition is often referred to as "honey bound" and is known to prompt a colony to swarm. A hive in the Champlain Valley of Vermont was once observed to draw

out 10 frames of foundation in a deep honey super, fill the combs with honey, and cap about half of the honey cells all within a period of one week. This illustrates how fast a strong, healthy colony can fill up a hive when conditions are ideal.

While it's important not to super a hive too late, it is also not good to super too early. One should strive to provide additional expansion room for the colony in a way that does not expand the hive cavity too much, too fast. The goal is to match the size of the hive cavity to the population of the bees. It may help to imagine the hive like a suit of clothing; you don't want it to be so large that you're swimming in the outfit, and you don't want it so small that it is too tight for you to move around in comfortably.

I try to time adding an additional honey super onto a Langstroth hive (or moving the follower board back and adding several empty top bars to a top bar hive) when all foundation is drawn out and seven or eight of the combs are full, leaving only two or three partially filled frames left in the hive. This is why regular hive inspections during spring and summer are so important if a beekeeper hopes to super a hive at just the right time.

Fight the temptation to cut corners and super the hive for the season all at once; it's a strategy that can be detrimental to the colony's development. In the old days when colonies were generally very strong and healthy, beekeepers could super up their hives in spring, ignore the bees all summer, and harvest a bountiful crop in autumn. Unfortunately, if our modern-day hives don't have a large-enough population of workers to be able to adequately patrol and maintain the additional space, problems can arise from wax moths, small hive beetles, mold, and a host of other issues. Avoid such problems by waiting to add more space until just before the colony actually needs it, rather than way ahead of time. Weather and foraging conditions, along with colony population, play a large role in determining how fast a colony will fill up the space within the hive and therefore how often the hive should be inspected and supered. Ideally, the beekeeper will keep just slightly ahead of the hive's growth.

If you are an experienced beekeeper, you are probably saying to yourself, "The bees will make a lot more honey if supers are added much sooner, such as when the top super is only half full." This is true; however, supering earlier on a regular basis can lead to a "chimney effect," where the bees fill all the center frames of the hive with honey and leave the outer frames of each box virtually empty. This can lead to starvation issues during long dearths, such as during the winter. Because hive survival is my top priority, I prefer to super colonies when about 80 percent of the combs have been filled. Also, if foundation is added to a hive too far in advance, the bees will tend to chew away at the foundation wax, thus disfiguring the foundation.

I would also like to address here the fact that there are beekeepers who teach others to leave the bees alone and not bother the hive with inspections. To inspect colonies regularly disturbs the bees too much and is unnecessary, they say. I believe that such advice, while well-meaning, is misguided and ignores our current real-world situation. Today's bees are up against a host of threats and stresses including the industrialized agricultural system we've created based upon monocultures and toxic pesticides, herbicides, fungicides, and insecticides; diseases and pests we export and import with help from the corporate globalized economy; and the human-induced rapid increase of greenhouse gases, which is leading to a destabilized global climate. Since we are collectively responsible for many of the challenges that the bees of the world face today, we have a responsibility to help the bees deal with these threats to their survival. Once we have gotten our act together and cleaned up the messes we have made, then we can leave the bees alone and let them "do their thing" without constant intrusions into their homes. In these modern times, though, it is unfortunately the rare hive that will survive for more than a year or two without some kind of human assistance.

A colony may lose its queen as a result of numerous events including disease, failure to mate, beekeeper carelessness, and old age. A hive that becomes queenless may exhibit a variety of symptoms within the brood nest depending upon how long the colony has been without a laying queen. It takes careful observation and analysis to determine whether the queen is in the process of being replaced, is missing altogether, or whether it still has a queen that has become infertile. The distinction is important in terms of the appropriate hive management response.

For the first three days following the sudden loss of the queen, there may be no apparent changes in the brood within the hive. The fourth day following the loss of the queen, no unhatched eggs will be found. As time progresses, the beginning of queen cells are likely to appear. As more time goes by, all

FIGURE 3-33. Allow the hive to become too crowded and the bees are liable to swarm. Swarms come in many shapes and sizes.

cells containing larvae will be capped, including any queen cells that were created. Since no eggs are being laid, an absence of uncapped cells containing larvae eventually results. In the most advanced stage of queenlessness only adult bees are present and there is a complete absence of brood in the hive.

Do not be fooled by hives that contain no brood: they are not necessarily queenless. Lack of brood can occur during November, December, and/or January—depending on your geographical location—simply because queens usually slow their egg laying and often stop laying altogether for a short period of time as the hive conserves food and energy during the dearth of winter.

The absence of brood in a colony can also result when there is not enough food (honey and/or pollen) available to accommodate brood rearing. Therefore, when diagnosing the cause of broodlessness within

a hive, consider the amount of honey and pollen stored within the hive as well as the time of year.

The easiest and clearest indication that things are amiss in a hive is when the only brood present in a hive is drone brood. This situation signals to the beekeeper that the colony lacks a fertile laying queen. There may be a queen in the colony that is infertile, but it's more likely that the complete absence of worker brood is an indication of queenlessness. In a healthy hive, the brood releases pheromones that inhibit the development of the worker bees' ovaries. In the absence of brood, the ovaries of some worker bees will typically develop to the point that the workers will be able to lay eggs. But because the worker is incapable of mating, the eggs laid by such a worker will always be unfertilized and tend to result in the production of drones. According to Tom Seeley's book, *Wisdom of the*

Hive: The Social Physiology of Honey Bee Colonies, laying workers exist in hives most of the time. In a queen-right colony, however, the number of laying workers is extremely low, and the nurse bees will cannibalize the unfertilized eggs the laying workers produce in order to recycle the protein they contain. This dynamic changes, however, in hives where a fertile laying queen is absent.

The drone brood produced by a laying worker will not appear as neat and tidy as the drone brood produced by a fertile queen. Rather than patches of comb that contain only or mostly developing drones of approximately the same age, there will be numerous empty cells mixed in among the cells containing drone brood. This brood has a spotty, shotgun-like pattern that features eggs, larvae, and sealed brood all mixed together in close proximity. This is because laying workers do not lay their eggs in a tight pattern like fertile queens do, and also because many of the eggs laid fail to hatch.

In addition, the eggs produced by laying workers will be laying on their sides and attached to the sides of the cell wall, instead of standing up on end at the back of the cell. The abdomen of the worker bee is not as long as the queen's, and thus the laying worker cannot easily reach the back end of the cell with the tip of her abdomen when laying eggs. More often than not, the laying worker will also place multiple eggs in each cell. If any queen cells are present they are likely to be small and not viable.

A laying worker hive will often be more aggressive or "jumpy" when being worked than its queen-right neighbors. Anther clue that the colony is queenless is that, compared to a normal hive, it emits a louder, higher-pitched sound produced by the workers fanning their wings when the hive cover is removed. This sound is often referred to as a loud "roar." The workers also have a tendency to run around more when the frames are being handled and therefore appear "nervous."

Once a hive has established one or more active laying workers and becomes a drone layer producing only drones, correcting the situation is difficult, if not impossible. Trying to identify the laying workers and remove them from the hive is extremely difficult; they don't look much different from the regular workers in the colony. A frame of brood and unhatched eggs taken from a queen-right colony rarely works in prompting a drone-layer colony to begin to raise a new queen from the newly supplied fertile eggs. Nevertheless, if a frame containing unhatched eggs is inserted into the hive once a week for three or four weeks in a row, the colony will often begin to eventually build queen cells from the eggs. An alternative is to introduce a queen cell or a newly mated fertile queen into a drone-layer hive. The addition of a frame of brood and eggs, a queen cell, or a new fertile mated queen works best when introduced to a colony that has only recently become queenless, before the workers' ovaries develop to the point that they begin to lay eggs in earnest. Once the workers start laying, pheromones are produced that fool the rest of the colony into thinking that they have a proper queen. This suppresses their desire to readily accept a newly introduced queen or to raise a new queen from eggs or young larvae supplied by the beekeeper. The addition of a ripe queen cell that is days from hatching is also likely to fail, even though of all the options it may have the best chance of successfully transitioning the drone-laying hive into a normally functioning colony. This tends to be true for hives in all states and conditions: they will accept a queen cell and the queen that hatches from the cell much more readily than they will accept a mated queen that the beekeeper tries to introduce into the hive.

There are reported cases of normal queens developing from unfertilized eggs, such as those produced by a drone-laying hive. This process, known as parthenogenesis, is extremely rare in European honey bees, but is apparently more common in Cape bees from South Africa. Unfortunately, parthenogenesis cannot be counted on to save a drone layer.

Once the laying workers are firmly established in a queenless hive, or the number of workers in the hive is too small for the colony to have a good chance of caring for a queen on their own, the most

FIGURE 3-34. Relatively flat capped worker brood intermingled with the bullet-shaped capped drone brood is a sign that the queen is running out of sperm needed to fertilize eggs and reaching the end of her usefulness to the colony.

effective and efficient way to deal with the hive is to combine it with one or more queen-right hives that are in need of more room or more bees, or both. It is especially helpful to give the full supers or frames of bees, brood, honey, and pollen from a drone-layer hive to queen-right colonies that could use a boost in food stores or worker population. The workers in the queen-right colony will protect their queen from the worker bees from the drone-laying colony. This will allow the workers from the drone-laying hive time to get used to the queen's pheromones and to become integrated into the queen-right hive. Of course this option is only available if you have more than one hive.

Beekeepers who own a single hive and find it has become a drone layer can sometimes remedy the situation by shaking all the bees out on the ground at least 100 yards from the hive and then introducing a new queen to the colony. The theory is that the laying workers will be unable to find their way back to the hive and so they will be removed from the colony and won't interfere with the queen's introduction.

Sometimes a queen is not completely infertile, but rather is getting old and beginning to fail. One sign of this is a brood pattern in which both worker and drone brood are intermingled instead of neatly separated (see figure 3-34). This happens when the limited quantity of sperm that the queen had stored in the organ called the spermatheca is running out. During this period the queen will often attempt to lay fertilized eggs, but fail to successfully fertilize them due to the shortage of sperm in her system. A queen typically lives four to six years before reaching this stage in her life cycle, although the timetable is dependent upon the number of drones that the queen was able to mate with during her mating flights. In a hive with an aging queen, queen cells may or may not be present. Such cells indicate that the workers are trying to raise a replacement queen.

When a queen is reaching the end of her useful life span and is running out of sperm to fertilize eggs, the workers within the colony will often raise a new queen to replace her, a process called supersedure. From a natural/organic perspective, it is best to allow this natural process to unfold on its own. Unfortunately, not all hives will develop queen cells in an effort to raise a replacement queen. Beekeepers who wish to help the process along may choose to introduce a ripe queen cell. A close eye needs to be kept on the hive, however, because the raising of a new queen either from a supersedure cell or an introduced queen cell is not always successful. Even though the workers may build queen cells, or allow an introduced cell to hatch, there is no guarantee the queens that hatch will successfully mate and make it back to the hive. The process of mating is the most dangerous time in the queen's life. She could get eaten by a bird, get caught in stormy weather, or succumb to a host of other disasters that would prevent her from successfully mating and returning to the hive. If the colony is unsuccessful in raising a new fertile queen, a ripe queen cell can be placed in the brood nest if there is still enough time left in the season to give the hive another chance to succeed.

For those who don't want to risk letting the hive replace the queen on their own, or when there is not enough time left in the season for such a slow process, the introduction of a fully mated and fertile queen to replace the failing one will resolve the situation. To help ensure the successful introduction of a fertile queen, you must first remove the failing queen from the hive. For best results, remove the old queen one or two days before introducing a new queen to the hive.

During the process of replacing the queen, a hive will undergo a period where brood is absent. This may be mistaken for a queenless condition unless one notices the clues that indicate a new queen is in the works. These include the presence of supersedure queen cells or queen cell remains (left after the developing queen vacates the cell). Another sign is large patches of comb where no

FIGURE 3-35. Vertically oriented queen cups are often found in hives during swarming season. They will serve as the foundation for queen cells once the queen decides to swarm and lays an egg in the cups.

eggs are present but the cells of empty comb are all cleaned and polished; the bees are keeping these cells clear of honey and pollen in anticipation of the queen beginning her egg laying. These clues indicate that the queenless hive is a transitory condition that usually leads to a hive that successfully raises a new fertile queen, or, if unsuccessful, that ends up containing laying workers.

The challenge is accurately determining whether or not the hive is really without a queen. Timing matters. If the condition of queenlessness is caught early enough, there will be no drone brood to reveal the presence of laying workers. When observed late in the process, the supersedure or emergency queen cells will have hatched out and the worker bees torn down the remains, destroying the evidence that would let you know that a new

FIGURE 3-36. Swarms are usually very docile and easy to work with even without a veil or smoke. As long as you have the queen, the rest of the bees will follow. PHOTO BY ALICE ECKLES.

queen is present (but has yet to start laying eggs). If you misinterpret the evidence, and attempt to introduce a new queen to such a colony, it will be a waste of time and effort. The introduced queen is liable to be rejected by the workers or end up fighting the hive's natural queen to the death. On the other hand, if it is impossible to determine the exact state of the hive, the safest route to restoring the hive to normalcy may be to introduce a queen, just in case the hive has not been able to raise a new one. Sometimes despite all our learned science, intense observation, and experience in beekeeping, decisions need to be based upon simple intuition and what feels right at the time.

Another situation that impacts the state of the queen and the condition of the brood nest is swarming activity. When a colony decides to swarm, the queen will lay an egg in one or more queen cups. As the new queens develop, the mother queen will slow down and eventually stop her egg laying in preparation for leaving the colony in search of a new home. The queen and approximately two-thirds of the workers will leave shortly before the new queens hatch. The symptoms visible in the brood nest of such a hive will be similar to those of a hive that is replacing or superseding its queen as described above. The primary difference is that queen cells in a swarming hive (aka swarm cells) are most often found along the outer edges of the combs or on the edges of damaged comb, whereas emergency or supersedure queen cells will be located in the center of a patch of comb of worker-sized cells.

Once a queen and the colony have decided to swarm it is not easy to change their mind. Swarm

FIGURE 3-37. An unmated virgin queen or newly mated queen who has not started laying eggs and whose abdomen has not swelled up, such as pictured here, can be difficult to differentiate from the surrounding worker bees.

cells can be cut out in an attempt to stop the swarming impulse, but if you miss a cell, you're out of luck. Plus, swarm cells can't simply be cut out of the hive once. This task has to be done regularly, about every five days or so, before a hive will give up trying to swarm. Even worse, if you cut all the swarm cells out of a hive too late in the process, the colony may still swarm, leaving behind a hopelessly queenless hive. This complication, combined with the brutal nature of cutting queen cells, is an indication to me that this method of swarm control is not the best one.

The easiest way to deal with a swarm is to let nature take its course. Unfortunately, in some cases, letting nature take its course is not a good option. For example, if you are an urban beekeeper with hundreds of neighbors, most of those neighbors won't appreciate seeing a huge cloud of bees flying past them on the street. And whether a hive is urban or rural, when it's late in the season, a swarm will have little chance of building up enough honey stores for winter, and the mother hive is unlikely to encounter weather favorable for a virgin queen to mate. For times when allowing a natural swarm isn't the right choice, an alternative approach that respects the hive's desire, recognizes

the beekeeper's needs, and often stops the swarm in its tracks is to manipulate the hive in a way that simulates the act of swarming without having the bees actually fly off.

To accomplish this, remove the queen along with a number of frames of brood, honey, pollen, and bees from the hive, place them into a new hive, and move this new hive to a new location. It may become necessary to handle the queen to move her into a new hive. I never like to grab the queen by anything but her wing. The wing—being extremely thin, flat, and flexible—is very difficult to injure if handled carefully. The same cannot be said for the rest of the queen bee's body parts. If you are squeamish about picking up your queens by a wing, try practicing on drones first in order to become more comfortable with the procedure. There's no need to be worried about being stung by the queen; it seems as though queen bees will not sting a person. I have handled many queen bees by picking them up by their wings and I have never been stung; nor has such treatment injured a queen as far as I can tell. Place the queen in the new hive, but leave all frames containing queen cells in the original hive, which is left in its original location. When timed correctly, this maneuver will eliminate the desire and potential for swarming on the part of the workers and the queen.

This is most effective when the queen's new location is close to the original location of the hive. This will allow all the older foraging bees to return to the original hive that is now queenless. With the loss of all the older foraging-age workers in the hive and a significant amount of brood, combined with the addition of new combs with space to lay eggs, the old queen will tend to lose interest in swarming. Meanwhile, the workers in the original hive are without a queen to swarm with, and so they continue to raise the newly forming queens in the queen cells until one of them takes over the hive. Be sure that you transfer plenty of young nurse bees (usually found on the frames of brood) into the new hive with the queen so there will be enough bees to take care of the queen and the brood once the older bees leave.

• URBAN BEEKEEPING •

Along with a phenomenal rise in beekeeping in general, there has been equally rapid growth and interest in urban beekeeping recently throughout North America. I believe this is partly due to a greater awareness of the extensive ecological degradation the world is experiencing and a growing desire to do something positive for the environment. I also think many city dwellers recognize, either consciously or unconsciously, that city life is unsustainable. Sure, per capita urban dwellers consume fewer energy resources then most suburbanites and rural folks on a daily basis, but when the energy spent on constructing and maintaining the city (along with the environmental costs that result) is taken into account, the energy saving benefits largely disappear. Additionally, a city can be defined as an area where the population of humans far outstrips the carrying capacity of the land to support them. Continuing to rely on distant lands for the vast majority of one's food and materials necessary for life during this time of failing economic, energy, and political systems is enough to give all but the most oblivious among us an uneasy feeling. Thus, the city public's interest in beekeeping has risen right along with the growing interest in urban farming, community gardening, farmers' markets, and community-supported agriculture.

Urban beekeepers have unique challenges that make the craft of keeping bees a bit more difficult than for their rural and suburban counterparts. Helping to offset this is the fact that bees seem to do very well in an urban environment. There is an abundance of forage in parks, along tree-lined streets, in backyard and rooftop gardens, and on balconies filled with potted plants. There are few if any pollinators to compete with the bees due to the intense human manipulation of the city environment, which has destroyed much of the natural habitat for most other insects. Add to this the fact that city bees face much less exposure to agricultural pesticides than rural bees do, and it's easy to understand why urban beekeepers are poised for success.

FIGURE 3-38. This bee is foraging on a bouquet of flowers at a sidewalk flower stand on the streets of Melbourne, the second-most populous city in Australia.

While there are cities throughout North America that ban or severely limit beekeeping, the number of cities that are legalizing it is growing, mostly due to strong demand from residents. Beekeeping can be conducted safely in cities of all sizes, and proactive beekeepers are developing city ordinances for municipalities that lack them rather than waiting for an "incident" to trigger the city to come up with its own beekeeping regulations.[18] As a general rule, city governments don't have a problem with beekeepers unless citizens complain. As a result, being considerate of your neighbors and keeping a low profile will go far in preventing potential backlash to beekeeping activities.

The location of the urban apiary plays a large role in helping to keep beekeeping discreet in the neighborhood. Making sure hives are located well away from property lines or are screened from adjacent properties can be important. By positioning hive entrances away from high-traffic areas, interactions between people and bees can be kept to a minimum. Some beekeepers go so far as to camouflage their hives so they look like a common urban element such as air conditioners, while others will paint their hives a neutral color. Rooftop

FIGURE 3-39. Bees swarm in an effort to give birth to a new colony, reducing this natural tendency is critical for successful urban beekeeping.

beekeepers should utilize buildings of no more than 12 stories in order to prevent colony stress from excessive winds. In general, it is a good idea to keep the urban apiary small: two or three hives, and no more than five.

The conscientious beekeeper will be considerate of neighbors by practicing management techniques such as not opening the hives when neighbors are around, using smoke or sugar syrup spray to prevent the bees from becoming defensive, requeening aggressive colonies, taking steps to avoid robbing, controlling swarming, and posting warning signs in or around the apiary. Most of the time people fear bees because they don't understand them. Take the time to educate your neighbors about the fact that European honey bees are actually gentle creatures that play a huge role as

pollinators in helping to maintain life on our planet, and that sting only when threatened. With this knowledge, folks will tend to be much more willing to live and let live with bees in their midst. To help pave the way for the bees, consider starting an educational campaign even before you establish your apiary. This can help get neighbors on board with your plans, and they will be unlikely to file complaints. Or, beekeepers may choose to quietly bring in their hives and set them up in an inconspicuous location. After a year or two, when a honey harvest has been completed, the gift of a jar of honey to the neighbors living in the immediate area can be a gentle way to break the news. Any resistance to your beekeeping will tend to vanish once neighbors find out that you have been keeping bees next door for over a year with no bad result and that you are willing to share some of the honey with them.

Another major concern that is unique to the urban beekeeper is the great importance placed on controlling swarming behavior in bee colonies. Most people don't understand that honey bees are at their most docile during the process of swarming, and they freak out at the sight of a 30- to 50-foot-long cloud of ten thousand bees making their way down the street. Swarm control techniques—such as supering often and reversing hives or making splits/nucs to reduce crowding, requeening regularly to be sure each hive is led by a young queen, and placing swarm traps out around apiary sites in order to catch the errant swarm that may issue forth despite your best efforts—are all important steps that should be taken by the urban beekeeper. Beekeepers should also provide water for their city bees year-round in order to keep them away from dripping faucets, fountains, or other water sources where people may congregate.

Despite your best efforts, you may have a neighbor who just likes to complain and make trouble for you and your bees. This is why registering your hives with the local or state bee inspector, if there is one in your city or state, is a good idea. If your city has adopted best management practices regarding beekeeping and you can demonstrate and document that you have been following them and your hives are properly registered, the city will be on your side no matter how much complaining the neighbor does.

One final suggestion for the urban beekeeper: get to know the businesses within about a three-mile radius of your apiary. Certain businesses may be especially attractive to a colony of bees. Consider the infamous example of Brooklyn beekeepers who discovered their bees producing combs of a beautiful neon-red, honeylike substance. It turned out that their bees had been visiting the dumpsters behind Dell's Maraschino Cherries Company in the nearby neighborhood of Red Hook. A preemptive visit to such businesses to let them know your plan to bring bees into the area and to be sure they take steps to prevent bees from creating potential problems can eliminate a lot of headaches. In the end, being considerate and discreet will go a long way in making your urban beekeeping adventures enjoyable for all.

CHAPTER 4

Genetics and Breeding

Make boot upon the Summer's velvet buds,
Which pillage they with merry march bring home
To the tent royal of their emperor:
Who, busied in his majesty, surveys
The singing masons, building roofs of gold.

—WILLIAM SHAKESPEARE

As with all forms of husbandry, when working with bees, "an ounce of prevention is worth a pound of cure." This is more than just a trite saying. Organic farmers know that the soil is the place to begin in their efforts to produce a successful crop and prevent crop failure from disease and plant malnutrition. Healthy soil is required to grow strong healthy plants. The more robust the plants, the higher the quality of the crop harvested from those plants. The beekeeper, however, is not tied to a single plot of land and the soil it contains, but instead must rely on the genetic "ground" within the hive—in other words, the queen, or mother bee. I firmly believe in the wisdom shared with me by Charles Mraz that the more genetic diversity that exists in your hives, the better. Charlie believed, as I do, that raising "mongrels" rather than "thoroughbreds" results in a greater depth of genetic diversity that may give the bees an edge when dealing with unexpected challenges. As a result, over the years I have purchased and introduced into my apiaries several strains of honey bees with mite- and disease-resistant tendencies. The apiculturist who maintains pure strains of bees in all of his or her hives can be compared to the farmer who grows only a single strain of one type of crop. This type of farming is known as a monoculture, and unless you are a researcher or a queen breeder dedicated to maintaining a specific breed of bee, keeping a pure strain of honey bee may leave your hives unnecessarily vulnerable.

The Irish Potato Famine is a poignant example of the folly of relying on a monoculture. During the course of the two centuries leading up to 1845, the poor families of Ireland had become increasingly dependent on the potato to feed themselves. Potatoes provided good nutrition; were easy to grow; required a minimum of knowledge, labor, and technology to produce; and they kept well when properly stored. All this made the spud a favorite of the lower classes in Ireland. Over time, Irish farmers came to rely increasingly on the most productive potato varieties available. It has been estimated that, in 1845, only two potato varieties accounted for the majority of the spuds planted in Ireland. Unfortunately, these varieties were highly susceptible to a blight that left acre upon acre of Irish farmland covered with blackened dead potato plants between 1845 and 1850. Over a million Irish people are said to have perished during the Irish Potato Famine of the 1840s.

The apiculturist can learn much from this tragedy. By keeping honey bees of varied genetic stock in their bee yards, beekeepers can help to ensure that at least some of their bees are likely to contain traits that will help them survive outside stresses that may come along to threaten the viability of the colony.

Although some queen breeders make use of isolated breeding yards, purebred honey bees are typically obtained through the practice of artificial insemination. Unfortunately, artificially inseminated queens exhibit more problems than naturally mated queens. It has been my experience that, shortly after an artificially inseminated queen has

FIGURE 4-1. Swarm cells that house a queen tend to be built along the bottom edge of the comb, from queen cups into which the queen has laid her eggs. Supersedure queen cells tend to be built around an egg that the queen has laid within worker comb. PHOTO BY HILDE WALLEY.

been successfully introduced into a queenless colony, a hive will often raise a new queen through the supersedure process to replace the artificially inseminated one. Thus, to help ensure the longevity of the genetic characteristics contained within an artificially inseminated queen that has been transferred into a bee population, it is wise to raise daughter queens from the eggs laid by these mother bees as early as possible. Such actions are not as urgent when dealing with naturally mated queens, who tend to lay more eggs on a daily basis and live longer than queens mated in a laboratory. Although at least some of the genetic traits of an artificially inseminated queen are likely to continue on in the hive that successfully replaces her through supersedure, such a loss can certainly be demoralizing to the beekeeper. This is especially true in the case of "breeder" queens, which cost $250 or more. In addition, supersedure may be accompanied by a

decrease in egg laying or an interruption in the brood cycle. This can slow the population growth of the hive and may adversely impact the colony's ability to store enough honey and provide for itself during the winter months. Common sense—which is not so common anymore—tells us that bees know how to inseminate and raise queen bees better than humans do. To me, it makes economical as well as ecological sense to let the bees raise their own mothers.

This is not to imply that artificially inseminated queens cannot be as good as, or even better than, open-mated queens. It's just that with artificial insemination, a human-imposed factor enters the equation, a factor that does not exist with queens that mate naturally. This additional factor, when added to the long list of small miracles that must take place for the successful enthronement of a queen mother, brings with it additional complications and the possibility of failure. The potential is created for something to go wrong, whether due to human error, poor-quality work during the insemination process, an oversight, or some unknown or poorly understood factor that silently makes its presence felt when the honey bee's natural mating process is bypassed.

As a great appreciator of honey bees, I like to consider myself a steward of the hives in my care, and I make an effort not only to keep hives alive and thriving, but also to avoid crushing or killing individual bees when working with them. By the same token, I do not believe in purposely killing and replacing aged or poorly performing queens. The act of requeening a hive by removing a colony's queen and replacing her with a new queen has long been a recommended practice in beekeeping circles. The primary reasons for this are that younger queens are less likely to swarm and that they tend to lay more eggs than older queens. The resulting larger colonies are more likely to produce bumper honey crops. By requeening a colony with a mated queen instead of allowing it to naturally supersede the old queen, the beekeeper will tend to have more control over the genetic makeup of the hive and will avoid mating issues due to inclement weather or an insufficient number of drones in the mating area.

On the other hand, beekeepers who requeen their hives find that it can be a laborious, time-consuming, and frequently expensive task. To make matters worse, requeening efforts will often fail, even when carried out by an experienced beekeeper. These disadvantages to requeening may encourage some beekeepers to pursue a different approach.

Rather than requeen, I prefer to allow each hive to live its natural life cycle and allow each queen to live out her life fully, in accordance with their own timetables. This may mean that hives will supersede or swarm more often, though there are management techniques that introduce space into the brood area and deter the swarming instinct without requeening (see "Identifying and Working with Queen Issues" in chapter 3 on page 78). And even if superseding and swarming are more frequent, the time and money saved by not having

to raise and/or purchase replacement queens and introduce them into the hive more than make up for such inconveniences in most cases. Unless you are doing research, keeping bees in Africanized bee territory, or trying to breed a pure strain of honey bee, requeening is an option, not a necessity.

These two divergent approaches tend to be guided by differing worldviews. At one end of the spectrum is a perspective that views life as a ladder or pyramid, with dirt at the bottom rung or level, humans at the top, and everything else filling in the hierarchy somewhere in between. From this perspective those at the top of the hierarchy have dominion over those below. This leads to the contemporary concept of ownership that allows one to do what one will with whatever one owns. Unfortunately this seems to perpetuate a system that encourages the use of force to obtain resources. Even the term "resources" is a result of this view. Referring to land, trees, minerals, and bees as "resources" assumes that they are here simply for our use, and that nonhumans don't have a right to freedom of expression that is uniquely their own for their own sakes. From this viewpoint, it is perfectly normal and expected that one would purposefully crush or "pinch" a queen bee that is not performing up to our human standards and replace her with one we expect will do better. In most cases, though, the bees are not trying to get rid of their queen, and in fact may try to protect her from our requeening efforts. When we purposely kill a hive's queen and replace her with a new one, we tend to be acting on behalf of our own interests and not necessarily those of the bees. What gives us the right to treat our hives this way? On the opposite end of the spectrum is a circular worldview in which there is no top or bottom. All beings coexist in a community of life in which everything supports everything else. In this worldview the tree, the blade of grass, or the honey bee colony is no more, and no less, important than the human being. From this perspective, everything in nature has its own right to life, liberty, and the pursuit of happiness (to use one of our foundational American expressions).

The majority of reasons that might cause a beekeeper to requeen a hive relate to the desires and needs of the beekeeper: a wish for more brood or honey production, a wish for a younger queen (to make the hive more productive), or a desire to change the temperament of the hive. Little if any consideration is given to the desires and needs of the bees. To be fair, this kind of attitude is typical throughout our agricultural industry. As part of the corporate industrial model of agriculture, we often impose our limited ideas of how things should be upon the animals, plants, and land we are responsible for rather than letting things unfold in a more natural holistic way. By acting this way as beekeepers we are implying that we know, better than the bees themselves, what is best for the hive. This interventionist way of thinking reveals itself when beekeepers clip the wings of a queen in an attempt to stop her from flying off with a swarm, install a queen excluder to limit the area in which the queen can lay eggs and raise brood, or remove honey and feed sugar syrup back to the hive as a replacement.

Now, I must confess that I have never requeened a hive, mainly because this is the way I was taught by the Mraz family of Champlain Valley Apiaries, who mentored me in my early years of beekeeping. My tendency, as was the tendency of the Mrazes, is to allow each hive to prosper or decline according to its own abilities and fate. Sure, I'll do what I can to try and assist the hive by attempting to keep it as pest- and disease-free as possible, and I'll take pains to see that each hive has enough honey to see it through the winter, and so on. The ultimate decision as to whether the queen lives or dies (and thus the ultimate fate of the hive), however, is left to the bees. The only time I have introduced a mated queen into a hive is when the hive has gone queenless (and I have caught it *before* it has become a drone layer), or if I have made a nucleus colony and don't want to wait for the hive to raise their own queen from an unhatched or very recently hatched egg.

Over time the wisdom of not requeening my hives has subtly influenced my beekeeping management style. I have noticed that sometimes a weak hive that had to be nursed along all year surprisingly survives the winter in good shape. The following season such hives will typically produce a respectable crop of honey for harvest. Why this happens I am not certain, but I have to admit the possibility that the initial poor performance may have been due to my management choices, and not any fault of the bees. If the trouble a hive is experiencing is caused by me, how does my killing the queen and replacing her improve things? Contemplating such things injects a sense of humility into my beekeeping.

In addition, sentencing a queen bee to death based solely upon her age, or the size and shape of her brood-laying pattern, ignores a host of other characteristics that may or may not be readily observable. Although a particular queen may not possess certain characteristics that are deemed important to us, such as high honey production, she may harbor other traits that are extremely beneficial, but not as obvious. Resistance to mites and diseases are two examples. Out of respect for the bees that provide so many benefits for the rest of us that share this wonderful blue-green planet, I find myself resistant to purposely killing a queen mother and replacing her. It's simply a matter of stewardship and an approach to apiculture that I prefer to practice. This also means that, for me, one of the primary causes of colony loss is queen failure in hives that are unable to successfully replace their queen through supersedure, and I am comfortable with this.

Luckily I keep bees in a rural setting. As mentioned earlier, if I kept bees in an urban area where swarm prevention was paramount in order to remain on good terms with neighbors and the municipality, then I would feel compelled to replace older queens every year, even though it would not sit well with me philosophically.

There are many beekeepers who would vehemently disagree with my philosophy, and who argue that requeening a colony that is struggling is not only more profitable in the long run, but the most humane thing you can do for the bees. It seems to me that this position assumes we know better than the bees what is best for the hive when it comes to raising and installing the colony's reproductive organ, and I am not willing to make that assumption. This same line of thinking is what prevents me from combining a weak hive with a strong hive. As long as the weak hive is queen-right and has enough bees to care for the queen, I will let the hive be. By cultivating an attitude that allows us to treat the honey bee with greater reverence and respect, we nurture the same mindset that provides hope for the future of human culture. In essence, if we want to create a different world, we have to start by changing our minds. In the universe of the honey bee, the workers will carry out the process of supersedure to replace an old, failing, or injured queen and raise a new queen that will inherit the throne. Should the hive fail to raise a new mother, the beekeeper can respond by simply combining the remaining bees with another hive. This provides the opportunity to help a hive that is weak and in need of a population boost. This approach can also be used with drone-laying queens late in the summer, when the chances of raising a new queen and building up a healthy population of bees large enough to make it through the winter are doubtful. (For more on drone layers, see "Identifying and Working with Queen Issues" in chapter 3 on page 78.)

In essence, we can treat the hive body and supers of the failed queen like empty equipment and use them for supering colonies in need of more room, while the attending bees within the equipment will be accepted by their foster mother and able to join forces with the hive the supers are placed on. Many books will tell you to find the queen within the failing hive and kill her before combining colonies. This can be difficult, if not impossible, in the case of a hive that has been taken over by one or more laying workers, because the laying workers are not that much larger than their sisters and are difficult to locate. Unfortunately, drone-laying hives rarely accept a new queen that is introduced into their hive directly. A queen-right colony (with a healthy, fertile laying queen) will, even when weak, typically be able to accept a hive body containing drone-laying workers and their attendant workers and drones with little difficulty. It is also possible to encourage the workers in a drone-laying colony to raise a new queen from freshly laid eggs taken from a queen-right colony and inserted into the hive. This procedure tends to work best when carried out early on, when the queen first begins to fail, rather than later, after some time has passed and drone-rearing ways have become firmly established within the hive. It is almost as if, after some time, the workers left within the colony give up trying to raise a new queen from the fresh eggs laid by their failing mother and are thus unable to capitalize upon the fertile eggs donated by another hive. Compared to killing the failing queen or laying workers and then requeening, allowing the hive to raise its own replacement queen, or at least combining the remaining bees with another hive, is in greater alignment with a basic respect for nature's biological processes and the bees themselves. Such respect has always existed at the heart of the organic movement.

· A CASE FOR ·
LOCALLY RAISED QUEENS

No matter what the qualities of a hive are, the beekeeper cannot fully benefit from its specific genetic traits unless the bees can thrive in the climate where the beekeeper has located them. Living in Vermont, about 75 miles from the Canadian border, I have found that bees raised in the western and southern United States, where the majority of American queen breeders operate, are not necessarily well suited to the northeast climate. The same may be said of bees and queens imported from other countries. This seems especially true in the effort to find strains of honey bees that contain a natural resistance to mites and diseases. Just because bees show some mite or disease resistance when bred in warm climates like California or the southern United States, where hives are active on a more-or-less year-round basis, does not necessarily mean that those same bees are capable of surviving side by side with mites or disease in northern climates. Rather than enter a deep sleep and hibernate like many other insects and animals, the honey bee becomes dormant to conserve resources until spring. A key survival trait for northern bees is their ability to raise a large population of healthy bees in autumn capable of surviving the winter's period of reduced brood production. Unlike its warm-weather cousin, the Africanized honey bee (*Apis scutellata*), the European honey bee (*A. mellifera*) has fully developed this mechanism for surviving the cold winter months in northern climes. These late-season bees, born in autumn, have the capacity to live for several months within the relatively inactive environment of the winter cluster. To stay warm, the bees eat the honey the hive has gathered during the summer and use these calories to flex their wing muscles and generate body heat. The bees literally snuggle with one another throughout the winter by huddling close and creating an insulating shell with their bodies, to the point where they form a warm ball surrounding the queen. As the bees on the exterior of the cluster cool down, they slowly work their way into the interior to warm themselves, while warmer bees nearer the center move to the outside to take their place. Although the cluster does not heat the unoccupied space within the hive to any great degree, it can regulate the temperature in the middle of the group—where the queen remains—by contracting or expanding the cluster size.

As the cold weather comes on, the cluster will contract, and the queen slows down her egg laying. She eventually may stop laying altogether for one or more periods before resuming her reproductive duties and slowly building the hive's population back up in preparation for the spring bloom. This reduction in egg laying serves two distinct purposes. The first is to prevent the possibility of chilled brood, which can be fatal for the baby bees. Chilled brood occurs when the hive does not have enough bees to cover the developing brood and keep it warm with their body heat. The reduced egg laying also conserves the hive's resources, because

it takes significant amounts of honey and pollen to raise baby bees into adults. Reducing the need to consume pollen and honey helps ensure that the colony will have enough provisions to carry it through the winter.

Honey bees born in late autumn have special attributes that make them much better suited to winter survival than their summer-born sisters, who literally wear themselves out by working so much that they may live for only five to six weeks. The fact that honey bees have a hotter sting in the fall than during other seasons indicates that the bees are indeed different at this time of year. Unfortunately, winter survival can be especially challenging for these bees when they have to share their hive with mites. Honey bees are weakened both by mites feeding on them during their development and in adulthood, and also by associated viruses and diseases the mites may spread through the hive's population. (I cover mites in much more depth in chapter 5.)

The best way to build *Apis mellifera* bee stock that is acclimated and well suited to survive the unique conditions of one's area is to raise succeeding generations of them in that same locale. Luckily a lot of specialized knowledge and equipment is not necessary to do this. Good results may be obtained by breeding from one's best hives and working with the honey bees' natural instinct to raise a queen when the hive finds itself without one. Reserving the option to rear queens that have performed well for you in your area by creating new colonies from existing beehives is yet another reason for not automatically requeening hives, as some authorities recommend. In the days before varroa mites, beekeepers would typically split the hives that made the most honey.

Another reason for splitting a colony would have been to reduce the hive's population during overcrowded conditions in an effort to avoid having the bees swarm. With the advent of varroa, however, high honey production ceased being the primary trait beekeepers focus on when deciding which hives to use as breeding stock. Hives that have survived the winter and have a relatively high population of bees in the spring are the colonies that are bred these days, in an effort to increase the number of hives with a genetic disposition for survival in spite of the varroa mite and its associated diseases.

When a hive is divided to create one or more new colonies, each new hive is given a small version, or nucleus, of what is found in a full-sized hive with which to build upon. Thus, the term for these hives is nucleus colony, or nuc for short. The beauty of using the existing hives in your apiary to create more colonies is that the potential of importing into your bee population diseases, pests, or unwanted genetic traits, such as those found in the Africanized honey bee, is greatly reduced. Some queen breeders are utilizing nuc making in combination with the elimination of mite treatments altogether in an effort to develop varroa-resistant strains of honey bees. It is clear to me that a strong nucleus colony program is integral to the success of this "tough love" approach. "Treatment-free" beekeeping is covered more fully under "Cultural Management Practices and Biological Controls for Varroa" in chapter 5 on page 121.

In this day and age, when varroa mites are so prevalent, making splits or nucleus colonies from surviving bees is one management technique that has definately increased the likelihood of honey bees surviving winter. Not only do these splits carry on the genetics of hives already proven to have a level of survivability, but the act of making nucleus colonies, in and of itself, consistently imparts a higher survival rate for the nucs over the winter than for the older, "full-grown" colonies. Splitting up a hive to make a nuc not only divides the number of bees in the hive, but also reduces the number of mites at the same time, because some are removed from the original hive when the nucleus colony is made. Splitting a hive and then allowing the bees to raise their own queen also dramatically interrupts the colony's natural brood cycle, upon which the varroa mites depend. The result is an interruption in the reproductive cycle of the mites as well. By interrupting the brood cycle of the mites through

the creation of a nuc that must raise its own queen from an egg, we slow down the exponential growth curve of the varroa population and reduce the likelihood that the number of mites in the hive will build to a level that will cause the colony to collapse later in the season. Indeed, a higher tendency to swarm is one of the characteristics attributable to the mite resistance found in both the Africanized honey bee and certain Russian breeds. Some beekeepers achieve similar results by caging the queen for a while, thus creating a short broodless period. Thus, the ultimate experiment may be to see whether a strong hive can survive for several years without mite treatments *and* without artificial broodless periods being created within the colony. This may require that today's beekeepers and breeders become more tolerant of the natural swarming instinct within their colonies, a trait that most have long tried to breed out of their bee stocks in an effort to maximize honey production and profits.

The naturally emitted swarm is *Apis mellifera*'s way of giving birth to a new colony, and it is one of the most fascinating activities in which honey bees engage. It stimulates the human imagination and stirs the soul to witness a swarm issuing forth from a hive in search of a new home. The queen mother and half to two-thirds of the workers within a colony will time their departure so that it takes place shortly before one or more new virgin queens, which have been lovingly raised and cared for, are set to hatch out of their birthing cells and take over leadership of the hive. The confidence the swarm shows in itself and the faith it has in the universe's ability to provide for its future well-being are inspiring. Certainly the mother and her daughters do not make the decision to swarm carelessly, typically timing their move to coincide with conditions when the hive is chock-full of bees, honey, pollen, and brood, and there is plenty of nectar and pollen still to be gathered. It makes good sense, after all, to wait until times are plentiful and prosperous before deciding to build a new home and start another family. However, when compared to human parents, the widowed mother bee shows unparalleled generosity of spirit in her actions.

HOMEGROWN QUEEN REARING

One queen breeder who is working with nucs to try and develop a mite-resistant bee is Kirk Webster of Champlain Valley Bees and Queens in New Haven, Vermont. Kirk's experience has proven that the "live and let die" method is not for the faint of heart. A full-time beekeeper with about a thousand hives at the peak of his season, Kirk has been slowly weaning his bees off varroa treatments, to the point where he doesn't treat his hives for *Varroa destructor* at all anymore. Without invasive mite treatments, the hives at Champlain Valley Bees and Queens experienced winter losses of 60 percent or more in some years, but losses have since leveled off to between 30 and 50 percent. Kirk's approach is to split and raise queens from his surviving colonies each spring and repeat the cycle over and over again. The hope is that, over enough generations, the bees in his care will self-select for varroa-resistant traits, to the point where a bee that is immune, or at least highly tolerant of mites, will result. Although this is not an attractive option for all beekeepers, this radical strategy certainly has the potential to separate the wannabees from those that are truly resistant to *Varroa destructor*. However, even if efforts such as Kirk's do not result in true immunity, the benefit of any incremental increases in varroa resistance that may be developed is likely to prove invaluable in the long run.

As with all things, the cold-turkey approach also has its shortcomings. Aside from the obvious emotional and economic stress that may be caused from the loss of a large percentage of hives each year, and the additional labor involved with cleaning out dead colonies each spring, there is no guarantee that this approach will ultimately prove successful. Being as much an art as it is a science, the craft of beekeeping involves a staggering number of variables. Honey bee genetics, atmospheric weather patterns, local flora and fauna, and the type and timing of hive manipulations are just a few of the vast number of influences that can affect the health and survival of a hive.

When humans establish a home, become prosperous, and have raised their offspring to adulthood, the tendency is to push the kids out of the nest, so they either learn to fly on their own or hit the ground trying. In contrast, by leaving behind the hive she has guided to maturation and prosperity, along with her most loyal family members, the queen bee exhibits courage truly worthy of royalty, transferring ownership of her entire estate to the daughters she has left and in the process ensuring their continued well-being. Flying away from the safety and security of the hive toward the unknown, the swarm leaves behind all the material possessions it can't carry with it. The honey bees' act of swarming inspires a sense of generosity at a level that is unmatched in the world of insects.

Although the idea of creating artificial swarms and rearing new colonies from your own bee stocks may seem intimidating, it is relatively easy provided you keep in mind the basic biology of *Apis mellifera*. When a hive finds itself without a mother, the workers will go to work and attempt to raise a new queen in short order. The primary requirements to accomplish such a task are a few freshly laid fertile eggs, ample supplies of honey and pollen, enough bees of all age groups to survive the roughly eight-week period before the freshly hatched brood from the newly mated queen emerges from its birthing cells, and enough good weather during the third and fourth weeks for the new queen to make one or more successful mating flights. With thoughtful planning and attention to detail, the beekeeper can often orchestrate the opportunity for the necessities of enough eggs, bees, and good weather to come together successfully.

A well-known example of a strain of bees bred for a specific trait is the Buckfast bee. These bees are genetically resistant to tracheal mites (*Acarapis woodi*) and were developed at England's Buckfast Abbey by the famous beekeeper Brother Adam (1898–1996). In recent years, apicultural researchers have made genetics the focus of much of their efforts to control the damage resulting from varroa mites. Hygienic behavior (increased grooming and hive cleaning) has become the most studied genetically determined characteristic of honey bees that exhibit varroa resistance. This in turn has resulted in the commercial availability of several bee strains with varying levels of resistance to varroa. These include bees that can trace their genetics to western Russia, near Vladivostok along the Pacific Coast (Primorski stock); the survivor stocks from across the United States that were collected and refined by Dr. John Harbo and Dr. Roger Hoopingarner; and the Minnesota Hygienic stock developed by Dr. Marla Spivak of the University of Minnesota. The Russian breed has been shown to control mites at least partly through increased grooming behavior. Grooming behaviors that fend off the mites include bees biting mites that are on either their own bodies or the bodies of other bees. Should a bee chew off a mite's leg, the break in the mite's exoskeleton will cause the parasite to quickly dehydrate and die.

It is similar hygienic characteristics that make these bees more resistant to other hive diseases as well. The Minnesota Hygienic stock and the bees that were bred from various survivor stocks—hives that were able to survive despite the onslaught of varroa—have developed a hygienic mechanism for removing diseases and mite-infested brood from their nest. When applied to mite-parasitized broods this behavior was initially labeled suppressed mite reproduction (SMR). This genetic trait has since been renamed varroa-sensitive hygiene (VSH). VSH bees exhibit the ability to detect varroa-infested brood within the sealed birthing cell and remove it. These bees are also more selective about removing brood infested with mites that are capable of reproducing, and passing over brood that share cells with mites that are unable to reproduce.[1] How the VSH bee is able to differentiate between mites that can reproduce and mites that can't is still the subject of study. Nevertheless, at this time it seems that the best chance of breeding a honey bee that is truly immune, rather than just being tolerant of or resistant to varroa, lies in combining these types of hygienic behaviors with a strong swarming instinct.

For those who don't have the stomach for the radical measures of the "live and let die" strategy, a more systematic approach has been developed to test for hygienic behavior. Basically, hives are tested to see how rapidly they are able to detect, uncap, and then remove dead or damaged honey bee brood. There are several variations of this test that have been developed over the years. One is to cut out a section of sealed brood, freeze it, and then replace it within the frame in the hive. The frame with the dead brood is then inspected 24 to 48 hours later, and the number of cells that have been opened and have had the dead brood removed are recorded. The downside to this technique is that it is laborious and very disruptive to the colony. One could even make the argument that any hive that didn't notice something wrong after going through this procedure is *really* deficient. A more refined approach has been developed by Marla Spivak using liquid nitrogen to freeze the sealed brood without removing it from the frame. A hollow cylinder, such as a soup can or PVC pipe open on both ends, is driven through the brood to the middle of the comb (midrib). The rim should be smooth, with all the rough edges filed down to provide a tight seal between the cylinder and the wax midrib of the frame. A good seal is essential for preventing leakage of liquid nitrogen from within the test area. The liquid nitrogen is then poured into the can or cylinder and onto the frame, instantly freezing and killing the brood. Again, the percentage of dead bee brood removed in a 24- to 48-hour period is observed. The major shortcomings to this approach are the difficulty in obtaining liquid nitrogen and the dangers associated with its handling.

A simpler technique to conduct the hygienic test uses a pin or needle to fatally injure the developing brood. After the needle is inserted into 100 cells of sealed brood, the results are evaluated 24 to 48 hours later. Although still labor-intensive, no special or hard-to-find equipment and materials are required with this method.

Hygienic behavior is determined to be present if 95 percent or more of the dead brood is removed within the initial 48-hour period. Limiting the sample area to 100 cells helps to simplify the calculation of percentages. It also helps if all the brood used for the test is grouped together, so that a clear delineation between the dead brood removed and the live unharmed brood on the same frame is evident. Some researchers believe that a hive should be categorized as hygienic only if the test is performed twice and over 95 percent of the dead or damaged brood is removed within the same time span on both occasions.

Either way, it is important to realize that the proportion of workers required for a hive to express hygienic behavior does not have to be exceedingly high to be obvious and effective. Some research has suggested that colonies with only 15 to 50 percent hygienic bees may be just as hygienic when confronted with large amounts of dead or damaged brood as colonies with 100 percent hygienic bees. This may prove important, because queens are believed to be able to mate with as many as 20 or more drones, and only a certain proportion of these drones carrying the hygienic gene are needed to produce a hygienic colony. By the same token, our lack of knowledge as to what degree of hygienic behavior is desirable means that it is possible that it may be overdone. Perhaps, in its extreme form, hygienic behavior can be detrimental to the colony. At this time, current thinking holds that selection for hygienic behavior by breeders will produce far greater resistance, or even true immunity, to *Varroa destructor*, along with diseases like American foulbrood, sacbrood, and chalkbrood.

Whether we take the radical "live and let die" approach or a more scientific one, once a hive with favorable genetic traits has been identified, we can breed the colony in an attempt to pass on and improve upon the hive's favorable characteristics. One way to accomplish this is to raise or graft queen cells and then transfer them into nucleus colonies. This can be a highly efficient method for producing many queens. It also often requires specialized equipment and skill. As a beekeeper who is continually trying to reduce the amount of equipment I must deal with, I have forgone this approach for a less efficient but simpler and less labor-intensive way to propagate a colony. The technique I prefer is to simply split a hive in order to create one or more nucleus colonies from the mother hive.

• CREATING THE SPLIT • OR NUCLEUS COLONY

If you already own a hive or two, you can make your own increase in order to establish new hives. This should be done using your strongest, most productive hives and is best carried out during the primary swarming season, which tends to coincide with local honey flows and will vary among geographic locations. By timing your breeding (split or nucleus creation) efforts to overlap with the natural swarming season of your area, you can take advantage, as the bees do, of the favorable conditions that typically exist during this time: warm average daytime temperatures and an abundance of blossoms for foraging. Plus, enough time is left in the season for the new hive to build plenty of combs, fill them with honey, and build up a large-enough bee population to survive the long dearth of winter.

In Vermont in an average to good honey year, a nucleus colony or split can be created in late April or as late as the end of June and still have a good chance of successfully surviving the winter without supplemental feeding. Late-season swarms, nucs, splits, and packages that are caught, created, or purchased in July or August almost always require additional feeding in the fall or during the winter and early spring to prevent starvation.

Once you have identified breeding stock within your established apiary and the timing is right in your geographic location, you can get down to the business of creating more colonies. Once the hive has been divided, additional labor on your part can be eliminated by allowing the bees to raise their own queen. Doing so will run the risk that the queen the hive raises may not mate successfully. If they are successful, however, the population of varroa mites in the colony will not increase as fast as it would have otherwise, due to the interruption of the brood cycle during their queen raising efforts (more on this in chapter 5). Control over the genetic makeup of the hive is reduced, since the queen raised by the nuc or split is free to mate with drones throughout the mating area, which is reported to range about 6 miles. On the plus side, though, queen introduction issues are eliminated.

For a relatively small cash investment, splits taken from a hive and made up of frames of bees, brood, honey, and pollen can have a queen introduced to the colony in much the same way as performed with a package of bees. If such a split also includes unhatched eggs, there is a chance that the colony may be able to fall back on raising their own queen should the queen introduction fail.

As with all aspects of beekeeping, there are numerous ways to successfully go about making a nucleus colony or split. For beekeepers whose management style is to overwinter hives in two deep hive bodies, the easiest method is to separate the hive bodies and provide both boxes with a bottom, an empty honey super or hive body for anticipated honey storage, an inner cover, an outer cover, and an entrance reducer. Make sure there are at least a few unhatched eggs in both of the hive bodies so the worker bees in the box without a queen will have the resources to raise a new mother bee. The eggs, which resemble miniature grains of rice, can be hard to see, especially when the comb is crawling with many bees. One easy way to encourage the bees to move off an area of comb so you can inspect the open cells below the bees for eggs is to *gently* blow on the area that you want the bees to move off of, as if you were blowing on a seed head of a mature dandelion.

Initially, you may not know which hive body contains the queen, so it is important that some sealed worker brood, honey, pollen, and plenty of bees are also included in both boxes. After creating the split, it is best to leave it alone for about 30 days before conducting an inspection for the presence of a queen. If you rush the inspection, any newly formed unhatched queen cells may be damaged while you are manipulating frames. Unfortunately, there is no guarantee that the newly created split will be able to raise a queen or, if it does, that the queen mother will be able to successfully mate and return to her hive without being eaten by a bird or falling victim to some other tragedy.

The hive containing the original queen will not experience a disruption of its brood-rearing cycle and should contain frames with sealed worker brood 30 days after formation. The split, on the other hand, will not contain sealed brood but should have a newly mated queen that is laying her first eggs. During the initial inspection of the split, I have observed that the first eggs laid by a new queen may look similar to those of a drone-laying queen or worker. Don't allow this to confuse you! Eggs flopped over on their sides or not positioned in the apex at the back of the cell, as well as more than one egg laid in a cell, are all part of the learning curve the new queen may go through as she grows into her role within the hive and figures out the nuances of her plumbing. It is the frequency at which such instances are discovered that helps the beekeeper distinguish between the hive with a mated queen and the one with a drone layer or laying worker. Cells with mislaid eggs will be the norm when dealing with the latter, rather than the exception occurring early in the egg-laying career of the former.

This "splitting" method of creating new colonies requires a minimum of hive manipulation and relieves the beekeeper from the work of having to locate the queen prior to dividing the hive. However, this should be tried only with a hive that has a large population of bees occupying both chambers of the original full-sized colony. This is because the older field bees that leave the split to gather nectar and pollen will follow recognizable landmarks and return to the location of the original hive after completing their foraging activities. Therefore, it is important to be sure there will be enough young bees left behind in the split to carry on the essential activities of the developing hive once all their older sisters have departed.

The entire procedure for making a nucleus colony is basically the same as for a split, only nucs are typically made up of only three to five frames. One way to prevent the reduction in population caused by the drifting of foraging-age worker bees from a nuc back to the mother colony is to move the nucleus colony to a new location over 2 miles away from the original hive (this can also be done with a split). This will force the foraging bees to reorient themselves and become familiar with the strange new territory surrounding their hive location, memorizing the landmarks that will lead them back to their newly created home location rather than their original hive. However, although this is a great way to improve the chances of a newly created hive's success, it is not always practical or possible to move a nuc to a new location 2 or more miles away.

Older established hives may be moved a short distance within an apiary without the loss of their foraging force by using one of two methods. The first is to move the hive over 2 miles away for a period of about two to three weeks, as described above. Then, once all the older bees of foraging age that would recognize the local landmarks have either died off or reprogrammed themselves to this new apiary, the hive may be moved back into the original apiary and placed in its new location to relearn the area once again. Alternatively, a hive may be moved a very short distance (one to two feet) away from its original location. Such a minor move will not cause foraging bees to become so disoriented that they are unable to find their way back home. Regularly moving a hive such short distances and providing a period of a day or two between each move can progressively transport a colony to a completely new location within a bee yard without a significant loss of foragers.

Fortunately, leaving the baby colony in the same bee yard as the mother hive without first moving it 2 or more miles away, and simply ensuring that there are plenty of young worker bees in the nuc to carry out the hive duties, can still obtain good results. Whenever you make a nucleus colony or a split, it is important to reduce the entrance down to a single small hole or two by blocking most of the open entrance areas with something like a small piece of wood or some wads of grass. This helps reduce the potential for robbing, especially if the honey flow ends or fails to materialize altogether. During years with a poor honey flow it is possible for a nuc to be totally robbed of all its honey stores due to the bees' weakened and disorganized state, all because the hive entrance was not reduced to allow the guard bees to easily defend the colony.

When you wish to bring specific genetic traits into your bee yard, a queen-right nucleus can be created quickly by introducing a queen instead of waiting a month or so for the bees to raise their own. This method has the benefit of gaining an extra four weeks of brood rearing, resulting in the faster buildup of the hive. Although it is possible to introduce a queen directly into a nucleus colony that has just been created, complications can be reduced and the chances of success improved if the hive is left queenless for at least a day or two before a new queen mother with desired genetics is placed into the fledgling colony.

The best way to introduce a queen into a colony is in a queen cage. The cage slows down the introductory process so that the bees have a chance to become familiar with, and accept, the queen prior to her attempt to integrate into the hive. To introduce a new mother into a hive or nuc, remove the cork from the candy end of the cage. I prefer to poke a small hole through the candy with a nail or some other similar object, like the reamer/threading tool on a Swiss Army knife. Although it is not necessary, and many beekeepers don't bother to do so, the hole encourages the bees to chew through the candy, thus releasing the queen more quickly.

I like to position the cage between two frames in the nuc, preferably with the screen side facing the bottom of the hive, and tilted so that the candy end is a little lower than the area occupied by the queen. This will make it easier for the bees to make contact with and feed the queen through the screen until the candy plug is eaten away. Unfortunately, the cage will not fit between the frames of a ten-frame hive body without damaging the comb. A frame (preferably one that does not contain brood) can be removed so the cage will fit without crushing any sections of comb. Be sure the cage is wedged in tightly between the frame's top bars because the queen may die if she falls to the bottom of the hive. It is preferable to position the cage between two frames of brood within the colony so that the queen will be located within the cluster, the warmest area of the hive. Once the queen cage is in position, close up the hive and wait three to four days. When you open up the colony again and find the cage

empty, look for freshly laid eggs in the combs as an indication that the queen has successfully been released and has been integrated into the hive. If the queen has not yet been freed from her cage, you should release her by removing the screen on the cage and letting her wander out into her new abode.

Once you have gained some experience and are feeling more confident in your abilities, you can make two or more nucs from a single hive body filled with bees, brood, honey, and pollen. This can be accomplished by removing three, four, or five frames that together contain at least a few unhatched eggs from the parent hive to make up each nuc. Properly laid honey bee eggs resemble miniature grains of rice standing up on end and are located way in the back of the cell at the apex or point where the six sides of the cell meet. This means that from a 10-frame hive body you will need at least two frames containing eggs to make up two 5-frame nucs, or three frames of eggs to make up two 3-frame nucs and one 4-frame nuc. Be sure that the frames that make up the nuc also contain worker brood in various stages of development, capped honey, pollen, and nurse bees who will care for the young. Position the frame(s) with eggs between the frames containing both uncapped and sealed brood. This will ensure that the eggs used to create queen cells will be located within the cluster of bees that are surrounding the brood and make it more

FIGURE 4-2. Positioning the queen cage with the screen side facing down and the candy end lower than the corked end takes advantage of the bees' natural tendency to move upward.

FIGURE 4-3. It takes about three days before a freshly laid honey bee egg will hatch. Notice the surrounding cells filled with the pollen necessary for feeding the growing young bees. PHOTO BY STEVE PARISE.

likely that the developing queen(s) will be well cared for. If the nuc is not going to be moved 2 or more miles from the mother hive, it is a good idea to shake additional bees from frames of brood taken from the mother hive into the hive body containing the nuc to ensure that an adequate supply of young nurse-age bees are transferred into each newly created colony. Special care must be taken to ensure that the queen is not among the bees adhering to the frames being transferred or the bees shaken into the nucleus colonies. When inspecting each frame for the queen it is best to hold the frame over the parent hive as you lift the frame from its resting place. This way, if the queen is on the frame and loses her grip on the comb, she will fall back into the hive and not on the ground, where you might accidentally step on her. Holding the frame over the open hive when it is initially removed is a practice that should be followed whenever frames are removed from a queen-right colony.

Inspecting the Hive and Finding the Queen

Bill Mraz of Champlain Valley Apiaries taught me a wonderful system for quickly and efficiently finding the queen, as her swollen abdomen full of eggs causes her to stand out from the rest of the individuals in a hive full of bees. Bill's approach is to first position yourself so that the sun is shining over your shoulder and then start the search, beginning with the inner cover. Upon removal of the inner cover, immediately check the underside for the queen. Although she usually remains within the confines of the brood nest, she sometimes wanders and could be anywhere. In fact, the queen mother is often found up on the inner cover on cold spring mornings in northern climates because this is often the warmest place in the hive at that time of year. If the queen is not on the cover, set it down by leaning it on end against the hive, and turn your attention to the second frame in from the outer edge of the hive. At this time of the year the second frame of comb is usually off to the side of the brood nest, which is typically located in the center of the

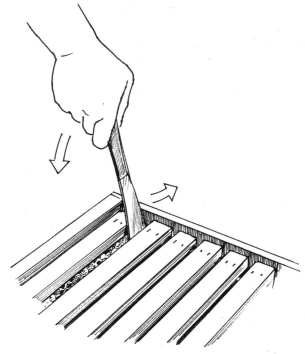

FIGURE 4-4. You may have to pry hard to force the frames over in your effort to create enough space to remove that first frame.

hive body and is where the queen is usually found. By first removing a frame off to the side of the brood nest, the chances are reduced that the queen will be on that particular frame and get injured as the frame is pulled from the hive, as compared to pulling the first frame directly out of the middle of the brood area. If this is the case, you might logically ask, "Then why not remove the outside frame first, since it is farthest from the brood area?" The reason it is preferable to remove the second frame rather than the first frame is that the center of the comb of the outside frame is often fastened to the side of the hive body with brace comb and will rip loose as the frame is lifted, spilling honey and crushing bees between the broken comb and the side of the hive in the process.

To remove the frame that is sitting in the second position, first slice through any propolis that may have built up between the end bars of the first and second frames and, using your hive tool as a lever, crowd the majority of the frames tightly together against the opposite side so that a little additional space is created between the first and second frames, which makes the removal process easier. The angled end of the standard hive tool can then be used to pry one end of the second frame away from the frame that is sitting in the third position, raising it up so that you can grab it with your free hand. Once you have a grasp of one end of the frame, pry the opposite end up in the same way and use your index finger to grab this end. The frame can now be lifted from its resting place, slowly and carefully, so that the comb does not rub against the frames on either side (see figure 4-6 on page 106). This will require a steady hand and a lifting motion that is straight up, all the while keeping the frame an equal distance from the frames on either side of it. Of course, the judicious use of smoke is recommended throughout this entire procedure to keep the bees in the hive body calm, to chase the bees away from the ends of the frame so that it may be pried up and lifted out of the hive, and to cool the tempers of any bees that get caught between the two frames and receive a rough ride as that first frame is being lifted from the hive.

As soon as the frame that occupied the second position in the hive is removed from the colony, peek down at the now exposed side of the frame that is sitting in the third position. The queen mother instinctively knows she is safest when she is deep within the dark recesses of the hive, and she has a natural tendency to run away from the light of the sun. If she is on the face of the comb that is exposed, she is relatively easy to spot as she runs down to the bottom edge of the frame and scrambles around to the other side to get out of the sun's glare. If you do not see her on the exposed comb within the hive, look for her on the frame that you have just removed and are holding over the open hive. When looking for the queen on a frame full of busy bees that are bustling to and fro, use a spiral search pattern, starting on the frame's outer edge and working your way inward toward the center of the comb. If the queen is on the frame that has just been removed, she will likely be trying to run from the light and will often be headed for the shaded side of the frame you are holding. By

searching the outer edges of the frame first, you are likely to spot her before she has a chance to duck around the corner to the other side. If you don't see the queen on the first side of the frame, slowly flip the frame over and repeat the spiral search pattern on the opposite side.

If Her Royal Highness is not found after searching both sides of the frame, set the frame down in an empty hive body, or stand it on end and lean it up against the hive where it will be out of the way, and remove the next frame, the one sitting in the third position. Pry it free from the fourth frame and make use of the space created by the missing second frame to lift out the third frame of comb without allowing it to brush close to the other frames, possibly rolling and injuring bees in the process. After taking a peek at the exposed side of the comb on the fourth frame still in the hive, run the spiral search pattern on the frame that was just removed. Keep repeating this search sequence, frame by frame, working your way through the entire brood nest until the queen is found. Remember, although the queen is typically found in the brood area, she could be anywhere within the colony. Therefore, it is important not to place the inspected frames back in the hive, since the sides of the hive body and the bottom board will have to be checked if the queen is not found on the frames. It can be convenient to have an extra empty hive body set up on a bottom board or cover to receive the frames as they are removed from the hive.

One last point to keep in mind when looking for the queen is not to use too much smoke. In fact, this is one of the few times in beekeeping that using too little smoke is preferable to using too much. The goal should be to use just enough smoke to keep the bees off balance so the inspection process can proceed smoothly. When too much smoke is used, the bees will tend to run around in circles, seemingly in a state of panic, which makes it much more difficult to locate the queen. By using just enough smoke to keep the bees simply distracted, they will tend to move more slowly and deliberately. By not causing a panic among the *mellifera*

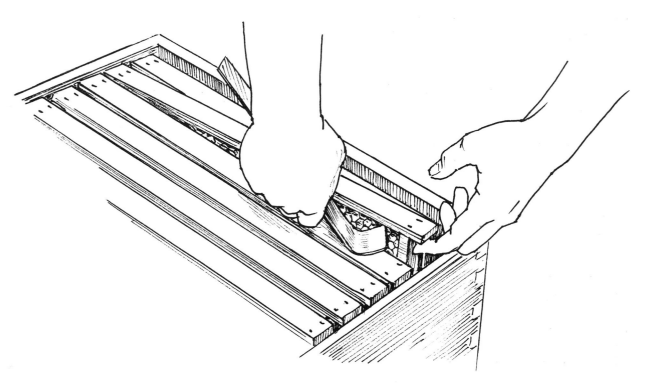

FIGURE 4-5. A little smoke can help to clear the area of bees before you go to work prying up the frames so you can lift them out of the hive.

masses, spotting the queen will become a lot less like trying to find a needle in a haystack. To make the process easier, some beekeepers have an empty hive body equipped with a queen excluder attached to its bottom into which they shake the bees and force them through the excluder with smoke. If a queen is present, she will be left on top of the excluder and easily located.

Once you know where the queen mother from the original hive is and have placed her out of harm's way, assembling the appropriate frames needed to make up the nucleus colony goes much faster. The ideal arrangement of frames when creating a nucleus colony will position the frame(s) containing unhatched eggs between combs filled with brood. This will help ensure that any queen cells that are built are located near the center of the cluster of bees warming the brood nest. Although the frames of eggs and brood will often have honey and pollen stored in the upper third of the comb, any frames consisting wholly of honey and pollen

that are included in the nuc should sit on the outer edges and sandwich the brood, acting as insulation against chilly nights and positioning nourishment near the brood nest, where the nurse bees can access it easily. After filling the space left in the hive body containing the nuc with additional combs or frames of foundation, several frames crawling with young nurse bees should be shaken into the box if the nuc is to be left in the same yard as the mother colony. A puff of smoke blown across the face of both sides of the frame before shaking will help prevent a harsh response from any shaken-up and irritated bees. Once completed, the fledgling colony can be closed up with the entrance reduced and left to its own devices to raise a new queen.

Some beekeepers recommend placing the newly created nuc in the location of the mother hive and moving the original hive still containing the old queen to a new location within the same bee yard. This is done so that all of the older foraging bees in the original hive will return to the queenless nuc

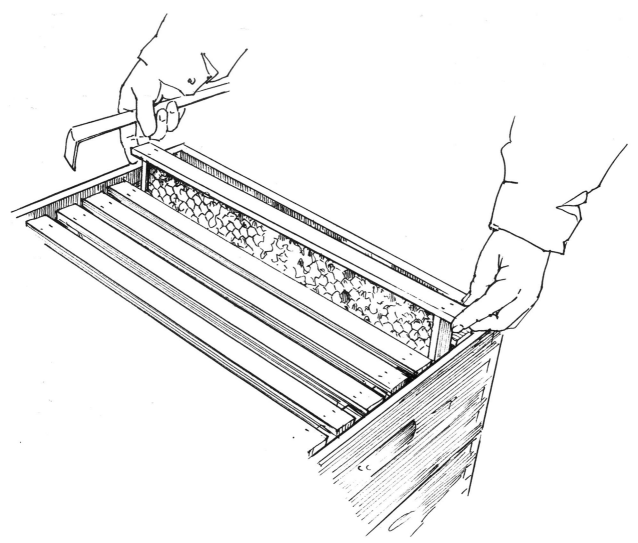

FIGURE 4-6. It takes some practice and dexterity to pry up and grasp the end of a frame while holding the other end of the frame in your other hand. Lift straight up slowly to avoid crushing bees.

that is now sitting where they have always known their hive to be. This will vastly increase the number of bees in the newly made split, greatly increasing the nucleus colony's chances of successfully raising a new queen. The downside of this technique is that it removes the majority of foraging bees from the mother colony and significantly reduces the honey-gathering potential of the nuc-generating hive even more at a time when nectar-gathering opportunities are often peaking. As a result, both the nuc and the mother hive are weakened and may need feeding to make it through the winter.

This problem can be mitigated by introducing a mated queen into the nuc. However, a mated queen does not have to be purchased, and the nuc should not be placed in the location of the original hive so long as there are eggs and sealed brood placed within the nucleus colony, a fair number of bees are on the frames that are transferred into the split, and several more frames of bees from the brood nest are shaken into the nuc. This will allow for an adequate number of bees within the new hive, allowing the colony to successfully raise a queen and get off to a good start without your having to resort to moving

FIGURE 4-7. For the double nuc system to work, there must be no way for the bees to be able to move from one side to the other. This modified bottom board helps accomplish this. PHOTO BY HILDE WHALLEY.

the mother colony and siting the nuc in its place, or moving the colony to a location over 2 miles away. If any supplemental feeding is required, only the newly made nucleus hive will usually need it, rather than both the nuc and the mother colony. (I want to stress again the importance of remembering to reduce the entrance of the nuc so that the hive can be defended by a minimum number of guard bees.)

A typical success rate of at least 80 percent can be expected when allowing the bees to raise their own queen from a nuc in a single ten-frame hive body during the natural swarming season, using four or five frames from the breeder colony and filling out the remaining space with frames of drawn comb or foundation. The success rate can be boosted well above 90 percent by placing two smaller nucs side by side in a single hive body. The

newly created nucleus colonies can be separated by inserting a divider such as a division board feeder with its openings sealed in the center of the hive body. This requires the modification of a bottom board so that it sports a raised section down the middle that is flush with its outer edges and rests against the bottom of the divider, eliminating any openings between the two sides of the chamber (see figure 4-7). Small openings must be created in the opposite ends of the bottom board, to provide separate entrances for the two hives that will be created. An empty poly grain bag makes an effective inner cover for this double nuc setup. The flexibility of the bag allows for one side of it to be peeled back to expose one nuc within the duplex hive with the residents of the other side being none the wiser. Duct tape can be used to seal the openings in the feeder to prevent commingling of the bees through the shared wall. To give each fledgling nucleus enough room to expand, a frame of foundation can be placed in the outside position farthest from the center of the brood nest.

The benefits of making double nuc colonies are that they make much more efficient use of available equipment, while relying on a minimum of specialized equipment, and they maximize the opportunities for success. Between the smaller space that the bees must maintain within the nuc and the fact that the shared wall between the two nascent colonies is kept warm by the fledgling hives on either side during cool weather, stress on the hives is decreased, which allows for greater success in raising a new queen when compared to a single nuc in a ten-frame hive body. This method of nuc making seems especially well adapted for use with Russian strains of bees, which seem to thrive in small spaces. All that is required of the beekeeper is to visit approximately 30 days after making the nucs and transfer each of the mini-hives out of the double nuc box and into its own ten-frame hive body. Positioning yourself with the sun shining over your shoulder should help provide adequate light to see if any eggs have been deposited in the back of the brood cells. On the rare occasion that you cannot find any eggs because one of the nucs has not successfully raised a queen, the frames of bees and honey from the queenless nuc can be combined with the colony that successfully raised a mother. Don't be fooled by the queen that, for some unknown reason, has delayed the start of her egg laying. Every once in a while I come across a hive that has no eggs present but does have large areas of empty comb with cleaned and polished cells in the center of the frames that the bees are keeping clear of honey and pollen in anticipation of the queen beginning her egg laying. In these cases, leaving the colony alone for another week will provide sufficient time for the queen to start laying.

If made in May or early June here in Vermont, successful nucs will often fill up their hive body and an additional super with enough honey, bees,

FIGURE 4-8. The easiest way to hive a swarm is to put the bees in the hive before they swarm. This double nuc uses a division board feeder to separate the box into two separate cavities.

and brood to get them through the winter in good shape. Nucs made later in the season, however, or those made with foundation rather than drawn comb almost always require additional feeding to prevent starvation from descending upon the young hive before the spring blooms herald the start of the next honey season.

Bee supply companies offer various styles of nuc-making boxes. Although they may work perfectly well, I have never tried them, preferring to utilize equipment that I already own rather than buying a new, specialized nuc box. As a result, I have always made nucs in standard ten-frame hive bodies, and occasionally I have used a modified bottom board and divider of some type to separate a hive body into two individual compartments with room enough for the two small nucs.

The big question for those beekeepers who want to increase the numbers of hives in their care through queen rearing or nuc propagation is: Which hives are the best ones to use for breeding purposes? The answer to this question is simplified for those with few hives, because their only option is to breed from the hives that are readily available. Operations containing large numbers of hives must decide from among their many colonies which of the hives are most worthy of being bred and encouraged to continue their particular genetic lineage. There are many traits that breeding can help bring forth in the bee yard. The typical characteristics that beekeepers have historically sought to encourage include high honey production, gentle demeanor, low swarming tendencies, fast spring buildup, low propolis use, and conservative honey usage during winter. However, the most important trait during these times of stress on the honey bee population is pest and disease resistance, with *Varroa destructor* by far being the biggest challenge facing the honey industry in the late twentieth and early twenty-first centuries. As a result, the breeding that is currently being done—by folks like Kirk Webster, Dr. Spivak, and the researchers at the USDA Agricultural Research Service laboratories, along with every backyard beekeeper breeding colonies with the evolution of the bee in mind—is among the most important work going on in apiculture today. The effort to help the bee reach a level of immunity against varroa and other pests and diseases is the long-term solution required to ensure the ultimate survival of the species. All the management techniques and methods that beekeepers use to help the honey bee survive (many of which are described in subsequent chapters) can be seen as stepping-stones—temporary Band-Aids that are applied in an effort to buy the honey bee enough time to do the work required for its evolutionary growth and survival.

The famous German teacher and philosopher Rudolf Steiner certainly touched on this subject over 80 years ago when he noted how the techniques and mindset of those who care for honey bees affect the evolutionary process and, thus, the ultimate survival of this fascinating insect. As a result, during a series of lectures, Steiner accurately predicted the difficult times the beekeeping industry faces today. Here are Steiner's own words from a lecture he gave on bees on November 26, 1923:

One is able to say—in the whole inter-relationship of the bee-colony—of this

organism—nature reveals something very wonderful to us. The bees are subject to forces of Nature which are truly wonderful and of great significance. One cannot but feel shy of fumbling among these forces of Nature. It is becoming increasingly obvious today that wherever man clumsily interferes with these forces he makes matters not better, but worse. He does not make them worse all at once, for it is really so that Nature is everywhere hindered, though notwithstanding these hindrances Nature works as best she may. Certain of these hindrances man can remove, and by doing away with them can make things easier for Nature. For example, he seems actually to be helping Nature when he makes use of beehives which are conveniently arranged, instead of using the old straw skeps.

But here we come to the whole question of artificial bee-keeping. You must not think that I am unable to see—even from a non-anthroposophical point of view—that modern bee-keeping methods seem at first very attractive, for certainly, it makes many things much easier. But the strong holding together—I should like to say—of *one* bee-generation, of one bee-family, will be impaired in the long run.

Speaking generally today, one cannot but praise modern bee-keeping; so long as we see all such precautions observed of which Herr Muller has told us, we must admire them in a certain sense. But we must wait and see how things will be in fifty to eighty years time, for by then certain forces which have hitherto been *organic* in the hive will be mechanized, will become mechanical. It is not possible to bring about that intimate relationship between the colony and a Queen that has been *bought*, which results naturally when a Queen comes into being in the natural way. Only, at first these things are not observed.[2]

Then, on December 5, 1923, Herr Muller asked Rudolf Steiner the following question:

I myself, cannot understand that within the next eighty to a hundred years the whole stock of bees will die out. I really cannot understand what Dr. Steiner means by saying that within eighty to a hundred years bee-keeping will be endangered.[3]

To which Steiner answered, in part:

When one makes these experiments one discovers the following:

One finds that calves bred from cows that have been brought to an excessive production of milk, are considerably weaker. You see it in the way the remedy affects them. The working or nonworking of the remedy, so to speak, can be tremendously increased in such cases. The calf grows up if it does not die of the disease, but the calf bred from a cow that has been over-stimulated to this over-production of milk, a calf of such breeding is weaker than calves bred from cows that have never been so forced. This change can be observed through the first, second, third, or fourth generations, but is then so slight that observation is not easy. This breeding for milk-production is still of short standing, but I know very well that if it continues, if a cow is forced to yield six gallons of milk a day, if you continue thus maltreating it, all breeding

of cows will after a time go absolutely to ruin. There is nothing to be done.

Well, in artificial bee-keeping things are, naturally, not fundamentally so bad, because the bee is a creature that can always help itself again, that is indeed, incredibly able to help itself because the bee lives so much nearer to Nature than the cow that is being bred in this fashion. It is not even quite so bad if cows so maltreated for milk-production are nevertheless at times taken out to pasture. But on the big dairy farms this is no longer done. These farms have nothing but stall-feeding; the cow is completely torn away from natural conditions.

You cannot afford to do this in bee-keeping. Thanks to its nature the bee remains united with external Nature; it helps itself again. And you see, gentlemen, this self-help in the bee-hive is something extremely wonderful.[4]

Our tendency to break things up into component parts rather than working with the entire system leads to the mechanized approach referred to by Steiner. It is important to keep in mind that, while we are dealing with individual honey bees, we must also respond to the hive in its entirety. Each individual honey bee can be likened to an organ or cell within the body of the colony, whereas the entire hive is the individual. Such a holistic vision has long been a primary component of truly organic agriculture.

CHAPTER 5

Parasitic Mites

If skies remain clear, the air warm, and pollen and nectar abound in the flowers, the workers, through a kind of forgetful indulgence, or over-scrupulous prudence perhaps, will for a short time longer endure the importunate, disastrous presence of the males.

—MAURICE MAETERLINCK

Adequate control of honey bee pests and diseases is recognized as an integral part of good beekeeping practices. Until the introduction of organic standards, the only federal laws affecting honey bees were regulations on importation, developed to reduce the spread of diseases and problem insects, and regulations on pesticide use, designed to reduce unintended bee mortality. State laws governing the interstate movement of bees and the registration of hive locations were also focused primarily on disease prevention. While many regulations were originally designed to control the spread of American foulbrood, the recent introduction of Africanized honey bees, small hive beetles, and varroa mites to the United States has increased the importance of such governmental oversight.

As we've discussed, the best approach to addressing disease and pest issues from both the bee's and the beekeeper's perspective is the propagation of resistant bees. If the bees have the ability to keep themselves healthy and strong, the apiculturist can focus less time and resources on such maintenance issues and instead direct energies toward growth and production concerns. Unfortunately, the development of truly disease- and pest-resistant bees is likely to take many years to

accomplish. In the meantime, what is the organically minded beekeeper to do?

The first thing I do is try to start with bees that already have some level of mite and disease tolerance. This means that when purchasing bees and queens, I choose only bees that have some mix of genetics from Russian, hygienic, and/or varroa-sensitive hygiene (VSH) bees. Otherwise I am simply purchasing bees that are highly vulnerable to these hazards. Why make my job as a keeper of bees more difficult than necessary?

Conventional disease and pest control procedures typically have the favorable attribute of significant labor savings over organic approaches. For the cost of a couple dollars or so, a prepared "dose" can be purchased and introduced to a colony quickly, with minimal effort, to achieve the desired result. In general, organic controls are more labor-intensive, even when they are comparable in cost. Some organically approved remedies also require additional equipment prior to use, which adds to their expense. Increased labor requirements combined with the additional expenses associated with organic certification result in organic hive products costing more to produce, and therefore commanding a premium price relative to their conventional counterparts.

In general, it is important to incorporate regular colony inspections into one's hive management program. Periodic inspections become even more critical when utilizing organic methods of disease and pest control. Recognizing the signs of disease or pest predation is crucial to timely intervention to prevent colony decline or collapse. The organically minded beekeeper must also learn to differentiate between, and prioritize among, the afflictions affecting a hive in order to choose the appropriate course of action for a given situation.

Efforts to detect diseases and pests begin as soon as one enters the apiary. Simple observations such as the amount and type of activity in front of a hive, or the presence of fecal matter or dead bees lying by the entrance, can reveal much about the condition of a colony before you even remove the outer cover to inspect the hive. Good beekeeping practices demand that the beekeeper remain alert to such clues regarding the hive's condition and learn to recognize their meaning. Working with honey bees teaches us the importance of paying attention to the fine points that are so easily overlooked. This attention to detail plays a crucial role in the survival of honey bee colonies, whether they are managed organically or otherwise. Noticing when the soothing hum that typically emanates from a hive has changed in pitch, when the faintly sweet scent that normally exudes from a hive becomes sour or foul, or when the movements and activities of the bees themselves are not in keeping with a happy and harmonious hive—these are the kinds of small observations that can provide clues to the status of a colony and how to assist the bees as their keeper. Of course, being able to identify problems within the hive requires that we first take the time to observe and experience the way things are when all is right within a colony. Once you become familiar and comfortable around the bees, you may even be able to use your intuition to sense when something is not right with a hive. Anyone who keeps bees long enough is likely to come to the realization that beekeeping is not an exact science. There is as much art as science involved in the practice of this ancient craft. All too often, just when we think we have the bees all figured out, a hive will do the unexpected. Beekeeping is one of those activities in life where the more you know, the more you realize how little you *truly* know.

When it comes to hive inspection and manipulations, there are two basic philosophies I have heard expounded over the years, and they are diametrically opposed to each other. The first contends that it is best to disturb the bees as little as possible. Honey bees know how to conduct their own affairs, it is argued, and the less disruptive we are of their organization and work, the better. The alternative view calls for regular hive manipulations and inspections, both for early detection of any potential problems and to cause the bees to constantly reorganize and redouble their efforts following each manipulation of the hive, further stimulating honey collection activities.

My preference is to follow the middle road. Striking a happy medium between the two extremes means that one incorporates a single thorough inspection of every hive early in spring, with regular cursory assessments approximately every 7 to 14 days during the rest of the active season (depending on how much forage is available and whether a honey flow is on). Such quick inspections typically consist of observing the bees at the entrance of the hive and peeking under the inner cover to make sure everything looks normal and that there is plenty of room for additional honey storage. With experience, these observations are all that should be needed to ascertain whether a colony is evolving in a normal and healthy manner. If during a quick check a problem is sensed or signs of trouble are noticed, a more thorough inspection is in order. Typical indications that a hive may be having difficulties and requires a closer look include weak hives that have an insufficient number of worker bees, colonies that sound or act queenless, and hives in which no worker brood is evident as one peeks down between the frames of the brood chamber. Otherwise, it is generally preferable to stay out of the brood nest and leave the bees to themselves to do what they do best.

How does one acquire such experience when just beginning with bees? Opening up the hive, removing frames, and inspecting them every week or two during your first year as a beekeeper is one way. As I mentioned in the beginner's section of the book, the only way to get experience working with bees and knowing what signs to look for is to do it.

• VARROA DESTRUCTOR •

Observing bees as they go about their business is truly fascinating. For me a large part of this fascination stems from the world honey bees create within the hive and its many similarities to human society. The constant activity and industriousness of the colony, and the clear delineation of individual responsibilities and jobs, with all activity geared toward the singular goal of the common good, inspire awe and inspiration. Unfortunately, observation can also cause worry and concern when disease or parasites such as varroa mites are detected (mites are arachnids and related to spiders). Spotting varroa within a colony is made easier by the fact that it is much larger than its cousin, the tiny tracheal mite (see "Tracheal Mites" later in this chapter on page 155). When you examine a frame of brood, varroa mites may be seen crawling around on the bees or, in cases of heavy infestation, even on the frame itself. Nevertheless, spotting a mite clinging to a frame or the exterior of a bee that is scampering around on a comb with hundreds of other bees can be a challenge. Good lighting and a sharp eye are necessary for detection. About the size of a pinhead, varroa take on a reddish-brown hue as they mature. This deep rustlike color makes the mite relatively easy to see when contrasted with the white of the developing pupa upon which

it feeds, sucking on its blood (hemolymph) prior to it developing into an adult bee and emerging from its birthing cell. The hole that the mite chews in the exoskeleton of the bee is slow to heal. Some researchers believe this is due to the mite's ability to somehow suppress the bee's immune system. This open wound in the bee becomes an entry point for bacteria, viruses, and other disease organisms.

Aside from the direct observation of varroa, a telltale sign that mites have infested a hive is the presence of bees with deformed wings and sometimes a shorter-than-normal abdomen. These sickly honey bees can typically be spotted in front of the hive trying in vain to fly, though usually they are weak and seem to be in the last throes of their lives, if they are alive at all. In addition, the hive exhibits a general weakness that expresses itself in deteriorated brood and declining worker populations. These symptoms are collectively known as parasitic mite syndrome (PMS).[1] Viruses that are spread by *Varroa destructor* are the primary cause of bee PMS symptoms.

Although varroa are able to reproduce on worker brood, research has shown that the mites prefer to raise their young on *Apis mellifera*'s drone brood. Thus, the regular inspection of drone brood can be a good barometer of the level of varroa within a hive. The easiest way to inspect the drone brood for the presence of varroa mites is to simply observe

FIGURE 5-1. Often the first chance we have to see the individual varroa mite is on drone brood. Brood exposed when supers are separated will typically be drone brood, which is the mite's favorite breeding ground. PHOTO BY TAD MERRICK.

FIGURE 5-2. Even if you didn't see the mite on the thorax of her sister nearby, bees with deformed wings are a typical sign that varroa mites are present in the hive. PHOTO BY STEVE PARISE.

brood that are exposed in the burr comb that the bees build in the space between the supers and hive bodies. These pieces of comb built outside of the frames typically occur when a honey flow is on and the colony runs out of room to store its bounty. The cells the workers build between the supers are most often the larger drone comb cells. This is because our modern Langstroth-style hives are most often fitted with worker foundation—most beekeepers want lots of workers because it's the workers that do all the work and make all the honey. To be healthy, however, hives need to raise drones during the times of the season when swarming may occur. In their desperation to find a place to raise drones, the bees build drone comb between the supers and

hive bodies even when this area respects their bee space. These drone cells are typically ripped apart and the enclosed larvae and pupae are exposed when the supers are separated. If mites are present, a quick glance at the exposed brood immediately after lifting the hive body or super will often reveal them scampering around on the developing larvae in an effort to get out of the sunlight, like rats abandoning a sinking ship. Since the mites may be hiding under the drone pupae, it is a good idea to remove the brood from their cells in order to fully inspect for mites.

If a hive has not been crowded enough to build burr comb between the supers, a frame containing sealed drone brood may be examined to check for

FIGURE 5-3. One way to ascertain how bad a mite infestation is involves examining immature drone brood exposed when supers are separated. Note the mite on the upper end of the larva to the left of center.

the presence of varroa. An ordinary capping scratcher (uncapping fork), inserted at a slight angle through the bullet-shaped caps of a section of sealed drone comb and pulled away from the frame, will carry with it the drone pupae within the cells. This method of detection allows both the brood and the now-empty brood cells to be easily inspected for the presence of mites. One should also keep an eye out for mite feces, which appear as little specks of white in the empty brood cells.

Another way of testing for the presence of varroa mites was developed early on and adopted by the beekeeping industry. It is known as the ether roll technique. This process requires a wide-mouthed canning jar and lid. Approximately 200 to 300 bees are collected in the jar (which roughly translates to about 1½ inches of bees in the bottom of the jar). The bees should be collected from the brood nest, and care should be taken not to accidentally include the queen in the jar. With the lid in place, a light rap of the container with your hand will knock the bees down to the bottom of the jar so a fairly accurate estimate of the number of bees the jar contains can be made. Once an adequate number of bees have been collected, the lid of the jar is cracked open and the honey bees are sprayed with ether from an aerosol can, such as those typically sold in auto supply stores for use in helping to start engines. A short, two-second blast of ether is enough to cover the bees. The lid is then replaced

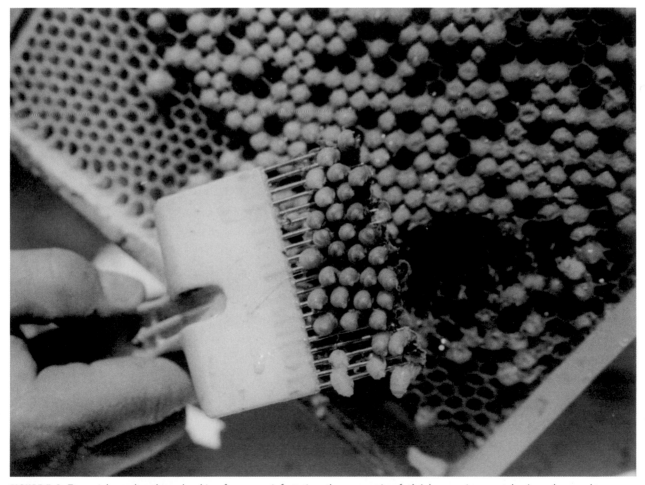

FIGURE 5-4. For quick results when checking for varroa infestation, the uncapping fork (aka capping scratcher) can be used on areas where sealed drone brood is concentrated. PHOTO BY HILDE WHALLEY.

and the jar is rotated so that the bees roll around on the side of the glass. Most of the mites that are present on the bees will fall off and stick to the side of the jar. Alternatively, ethyl alcohol or rubbing alcohol can be poured into the jar and the bee mixture shaken for a minute or two. When alcohol is used, the bees are separated from the alcohol by pouring the mixture over a wire mesh. The alcohol is then strained and the number of mites examined.

Unfortunately, all the bees within the jar are killed using these methods, although the use of ethyl alcohol may preserve the bees for other purposes such as dissecting them to inspect for disease or the presence of tracheal mites. In addition, the use of a toxic engine-starting spray does not really resonate with the natural beekeeping philosophy. An approach that is more in line with the organic state of mind consists of collecting some bees in the manner described above, but instead of adding ether or alcohol, drop in a tablespoon of powdered confectioners' sugar. The powdered sugar should be fresh and flow freely with no clumps. Shaking and rolling the sealed jar will coat the bees with sugar and make it difficult for the mites to maintain their grip on their hosts. For best results, set the jar down for about two minutes after shaking. The bees in the jar will raise their body temperature immediately following the dusting and the increased body temperature will further act to dislodge mites from their bodies. The lid of the jar can then be replaced with a piece of ⅛-inch hardware cloth, and the sugar, along with the mites, can be dumped into an empty jar, leaving the bees behind. The lid of the canning jar makes a good template when cutting a piece of hardware cloth to act as a screen for separating the bees from the mites and loose sugar. A fine nylon mesh can then be placed over the mouth of this second container that will allow the sugar to be separated from the mites in order to make the varroa more visible for counting. Alternatively, water can be added to the mite and sugar mixture, which will dissolve the sugar and make the mites visible. The bees, still covered with a dusting of powdered sugar, can be returned to

their hive, none the worse for wear. Research seems to indicate that varroa monitoring with powdered sugar is just as effective as using ether when mite levels are low and is more effective than ether when mite levels are high.[2]

A less labor-intensive and intrusive method of determining varroa levels within a hive relies on the fact that a small percentage of mites will naturally fall off their hosts during a given period. Researchers running trials on the toxic chemical strips sold under the names Apistan and CheckMite+ (see chapter 1) had the idea of placing a screen on the bottom of the test hives. This would allow dead mites that fell to the bottom of the hive to be collected and the various chemicals being tested could be evaluated for their efficacy. Of course, control hives had to be included to make the trials valid. The control hives were treated in the same way as the test hives except that the chemicals were not introduced into the hive. During these trials, researchers found that, over time, about 10 to 20 percent of the mites in a hive will naturally lose their grip and fall to the bottom of the hive. Replacing the solid bottom with a screen allows for the use of a sticky board trap to catch the varroa that drop to the bottom of the hive. Although I have yet to use a sticky board myself, it has become a common technique for evaluating the level of mite infestation within a hive. Often a sheet of white cardboard or heavy construction paper is used for this purpose. The face of the board is typically coated with vegetable oil or Vaseline so that mites coming in contact with its surface will stick to it and be unable to crawl away. (A paint roller soaked in oil works well for applying an even coat.) The sticky board is then placed on the bottom board of the hive. With ⅛-inch wire mesh over the sticky board, mites that lose their grip and fall off the bees will fall through the screen and stick to the board, and the bees are prevented from becoming entangled themselves. A single sticky board is best used for a short duration of one to seven days at most. Leaving the board in the hive for longer periods may allow it to collect significant amounts of hive debris, making examination for mites difficult. Lines drawn on the face of the board dividing it into sections about 2 inches square can greatly assist in counting the mites captured. A magnifying glass and good lighting are also very helpful.

Depending on the time of year, a 24-hour natural mite fall of 8 to 10 mites or more is considered by many to be the threshold for treatment. If you are unable to return to the hive and check the sticky board exactly one day after it is installed, and you instead pull it out after three days, just count up the total number of mites and divide by three to get your 24-hour mite count. Waiting three to seven days and then dividing the number of mites found by the number of days will provide a more accurate measure of the number of mites in a hive than relying on a single day. The higher mite-drop level of 10 mites is often used as the threshold for treatment during the spring when the colony's population is growing rapidly, while the lower mite-drop count (8 mites) is often used as the threshold in the autumn, when the queen is slowing her egg laying and reducing the population growth of the hive in preparation for the dearth of winter. It should be noted that this or any other method of evaluating the population of varroa mites in a hive may not always be accurate. Therefore, if you are going to monitor mite levels, it is a good idea to test mite levels more than once in order to confirm initial findings.

I've provided this information on techniques to evaluate varroa levels so the reader will have some background and knowledge on what is considered to be standard operating procedure within the beekeeping industry. In actual practice, however, many beekeepers, including myself, find that we just don't have the time or inclination to carry out these laborious procedures. Not only does monitoring for mites take time, but the various sampling methods can provide false readings on the population level of mites in the hive. I am constantly working to simplify my operation by reducing the amounts and types of equipment I must rely on. Because the regions where bees are

isolated enough not to be exposed to mites are few and far between, I simply always assume that my hives always contain mites. By regularly inspecting and being on the lookout for the basic indicators of varroa presence within the hive—mites in burr comb brood, mites on bees, sealed brood whose cappings have small holes in them, and bees with deformed wings and/or shrunken abdomens—a good feel for the level of mite infestation within a hive can be obtained without the use of sampling methods. Keeping an eye out for indications of varroa's presence and hives that are not building up normally while performing regular checkups for supering purposes every week or two between May and August allows me to make sure that the varroa mite levels within my hives are not getting out of hand.

Historically there has been good reason to take the time to monitor varroa population levels in hives, because the use of toxic chemicals to control pests has consistently resulted in the development of pests with resistance to chemical treatment. By avoiding treatments that are unnecessary when pest population levels are low, the speed with which chemical resistance will develop is reduced. In addition, most treatments are not cheap. Significant savings can be realized by applying treatments only when necessary, especially in commercial or sideline operations with numerous hives. However, times have changed, and these reasons to monitor for varroa may not be all that applicable any more.

Thankfully, these days it is the rare beekeeper that is still using the approved "hard" chemical treatments for varroa: Apistan (fluvalinate), CheckMite+ (coumaphos), and Apivar (amitraz). Because these products utilize a single mechanism to short-circuit the mite's biological system, resulting in death, varroa developed resistance to these compounds within a relatively short period of time (three to five years). This, combined with research that revealed that these chemicals build up in beeswax and can have sublethal impacts on colonies, has resulted in the majority of beekeepers now using "soft" chemicals to treat for mites, if they treat at all. These alternative treatment options work in ways that make the likelihood of the mite's developing resistance remote.

Unlike manufactured chemical compounds, treatments derived from natural ingredients tend to be composed of a complex mixture of compounds that act synergistically and often create a multi-pronged threat to pests. Materials that are toxic to pests through a variety of pathways are much less likely to lead to pesticide resistance within a few years, if at all. The mite treatments Apiguard, Api Life VAR, and HopGuard all fall into this category. Other treatment options such as powdered sugar dusting, Sucrocide, and organic acid treatments such as the Mite Away Quick Strip (MAQS) work physically, which, at least in theory, prevents varroa from being able to develop resistance of any kind. These varroa control options are explained more thoroughly later in this chapter.

The recommendation to monitor mite populations in hives is partly driven by a desire to curb the speed with which varroa will build up resistance to treatments, and thus the need to monitor when using treatments that are unlikely to lead to resistance makes the need for monitoring questionable. Beekeepers may be better off assuming that mites are present (as their ubiquitous nature undoubtedly ensures that they are) and automatically treat at the appropriate time of year.

There is also the economic argument that if treatments can be avoided when they are not needed, costs will be reduced and profitability increased. In commercial beekeeping operations, however, it is the rare business that will absorb the labor cost associated with monitoring for mites, and, if they do, they sample only a handful of colonies in each yard. It may be cheaper and safer for beekeeping outfits to simply treat all their hives "just to be sure." Once I began keeping more than a dozen or so hives, I gave up on monitoring and just treated all my colonies as a matter of course, usually once in autumn. Given the propensity for the various monitoring techniques to provide false

readings, it may be prudent for small-scale and backyard beekeepers to do the same. Sure, some money may be spent unnecessarily on occasion, but such an expense is minor compared to the cost of replacing a colony of bees that have died because a false reading indicated that a treatment was not necessary. Even the most costly mite treatment on the market is fairly inexpensive when you only have one or two hives that need treating. If you are among the minority of beekeepers that still use the hard chemicals for mites, monitoring makes a lot of sense; but it seems that the value of monitoring for beekeepers who use soft chemicals is mainly as a post-treatment confirmation that the treatment was effective.

When I see no deformed bees, or I observe few or no mites in burr comb and on bees during inspections, I assume that mites are present, but in such numbers that the low-impact passive controls I constantly keep in place are adequate to maintain a healthy hive for the time being. When I notice several mites within a single brood cell, and there are numerous brood cells so infested, this is a sign that mite levels are starting to build, and I keep a close eye on the hive to ensure that the mites do not become too populous too quickly. When I spot bees with deformed wings and numerous mites are reproducing in many brood cells, then I know emergency high-impact controls are called for in the near future to avoid the collapse of the hive within the next couple of brood cycles.

Now, looking for indications of mite damage does not sound very scientific—and it isn't. Nevertheless, it works for me given the time and resources I have available. So far, using this basic observational method, I have kept my average yearly colony losses under 20 percent since 2001, whereas most conventional beekeepers have been reporting regular losses of 40 percent or more. Monitoring for varroa regularly would make more sense to me if I were using a mite control method for which the mites are likely to develop genetic resistance. It might also be appropriate if I were extremely concerned about the cost of the additional labor and materials involved with mite treatments and felt that the additional cost of monitoring mite levels outweighed the cost of potentially applying unnecessary treatments. Most organic mite treatments do not allow varroa to develop resistance easily, and since I keep the number of hives under my care within a human scale so that the time and money invested in prophylactic treatments are not prohibitive, the savings in labor and time from forgoing mite sampling and simply assuming that all hives are infested with varroa and require some form of treatment is worth it for me. By forgoing the purchase of labor-intensive mite-monitoring equipment, my constant goal of simplifying and minimizing the amount of beekeeping equipment I rely on is more fully realized.

This approach could be considered a weakness in my beekeeping philosophy, especially by the large beekeeping operations that must work hard to eliminate all unnecessary labor and input costs to stay profitable. However, in the past I have been able to keep my bees quite healthy throughout the seasons here in Vermont without sampling for mite loads by continually using a variety of low-level mechanical controls from early spring through the middle of August. I find that an effective, high-impact active varroa treatment of a nontoxic nature should be applied in late summer, after the honey harvest, to minimize winter losses and help guarantee strong hives in spring. This means that the work to produce next year's honey crop actually begins in August, and it requires that if I want to treat my bees for varroa, I schedule my honey harvest earlier in the season than has traditionally been the case. Early harvesting allows me to treat each hive before the mites reach the level at which they cause the hive to collapse. As a result, I have virtually eliminated the loss of hives to mites during the spring, summer, and autumn, limiting any colonies that are killed by varroa to the winter and early spring. In my climate region, which is zone 4b on the USDA plant hardiness zone map (formerly zone 4), I find that an early treatment in mid- to late August also helps to ensure that the favorable

temperatures necessary for the effective application of many temperature-sensitive nontoxic controls will be available. If I wait until late September in order to maximize the honey crop, there is a greater risk that it will be too cool for some of the temperature-sensitive treatments to be effective, and the hives are likely to collapse from their mite load by midwinter. By keeping the varroa population as low as possible during the time when the hive is raising the majority of bees that will overwinter within the hive—September and October—we can help ensure a healthy winter honey bee population to carry the hive into spring. This is because sick bees that are stressed by high varroa levels are unable to raise healthy winter bees, and unhealthy winter bees do not survive the winter. It is important to treat bees early enough so that they can become healthy themselves and raise healthy winter bees. An early honey harvest also gives the colony the opportunity to gather additional honey stores and helps guarantee access to plenty of food for the hive over the long winter season. A drawback to this approach is that the hives may become so full of honey in September that colonies will throw off late-season swarms due to the overcrowded conditions. It is disheartening to know that these swarms, if not captured and fed, will not have enough time to gather and store enough honey on their own to make it through the winter.

This early-season approach to mite treatment requires that the beekeeper be willing to sacrifice what can sometimes be a significant portion of the honey harvest. Late August and September is the time when the bees gather rich, dark, and flavorful goldenrod honey in Vermont. Even though the hive may not technically need it, by treating the hive for varroa and leaving this additional honey for the colony, the ability of the hive to survive the winter and find itself with a healthy population of workers come spring is greatly improved. I have also discovered that without the darker, more robust flavor of goldenrod, the honey that is harvested and bottled has the consistently lighter color and more delicate flavor that most of my customers prefer.

• TREATING FOR • VARROA MITES

The best approach for treating varroa will vary depending on the level of infestation within a colony, the time of year, and the resources available. For example, certain high-impact treatments that may be used to prevent the collapse of a hive with high levels of mites may not be possible if honey supers are on the colony. In contrast, relatively few mites within a hive during the start of a major honey flow may be adequately addressed with low-impact mechanical mite control procedures that remove only a small percentage of mites but have no effect on honey that is to be harvested and sold. Conventional wisdom dictates that periodic sampling is required to estimate the population of varroa within a colony and to determine what type of mite control methods to use and when to use them. However, determining which level of infestation requires which type of treatment can be a challenge due to the large number of variables involved. When considering mite levels, it would be much simpler if there were a hard-and-fast number that a beekeeper could use to know when and how to treat for varroa. Some people have attempted to develop just such benchmarks; however, these efforts are complicated by the fact that the same mite populations within different colonies may result in a variety of sampling results, depending on whether a sticky board or one of the roll methods is employed. Thus, multiple samples are sometimes required to get an accurate result. To be sure a treatment is effective it is wise to sample for mites shortly after a treatment has been completed. To complicate matters even more, there is some disagreement over what level of mites is a cause for concern. Are chemotherapy treatments indicated when the level of mites will adversely affect the honey harvest and cause economic damage, or only when the mites reach the point where the hive's survival is at risk? Additionally, the threshold indicating the need for mite treatment will be lower if the hive's immunity is already weakened by the presence of other diseases or pests. Ultimately, the beekeeper must make the final decision.

As indicated above, I tend to divide organic varroa treatments into two general categories: active, high-impact, emergency treatments that will remove over 70 percent of mites within a hive during a single treatment or treatment period, and will do so within a short amount of time; and passive, low-impact, mechanical or cultural actions that will affect only 5 to 30 percent of mites, but that can be used over long periods and that are useful in helping to keep mite loads low between high-impact treatments without fear of contaminating honey gathered during major honey flows. Two high-impact treatments that remove the majority of mites in a hive are often required each year, once in spring and once in fall, to ensure the bees do not succumb to mites. Nevertheless, a single high-impact approach can be utilized with success when several low-impact varroa management activities are carried out during the rest of the year, serving to replace one of the high-impact treatments. For those who want to be aggressive and use a combination of four or more low-impact passive mite control activities throughout the year to replace high-impact treatments, colonies can be kept alive using low-impact mite treatments alone without resorting to any major chemotherapy methods.

I used to think that bees required two high-impact treatments during the year or its equivalent: one high-impact method in late summer, combined with several low-impact controls throughout the rest of the season. As we will see, however, my experience and that of numerous other beekeepers seems to indicate that this view may be overly conservative.

Besides having the ability to be utilized while a honey flow is in progress and honey supers are on the hive, low-impact passive varroa controls typically offer the added benefit of being inexpensive. (Some low-impact controls can be time-consuming, however.) Active high-impact varroa controls rely on introducing a foreign substance into the hive, often with the assistance of specialized equipment, and as a result these methods of mite control tend to be more costly, but are often easier and good for beginners. It is important to remember, however, that with few exceptions, no foreign substance should be placed in hives whenever honey supers are on that are intended to be harvested, in order to reduce the chances of honey contamination.

As mentioned earlier, the number-one biological threat to honey bees and the beekeeping industry in the late twentieth and early twenty-first centuries has been *Varroa destructor*. The reaction by the US beekeeping industry to the introduction of this mite into North America has been zero tolerance. The chemical controls Miticur, Apistan, CheckMite+, and Apivar were introduced with the expectation of a kill rate of 97 percent or better. From an organic perspective, the effort to eliminate every mite within a colony is misguided. In many instances, the bees can sustain an infestation rate of 10 to 15 percent with little harm. Allowing hives to be exposed to low levels of varroa on a constant basis will speed up the development of the honey bee's natural genetic resistance to the mite. Pesticides, on the other hand, act like a crutch and have the effect of coddling the bees and preventing them from using their own innate defense mechanisms to deal with parasitic invaders. Leaving some mites in the hive on a regular basis is the only way to give the bees a fighting chance of dealing effectively with varroa on their own terms. We should be encouraging the honey bee's natural genetic resistance by working with the bee's evolutionary process, for this is ultimately the best long-term solution to the mite problem.

• ORGANIC VARROA • CONTROL MEASURES AND METHODS

Unfortunately, simply ignoring the mites will, with the rare exception, result in the loss of the hive within one to two years. I therefore recommend that, in addition to starting with naturally mite-resistant stock, organic beekeepers adopt some combination of the following mite control approaches. Although not all of these varroa control techniques have been "scientifically proven" as yet, they have nonetheless been utilized in the field with varying degrees of reported success.

Cultural Management Practices and Biological Controls for Varroa

A cultural management practice refers to an activity that is a built-in part of one's hive management system. It can be represented by a piece of equipment that is part of the hive's standard architecture at some point during the season or a hive manipulation that is a regular part of the management cycle carried out by the apiculturist. As such, cultural management practices tend to be physical in nature and act as passive control measures.

Biological control of varroa relies on life itself to deal effectively with the destructive nature of this pest, thereby bringing a sense of natural balance back into the hive's ecosystem. Efforts to develop hygienic bees that are naturally resistant to mites have been the focus of the majority of the biological control efforts within the beekeeping community. Nevertheless, other examples of biological controls are being worked on and hold great promise.

Nucleus Colonies

Early on, when varroa first became established in the United States and no treatments had yet been developed, it became obvious that newly made splits from the previous year had a much higher survival rate than the older, stronger hives that had already overwintered a season or two. As a result, one of the low-impact practices that beekeepers should incorporate into their yearly management cycle is the regular production of nucleus colonies as described in chapter 4. This allows the apiculturist to actively expand the number of hives by splitting up those with a proven propensity for survival, increasing the hives' chances of surviving the mite at the same time. Why newly created colonies typically have a better survival rate when faced with varroa infestation than older established hives is a good topic for future research. I believe at least four factors may be at work here. First, by removing bees and brood from a strong hive to make a nuc, we are also removing many of the mites at the same time and interrupting the mite reproductive cycle enough to keep their numbers below the threshold where they will overcome a colony. Second, large families of bees that are represented by strong colonies have larger brood nests to support mite reproduction, and they have many more foragers in the field, which results in increased opportunities for mites to hitch a ride and find their way into the hive. By decreasing the size of the brood nest we reduce the incubation opportunities for the mite. Third, young, vigorous queens in their first year may be able to fight off the effects of the mites in a way that their older sisters are unable to match. And fourth, by replacing older combs in the hive with new foundation or comb that is less than five years old, we reduce the tendency for wax comb to build up pathogen loads and chemical contaminants.

The existence of viruses that are carried and spread by the mite is well established. The deformed-wing virus (DWV) has become a telltale sign of a strong varroa presence. A symptom of parasitic mite syndrome, deformed wings are a typical sign that varroa are lurking somewhere nearby, because the mite has been proven to be the vector that transmits and/or makes the bees more vulnerable to the disease. DWV is often found in hives, but seldom becomes visibly identifiable by the presence of bees with deformed wings unless exposed to varroa-induced stress. The visible presence of this virus is not so much a death sentence as it is a warning that unless something more is done soon, the colony in question is likely to collapse. These days, the health of a hive is strongly tied to the population of varroa found within the colony, and this has led to the establishment of an economic injury level (EIL), which is the level of mites a hive can tolerate without the significant loss of honey production and income for the keeper of the bees. This approach puts the emphasis on the hive as an industrial/economic system whose primary purpose is production, rather than a biological system for which the health and integrity of the bee colony is of primary importance. Clearly the biological perspective is more closely aligned with organic methods and natural ecological agriculture.

Varroa destructor does not have a high reproduction rate and typically lays only between two and six eggs, approximately one every 30 hours during each reproductive cycle.[3] Some 20 percent of mites will enter a second cell and lay again, and, depending on whether the mite has entered a drone brood cell or a worker cell, an average of only about 1.8 adult female mites on worker brood to 2.7 adult female mites on drone brood will manage to reach maturity during the course of each cycle to carry on the process of reproduction.[4] Though few in number, the young female varroa that are able to reach maturity have a high survival rate. This creates an exponential growth pattern in the number of mites found in a hive over the course of a year. Although a hive may start the year off with only a dozen mites in the colony, following the first brood cycle their number will jump to approximately 26 individuals engaged in parasitism. After the next generation of brood hatches, about 57 mites may be found inhabiting the hive. After the emergence of the third brood cycle, well over 125 mites may reside in the hive. The next brood cycle—about three months after the start of the year—may result in over 275 mites present in the hive, and so on. Of course, these numbers all presuppose that foraging workers or drifting drones bring no additional mites into the hive, which, in fact, they typically do. It is this exponential progression, combined with the weakening effect of the young developing mites feeding on the honey bee brood, plus the viruses carried by the mites (combined with the viruses that are latent within the colony) that allows the increasing presence of mites in a colony to eventually overwhelm the hive and cause its demise before autumn rolls around—unless the beekeeper, or the honey bees themselves, are able to take steps to keep the varroa population below the threshold that will precipitate the colony's collapse.

Although varroa prefer to raise their young on drone brood, they can be successfully raised on the worker brood, even though it has a shorter reproductive cycle. Any varroa egg laid within the standard European honey bee colony within the first 12 days of a worker bee's development, or the first 15 days of a maturing drone's, will typically have time to develop fully to adulthood. Varroa eggs laid after these times will produce mites that will fail to mature in time and will not survive.[5] Hence, a large population of mites can be raised faster on drone brood than on worker brood, and, in general, the more brood in the hive, the more opportunities for the mites to sustain prolonged population growth. Hives governed by a queen with varroa-sensitive hygiene (VSH) genes have the ability to sense which mites are able to successfully reproduce and can interfere with the varroa's reproduction cycle. They accomplish this through hygienic behavior, which motivates the bees to uncap and remove the developing bee larvae that are infested with reproducing mites to a much greater extent than the typical hygienic behavior of other honey bee strains.

Over time, organically minded beekeepers can develop mite-resistant traits in their hives through the regular establishment of nucleus colonies. The simple way to do this is to breed hives that enter the spring with the largest and healthiest bee populations. Hives that can survive and thrive in the presence of varroa should be the primary focus of breeders. Once colonies tolerant of varroa make up the majority of one's hives, breeding efforts can then focus on the development of other traits such as gentleness or high honey production. It doesn't matter how much honey a colony can produce, or how easy it is to work with, if its bees are dead from varroa infestation.

Regular Comb Replacement

As mentioned above, the regular removal of old comb goes hand in hand with the making of splits or nucleus colonies. Removing old comb contaminated with pathogens and toxic chemicals can aid in the ability of a colony to survive and tolerate the stress of varroa mites. [6] Research has shown that when the level of disease organisms in a colony is low, the colony is able to withstand a larger population of mites than when pathogen levels are high.[7] Varroa mites weaken the bees' immune systems to the point that the bees then become susceptible to the various viruses and fungi that exist within the hive. Thus, we now see that varroa don't actually kill the bees—diseases do. As a result, it is advisable to ensure that none of the combs in our hives (especially the brood nest) is older than about five years, particularly if we are using foreign substances in the hive to control mites or have the bees located in areas where chemical or pathogen contamination is likely.

Numerous beekeepers around the United States advertise "treatment-free bees" for sale. Most beginner beekeepers don't understand that while these bee suppliers are not lying (they don't treat their bees by introducing foreign substances into the hive), it is not as if they are doing nothing to help the bees deal with varroa. These bee suppliers are typically selling nucleus colonies that they make from hives of bees that have developed some level of mite resistance. By splitting up hives to make nucs, they are interrupting the brood cycles of both bees and mites, while at the same time removing old frames from hives which, all too often, are sold along with their nucs. Thus, unsuspecting newbies purchase the "treatment-free" bees thinking that they don't have to do anything to help the bees deal with the mites. Sure, colonies can be kept alive without treatments, but when beekeepers don't manage the hives in a way that suppresses mite populations and stresses on the colony, like the sellers of nucleus colonies do, the hives will typically die out within a couple years.

Screened Bottom Boards

As an integral part of the hive, the screened bottom board helps to control varroa mites with the least amount of money and labor. Nevertheless, as the saying goes, "You get what you pay for," because it provides year-round removal of about only 10 to 20 percent of the mites in a hive. The screened bottom board's effectiveness is based on the observation that a certain percentage of varroa will lose their footing and fall off their honey bee hosts during any given period. When a mite falls off a bee in a hive outfitted with a traditional solid wood bottom, the mite will land on the bottom board and have the opportunity to hitch a ride back up to the brood nest by attaching itself to a passing bee. It is believed that when the mites are over an inch and a half away from honey bee activity, they have a difficult time sensing where the bees are, and thus their ability to find their way back to the brood nest is impaired. When a screened bottom replaces the solid wood bottom, the mites that fall through the screen are in effect permanently removed from the hive.

I have experimented with two types of screened bottom boards. The first design was a screened frame that resembles the basic piece of equipment available through many bee supply companies today. In fact, when the first bottom board screen became available, I purchased one and then used it as a template to manufacture more of my own. This design placed the screened bottom on top of the solid wood bottom board so that the surface of the screen sat about an inch and a half away from the surface of the bottom board below. The screen in effect became the bottom of the hive for the bees. The former space that was used by the bees as the entrance to the hive just above the solid bottom board was blocked off with a piece of wood, which acted as a closure so the bees wouldn't confuse it with their new entrance that now existed between the screen and the first brood box.

Then someone figured out that by turning the bottom board around, the opening beneath the screen faced the back of the hive. Today's bottom boards are now made as a single unit with the screen and bottom board attached to each other and the opening for the catch tray facing the back of the hive. Often adhesive glue, a thin layer of petroleum jelly, or oil is applied to a catch tray set on the bottom board beneath the screen. This ensures that mites will become stuck after falling through the screen, thus removing any chance that they will somehow be able to find their way back into the hive.

Although the screened bottom above a solid wood bottom board works well in removing fallen mites from the hive, it tends to also accumulate wax and hive debris and requires regular cleaning to maintain the effectiveness of its mite-trapping abilities and to prevent wax moths from laying their eggs in the area. The additional equipment (bottom board screen and closure for the false entrance), the additional labor involved in cleaning out the area between the screen and the bottom board (and changing the sticky board or the sticky material on the catch tray), and the fact that this space below the hive is an attractive home for wax moth larvae and other insects prompted me to remove the solid bottom board. Now all my hives have a screened bottom board that's open to the ground.

For those who really want to save on equipment handling, cost, and storage, get rid of the bottom altogether. Some beekeepers have reported success by doing just that. However, forgoing the floor of the hive is not practical if you intend to move the hive any significant distance at some point. Because either solid or screened bottom boards can help keep the larger predatory insects and animals out, as well as keeping the bees in, removing the bottom is not recommended if such critters are active in and around your bee yard.

To create a screened bottom board, simply cut out the center of your existing boards and staple a piece of ⅛-inch hardware cloth, available at your local hardware store, over the opening. Then rejoice in the fact that this solution reduces your equipment requirements for each hive by two items. Now that the bottom of the hive consists of a screen, mites and hive debris that fall to the bottom of the colony will fall through the screen and land on the ground below. Cleaning out the area under the screen is no longer necessary, and even the extremely healthy and active mites that fall through are likely to become a meal for insects that live under the hive, rather than making their way back up into the hive. Replacing the solid wood

FIGURE 5-5. Turn your solid bottom boards into screened bottom boards by simply cutting out the center, leaving a 2-inch rim around the sides and back (leave about 4 inches in front) and stapling a piece of ⅛-inch hardware cloth over the opening.

bottom of the hive with a screen also improves the air circulation within the hive, making it easier for the colony to cool itself during the hot summer months. The increased ventilation also comes in handy when the hive entrances are sealed up, such as when the bees are contained while the colony is being transported or relocated.

Initially I was concerned about how the hives would survive the winter with a screened bottom board. I was worried about the possibility of increased winterkill due to the amplified draft the screen bottom would create. However, my concerns were based on anthropomorphic thinking and have proved to be quite unfounded. I am pleased to report that opening up the bottom of the hive, as opposed to

reducing the openings during the winter months, as is often done in northern climates, has not adversely affected my colonies' ability to survive during the long, cold Vermont winters. Unlike our own winter homes, which must be tightly sealed and heated to keep us comfortable, the bees build their winter accommodations with their bodies by creating a cluster. By snuggling together and using their body heat to warm one another during cold weather, the bees are able to regulate the temperature of their living space without much regard for the size of the opening built into the floor of their abode. Honey bees are similar to humans in one respect, however: if they become wet and are exposed to prolonged cold temperatures, their chances of survival become tenuous. Over the years the bees have proven to me that, here in Vermont, as long as they have plenty of high-quality honey available and are kept dry, they can typically handle whatever cold temperatures winter wants to throw their way.

One criticism of the screened bottom approach is that, because honey bee brood takes longer to fully mature in cool temperatures, the increased ventilation created by the screened bottom may lower the temperature of the hive and increase the time that the developing varroa mites have to mature within their birthing cells. This in turn, it is argued, helps to actually increase the mite population within the colony. This assumes, however, that the reduced warmth does not similarly slow the speed of varroa development. Although there may be a slight increase in the number of mites that are able to reach maturity due to the longer length of the honey bee brood cycle, I sense that the screened bottom board removes far more mites than it adds, and it acts as a net gain for the bees at the mites' expense, especially when the benefits of increased ventilation are factored in. Nevertheless, this would be a good question for researchers to answer definitively through further study.

Another criticism of using a screened bottom open to the ground is that there is evidence that it reduces the size of the brood area in early spring when temperatures are still rather cool. Experience, however, shows that colony populations quickly rebound from any brood reduction and brood nest sizes increase rapidly once the cool weather passes. For me the benefit of eliminating hive debris buildup beneath the screened bottom board more than makes up for any delay in population buildup that the open bottom causes.

Another potential drawback of using a screened bottom board is that colonies that are predisposed to using copious amounts of propolis within the hive are likely to try to seal up the opening created by the screen in the bottom of the hive. Over the course of a year, however, I have yet to see a hive propolize more than about a ¼-inch border around the outer edge of the screening. In warmer climates where the bees have the opportunity to collect propolis over a longer period, the sealing up of the screen may become more of an issue, and the screen may require cleaning during the course of the year to keep the bottom opening clear.

The major disadvantage of using a screened bottom that is open to the ground is that it can be difficult to use a sticky board beneath the hive should you wish to monitor the natural mite drop, to evaluate the population of mites within the colony, or test the effectiveness of an active high-impact treatment. Whether you decide to use a screened bottom over a conventional wooden bottom board or a screened bottom board that is open to the ground, a screen on the bottom should become standard equipment on all hives.

The effectiveness of a screened bottom or screened bottom board in removing varroa mites from the hive can be increased dramatically with the use of various techniques designed to increase natural mite fall. The use of these additional management tools can improve the effectiveness of a bottom board screen, transforming it into an important part of a high-impact emergency control measure rather than just a low-impact tool.

One method that encourages mites to fall and has met with success is the use of smoke. Like all living creatures, mites do not like smoke and drop off the bees in higher numbers when this irritant is applied. Although I have yet to try this approach myself in a systematic and regular manner, smoke from burning the leaves of various plants (tobacco, black walnut, cedar, grapefruit, and creosote bush), as well as mature sumac seed heads, has been reported to be effective in this regard. Depending on where you live, some of these materials may be available to you and some may not; however, pure tobacco that does not have fillers and additives may be obtained nationally through companies such as American Spirit Tobacco Company. Research conducted by Dr. Frank A. Eischen of the USDA-ARS Honey Bee Research Laboratory in Weslaco, Texas, has indicated that smoke weakens varroa to the point where they are likely to lose their footing and fall off their bee hosts. This action is brought about through the incomplete combustion that forms the toxic substances and particulates found in smoke. However, it has yet to be established whether specific substances produced by the burning of certain plant materials are toxic to mites, whether smoke itself acts to confuse the mites, or whether the mites fall off simply in an effort to get away from the smoke in much the same way that bees react to smoke. Perhaps all of these actions come into play. It is interesting to note that Dr. Eischen has discovered that some plants, such as the creosote bush, when burned can have a negative effect on the honey bees as well as the mites, whereas others, such as grapefruit leaves, do not seem to harm the bees at all—and yet mites are still observed to drop in high numbers when exposed to the smoke. This observation indicates that it is possible for smoke to act physically to confuse and irritate the mites and at the same time to carry compounds that may potentially act as a parasiticide. It has already been established that smoke from a simple wood fire contains numerous toxic compounds, including dioxin, benzene, and heavy metals.[8] It is not a good idea to treat hives with smoke for mite control when honey supers intended for harvest are in place, because heavy smoke can impart an off-flavor to the honey stored in the comb, indicating that chemical constituents within the smoke are migrating into the honey.

The application of smoke for mite control consists of filling a standard smoker with burning material and directing copious amounts of smoke into an opening of the hive. Because smoke naturally rises, the most effective approach is to blow the smoke through the bottom entrance of the hive and to close off all other openings that the bees could use for ventilation so that a heavy accumulation of smoke will take place. According to the literature and most testimonial reports, the ideal period for smoke treatments ranges from 30 to 60 seconds. Cool, thick clouds of puffy white smoke is the type required for mite treatments, just as it is when working with bees. When the smoke

becomes hot and gray, it is a sign that the smoker needs refueling. Because this control method will not affect mites that are sealed within brood cells or piggybacking on foragers in the field, two or three treatments spaced three to five days apart are indicated. In areas where the types of materials listed above (leaves of tobacco, black walnut, and so forth) are common, their regular use as smoker fuel during normal hive inspections and manipulations may offer a degree of varroa control, provided a screened bottom board is in use.

The biggest drawback to the use of smoke for mite control is the deleterious effect that some of the stronger smokes, such as those produced by tobacco and the creosote bush, may have on the bees if they are exposed to too much smoke over too long a period. Issues regarding the effect of smoke applications on the queen and uncapped developing brood, the chance that constituents of the smoke may be absorbed by honey or wax, and the possibility of harm to those who inhale the smoke during its application should also be considered.

Another technique that can be utilized to increase the number of mites that drop to the bottom of the hive and are removed by a bottom board screen is to coat the bees with confectioners' sugar or some other benign substance such as wheat flour. The use of inert, powdery substances that have a particle size between 5 and 15 microns has become known as the Dowda method. This approach is akin to throwing a bunch of marbles underneath each mite's eight little feet and causing them to take a spill. It also increases the bees' grooming behavior. and as the honey bees clean the dust off their bodies, mites are likely to be dislodged in the process.

I first became aware of the benefits of dusting the bees after reading about the experiences of beekeepers who were getting good results controlling varroa using a 4:1 ratio of confectioners' sugar mixed with garlic powder, which was added to help kill bacteria and fungi within the colony. (It is important to note, however, that dusting a hive with a powder will work only when combined with the use of a

FIGURE 5-6. Brushing confectioners' sugar through the window screen ensures that it will be in a fine, powdery form and not clumpy.

screened bottom or a screened bottom board.) It was reported that the powdered sugar/garlic powder mix did not affect the taste of the honey left on the hive for the bees, and that the colonies of Italian bees really enjoyed the garlic powder.

The typical treatment protocol for an average hive calls for about ¼ pound (4 ounces) of confectioners' sugar to be applied evenly to the top bars of the hive. The sugar is then distributed down between the frames with a hive tool or brush, covering the frames and the bees in the process. In actual practice, it is much quicker to just keep shaking sugar between the frames until all the bees seem to be coated. For best results, the hive should be taken apart and each hive body and super treated individually. When treating large numbers of hives, a container such as a plastic 5-pound honey jar with

holes drilled into the cover makes a great sugar shaker. Alternatively, some people like to use a flour sifter to apply the powdered sugar. In recent years a plastic sugar-dusting tool with attached bellows has become available commercially and is nicknamed the varroa "Dustructor." One places the end of the tool in the hive entrance with the opening pointing up at the frames above and blows the confectioners' sugar up between the frames. I have never used the device and have heard mixed reviews about its ability to blow sugar effectively. Such a tool could be quite helpful in dealing with mites, especially for beekeepers who have top bar hives that feature a screened bottom board, since the only commercially available mite treatment that is easy to use in a TBH is HopGuard (see "An Herbal Treatment" on page 154 later in this chapter).

Unfortunately, all of the dusting application methods described above are highly labor-intensive or are of questionable effectiveness when used on large numbers of hives. A faster method for dusting colonies was developed by commercial beekeeper Randy Oliver in California, who built a frame that fits over the top of a hive body and is fitted with a piece of window screen. With the hive uncovered and the screen in place, the powdered sugar is dumped on top of the screen and a bee brush is used to work the sugar down through the screen. The screen is then removed and the powder that accumulated on the top bars is brushed down between the frames. Randy reports that using this method he can treat hives in less than 20 seconds each.[9]

The amount of sugar needed for treatments will vary depending on the bee population, with less sugar required for weak hives and more needed for strong colonies. The goal is to cover every bee in the colony with a light coating of powder, so hives with fewer bees require less powdered sugar. As with the use of smoke, mites that are sealed within brood cells or piggybacking on foragers in the field will not be affected. Therefore, hives require four or five treatments spaced three to five days apart to provide adequate mite control with this method. A number of beekeepers in Florida have reported keeping varroa-infested colonies alive for several years in a row with no other mite treatment than powdered sugar combined with a screened bottom board.[10]

When powdered sugar is used rather than other benign powders like wheat flour or talc, the bees may make use of it as a small source of feed. Because it is unethical and a violation of organic standards to feed sugar to honey bees when it could be stored in honey supers and end up being harvested and sold as honey, the Dowda method should not be utilized at times when honey supers are on the hive. I used to believe that it was not advisable to dust bees when they could be confined to the hive for extended periods due to cold temperatures, because confectioners' sugar typically contains a small amount of indigestible cornstarch that acts as an anticaking agent and must be eliminated

after being ingested by the bees. Nevertheless, beekeeper James Gabriel of Lapham's Bay Apiaries in Shoreham, Vermont, has used powdered confectioners' sugar containing cornstarch to make his sugar syrup when mixing up bee feed and he has not found it to be detrimental to the bees. So, while concern about the indigestible material in the powdered sugar make sense logically, it does not seem to play out as a real issue in the bee yard.

The use of powdered sugar has great potential as a chemical-free control for varroa, but because small amounts of sugar may be taken in as food by the bees, organic certification committees may be conflicted as to its use. Organic standards prescribe that all sugar fed to the bees must be from certified organic sources. The certified organic confectioners sugar on the market is expensive, so you may want to consider making your own by processing granulated sugar crystals in a kitchen coffee or spice grinder.

Traps

The use of traps to capture creatures has a long tradition in human history. Mechanical in nature, trapping is unlikely to allow for the development of resistance in the traditional sense. I suppose that some resistance may develop when trapping higher-thinking organisms, in that the population targeted for trapping may be reduced to those that are smart enough to learn how to avoid a trap. Nevertheless, the lack of pressure for the development of genetic resistance in *Varroa destructor* makes trapping a very attractive option to the beekeeper concerned with sustainability in the battle against mites.

In an article in *Science News* from 1989, French researchers reported the results of a study indicating that when mating, the female varroa mite found the European honey bee male drone brood much more attractive than female worker brood. According to the authors of the research, the female mite relies on the pheromones given off by the developing larvae to guide her to the brood cell, and drone larvae produce more of this attractant than the worker larvae do.[11] Once in the cell, the mite feeds

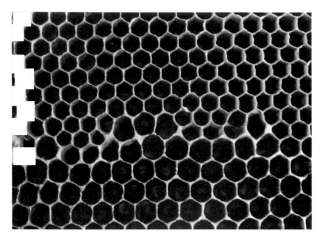

FIGURE 5-7. Due to their larger body structure, drones are provided with larger cells in the combs that they are raised in as compared to worker bees. Note the varied size and shape of the transition cells between the worker cells above and the larger drone cells below.

on the developing bee and then lays her eggs. The knowledge of this preference shown by the female varroa mite has been instrumental in the development of trapping devices.

The earliest trapping mechanism for mites, which technically is the culling of brood rather than trapping, consists of a brood frame fitted with a sheet of large-cell foundation that the bees can use to draw out into drone comb. Inserting a frame of drone comb into the hive's brood nest entices the queen to fill the comb with unfertilized eggs. As the eggs hatch and grow into drone larvae, varroa within the hive are attracted to the developing brood. Approximately 26 to 30 days after the comb is first introduced into the hive, the frame of sealed brood can be removed along with the reproducing varroa contained within the drone cells. Placing this frame in a freezer for 24 hours will kill the mites and brood without damaging the comb. Reinserting the frame back into the hive after it has thawed will then give the bees the opportunity to clean out the cells and allow the trap to be reused. (Some beekeepers will have a spare frame of drone comb that they can immediately exchange with the frame full of brood.) Even though the development period for drones averages 24 days, a maximum 30-day interval for replacing combs is

recommended to allow the workers time to clean out the drone cells and provide the queen with time to lay her eggs. An alternative approach to freezing that has been reported to be successful is to direct a forceful spray of water from a garden hose at the sealed comb, making sure that the water strikes the comb at a slight angle. The water will shear off the cappings and then flush away the pupae along with the developing mites. A much better use of the culled drone brood, however, is to feed it to chickens. Over time the chickens will learn to peck the drone larvae out of each cell without doing much damage to the comb.

Although I have not read any scientific research in this regard, I believe that varroa traps are likely to attract only the mites that are found within the general vicinity of the trap's location. As such, it is advisable to periodically vary the location of the trap in order to cover as much of the hive in as equal a manner as possible. For example, if a drone brood trap is placed on the left side of the hive body in position 3 (between frames 2 and 4), a refreshed trap, if used, should be placed in position 7 (between frames 6 and 8). If the brood nest consists of two hive bodies, then one trap should go in the lower hive body while the other trap is positioned in the upper chamber, each on opposite sides of the hive, to try to draw in mites from as wide an area within the colony as possible. Placing traps next to frames of uncapped worker cells that contain eggs will help keep the mites from infesting the worker brood.

A primary difficulty with culling frames of drone comb and then freezing them is that the beekeeper must have access to freezer space large enough to handle the frames. For those with numerous hives, this can prove to be a real challenge, and it is complicated by the fact that storing the frames outside the freezer for any length of time in a warm environment can encourage wax worms or small hive beetles (see chapter 6). Another potential drawback of using frames of drone comb to cull the brood and remove mites is that, if your timing is off and you don't pull out the drone brood frame at

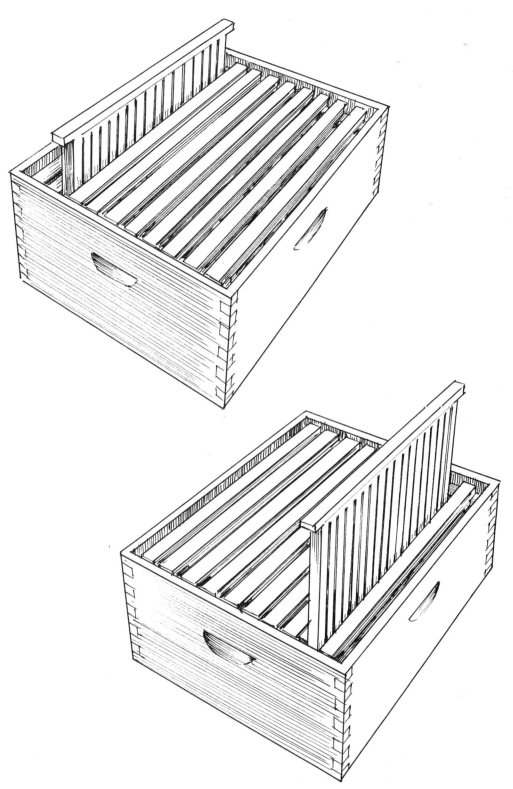

FIGURE 5-8. Regular rotation of the position of the traps within your hives will increase their effectiveness.

the right time, the brood will hatch and you will in effect be helping the varroa mites breed instead of suppressing their population growth.

Organic beekeeper Joshua White of Northwoods Apiaries in Westfield, Vermont, uses a modification of this technique to trap varroa. Joshua places a shorter frame, such as those found in medium or shallow supers, into a hive body containing the brood nest made up of deep frames. Faced with an abundance of space between the bottom of this short frame and the bottom of the hive body, the bees will naturally fill it with a large section of comb. More often than not, this comb tends to be composed primarily of drone-sized cells into which the queen will lay eggs. This is primarily due to the beekeeper having filled the rest of the hive with worker foundation and comb and not providing the hive with enough space to raise drones, as mentioned earlier in this chapter. When this drone brood is sealed, Joshua cuts the excess comb from the bottom of the short frame and renders the wax. Not only does this remove the mites that are developing within the drone brood and provide the beekeeper with a source of wax, but also the simplicity of a technique that does not require specialized or additional equipment appeals to my sense of Yankee frugality.

One drawback, however, is that the beekeeper is unable to evaluate the immediate effectiveness of this approach to culling the drone brood from the hive. Other than picking the dead brood out of each cell in the frame of drone comb and counting the mites present, there is no way to tell how many varroa have been caught within the drone brood and how effectively the trap is performing. In addition, this method results in the destruction of significant amounts of drone brood. This can be of particular concern for those involved with queen rearing, who are interested in flooding an area with drones available for mating rather than reducing the number of males. As a steward of my hives, I always try to base my management decisions on what is in the best interest of the bees and how I can boost their health and overall immunity, thereby making them

stronger. I find that I am personally uncomfortable deliberately killing bees, be they drones or otherwise. Even though they do not directly produce honey or seem to contribute in any active way to the day-to-day activity of the colony, drones are an essential part of a healthy hive.

I don't know how it is for hives down south, where winter barely makes its presence felt, but here in the Northeast winter is not a good season for the drone. For most of the year the drone leads a happy-go-lucky lifestyle when viewed from the human perspective. He does no work whatsoever within the hive, acting as if he were too sick to work and paying no attention to the abundance of chores that require the attention of the colony on a daily basis. With no stinger, the drone does not concern himself with the defense of the nest, preferring to spend his summer days drifting from hive to hive, begging the ever-active females of his species to feed him. Occasionally he'll get up off the couch and fetch his meals himself. The one task the colony calls upon the drone to perform is the act of mating with the queen. This is the one occasion when the drone rises to the call of duty. It would seem that the males of the bee world have an idyllic, carefree life. Nevertheless, their fortunes turn with the seasons.

As the cooler days of autumn herald the coming of winter, the workers' instinctive frugality and lack of tolerance for the drones' malingering become evident. They seem to sense that the abundance of nectar-bearing blossoms will not return anytime soon, and that a long dearth of honey will befall the colony during the frozen months ahead. Practical gals as they are, they realize they have to make the most of the honey they have gathered during the spring, summer, and fall, stretching it out until the following spring. At the same time, they recognize the drain on their collective efforts that the drones represent—and thus the worker bees can be seen every October preventing the males from gaining access to their honey stores, driving the drones from the hive, and blocking their reentry. Between the lack of food and the energy expended in vain trying to resist the female workers from throwing them out, the poor drone in its weakened state gets tossed out into the first snow of the season, only to land on its back and buzz dejectedly as if to say, "But sis, what about all the good times we had?" Unfortunately, the drones, like all honey bees, cannot survive for long on their own outside the hive, especially in cool weather. Thus, the lesson we can learn from the drones is to consistently carry our own weight and always contribute in whatever way we can to our family and community, so that we are not looked upon as expendable dead weight during difficult times.

Okay, so perhaps as a man, I'm a little overly sensitive to the destiny of the males within this female-dominated society. Despite the ultimate fate of these bees that may seem to have no obvious function other than to mate with the queen, I found that I had an aversion to intentionally killing drone brood. This got me to thinking of alternatives to the drone brood trap. What was needed was a trap design that would capture mites without harming the male bees. To accomplish this most effectively, the trap would need to be situated within the brood nest where the majority of mites in the hive are located. With these basic criteria in mind, and some insightful suggestions from Bill Mraz, I came up with a frame-shaped trap design with slotted sides. This trap replaces one of the frames in the brood area. The slots in the sides of the frame are large enough for a mite to fit through, but too small to allow the honey bee to pass. Positioning the entrance slits vertically allows the mites that enter to make their way down to the bottom of the apparatus unhindered. A strip of paper strategically placed on the floor of the trap and coated with a sticky substance allows the mites to "check in" but not to leave. This form of mite trap has the same localized efficacy as drone brood culling, but has the potential to capture far more mites. The trap should be staggered throughout the hive at various intervals as described above for maximum effectiveness.

The sticky paper within the trap can be made attractive to mites by adding methyl palmitate to its surface at the bottom of the trap. Methyl

FIGURE 5-9. A trap that utilizes sticky paper is the alternative to sacrificing drones in the quest to trap varroa.

palmitate is the pheromone given off by the developing drone larvae and was the substance found to be most appealing to female mites in the French studies mentioned earlier.[12] It can be purchased from chemical supply companies. At temperatures above 85°F (29°C), however, methyl palmitate crystals will melt. By folding the sticky paper so that it will hold the pheromone in its liquid form, the desired results may be obtained. The paper that sits in the bottom of the trap should also be folded in such a way as to ensure that the edges are flush with the sides of the trap, so that there is no room for mites to crawl underneath the paper rather than onto its sticky surface.

A significant benefit that a sticky trap offers over the culling of drone brood is that the number of mites caught within the trap is readily visible. This makes the sticky paper mite trap very effective as a monitoring device. Because methyl palmitate is naturally present within a colony, is approved by the USDA as an animal feed additive, and is used by industry to reduce foaming during cardboard manufacturing, trapping with pheromone bait should be safe to implement while honey supers are on the hive without the risk that toxic compounds will contaminate the crop.

Unfortunately, the bees have a tendency to fill in the slits in the trap with propolis when the trap is left in a hive for long periods. Cleaning the slots is simply a matter of sliding a hive tool along the propolis-filled openings to dislodge any buildup that may have accumulated. This job is made easier when the trap is exposed to cold temperatures first, causing the propolis to become hard and brittle.

When I first began using this method, I removed the traps, cleaning and replacing them with fresh traps, once every 30 days or so, similar to the schedule for removing drone comb. However, I found this to be unnecessary and extremely time-consuming, especially as the number of hives in my bee yards grew, and more and more of them survived through the winter. I eventually began changing the trap only once, about halfway through the active season, installing the first trap in early May, replacing it with a fresh trap and bait strip around the end of June, and then removing it in August. This has worked well for me, and I believe it is the ideal schedule. Of course there have been years when the traps were never removed, changed, or cleaned until August, during or just after the honey harvest, and just prior to applying mite treatments and tucking the bees in for the winter. When I don't have time to clean out the traps during the season, 15 to 20 percent of my hives will propolize well over half of the openings in the trap. This decreases the trap's effectiveness, but I rationalize that it is the trade-off for eliminating the additional labor of cleaning and replacing the traps during the course of the year. Because I use a number of other low-impact mite control techniques throughout the season in my hives, I have been able to get away with using the traps in this less efficient manner. If I did not have other effective mite control measures in use simultaneously, I would be much more diligent in refreshing the traps within the hives on a regular basis.

As you can see in figure 5-11, these traps are constructed utilizing a standard top bar combined with a fabricated bottom bar and end bars, so that when the slotted sides are attached the total width of the trap is ¼-inch thinner than a standard frame. Old-fashioned metal spacers on the end of the trap provide structural support and ensure that the final dimensions closely resemble those of a standard

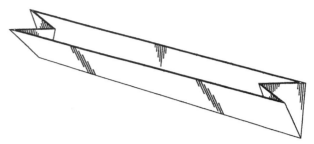

FIGURE 5-10. The sticky paper trap allows you to monitor the mite load while at the same time reducing mite populations within your hives. Be sure to fold up the ends if you plan on using methyl palmitate as bait.

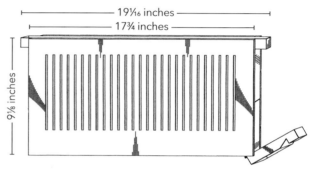

FIGURE 5-11. This trap is easy to make with basic carpentry tools and skills.

deep frame, thereby preserving the integrity of the bee space within the colony. Unfortunately, these metal spacers are becoming difficult to find as more and more bee supply companies discontinue their manufacture and availability. However, I'm sure someone with greater carpentry skills than I can come up with a design that precludes the need for the metal spacer and yet maintains structural integrity and proper spacing.

Another drawback of this trap design that I have not had time to improve upon is the use of Masonite board in constructing the sides of the trap. The small particles that make up the Masonite become exposed upon cutting the slots, and, as a result, the small pieces that are exposed along the ragged edges of the slits are chewed off by the bees and end up stuck to the paper in the bottom of the trap. Although initially this reduces the effectiveness of the trap, I have found that the accumulation of debris lessens dramatically after the first season, to the point where it eventually ceases to be an issue. Sanding these areas down after they have been cut may help reduce this tendency, but I have not tried it.

In 1999 I had the idea of applying for a patent for this mite trap and starting a business manufacturing and distributing it. However, after noticing how much more effective the trap was in catching *Varroa destructor* when methyl palmitate was used as bait, I sought a licensing agreement to market the trap, with the pheromone lure as an integral part of the trapping device. Shortly after starting the process to get methyl palmitate approved by the FDA for use as a honey bee miticide, I sent a letter to the French institute where the researchers had discovered methyl palmitate's properties and requested permission to use their intellectual property, for a fee of course. A couple months passed, and after not hearing back from them I wrote again, this time in English. Again I received no reply. Since the effectiveness of the trap is increased dramatically with the use of the pheromone, it did not seem wise to try to build a business promoting a trapping device without it, so I dropped the patenting and marketing idea and instead am satisfying myself with the opportunity to pass along the design of this device so that do-it-yourselfers have the chance to make use of it. Methyl palmitate may be used as bait within such a trap without infringing upon intellectual property rights, or the FDA's pesticide regulations, as long as the honey harvested is not sold but kept for personal use and you are not selling the pheromone to others as a mite control agent.

Obtaining methyl palmitate can be difficult, because it is not approved for use as a varroa mite attractant by the EPA, and because most chemical supply companies that sell it make it available in 55-gallon drums rather than selling a pound of the material here and there. I was able to obtain a 1-pound sample of methyl palmitate (enough to last many years) during the time I was pursuing the idea of starting a business manufacturing and selling the traps. The company I was dealing with was happy to sell me a small sample for testing because they hoped it would translate into a much larger order down the road.

Another variant on the drone brood trapping and drone comb culling idea comes in the form of a commercially available product called the MiteZapper. Invented by Dr. Zachary Huang, of Michigan State University, the MiteZapper is purported to be 85 to 95 percent effective in controlling varroa mites without chemicals. The zapper consists of a plastic frame with a heating element embedded within. The frame contains drone foundation and is placed within the brood nest. After the bees have drawn out the foundation into drone comb, the queen has laid eggs in it, and the workers have sealed the brood cells, the zapper is hooked up to a battery for about eight minutes. The electricity produces enough heat to kill both the pupae and mites. Within a couple of days the dead and dying pupae and mites will be removed by the workers, and the queen can once again lay eggs in the drone comb. It is recommended that the zapping process be repeated every 21 to 25 days during the drone-rearing season.

One argument that has been made against the idea of using drone brood culling, or traps that utilize either natural or synthetic versions of the pheromones given off by drone brood as an attractant to lure mites into the trap, is that by consistently removing mites that have a disposition for reproducing on developing drones, are we not inadvertently selecting for mites that will reproduce on worker brood? At first glance this sounds like a valid concern. However, due to the shorter brood cycle of the worker bee as compared to the drone, the reduction in mites successfully able to reproduce in worker cells could have a significant effect on their overall population buildup within the hive. So far, all efforts to remedy the varroa mite problem through a genetic solution have been focused on the honey bee. If drone brood trapping places selective pressure on the mite by increasing the number of mites that prefer to reproduce on faster-developing workers, such an event could serve to improve upon the ability of the mite and the European honey bee to coexist without the population of mites, and their attending diseases, overwhelming the host colony.

Whether or not it selects for mites that prefer to reproduce on worker brood, trapping, zapping, or culling drone comb filled with sealed brood can be an effective way to remove significant numbers of mites from a hive, and because they work physically, these methods can be utilized throughout the active season without fear of contaminating the harvest. In my experience, traps and culling alone have not proven to be enough to prevent a colony from crashing due to an overabundance of varroa. However, when used effectively and in conjunction with other low-level control devices such as the bees' natural mite resistance and a screened bottom, the level of mites can be kept low enough throughout the spring and summer so that a single high-impact treatment in the fall is all that is needed to ensure that the bees will survive the winter.

Over the years since I wrote the above section on drone brood culling and varroa mite trapping, I have changed my thinking about purposely killing drone brood to suppress mite populations. The "massacre" that occurs in each beehive just prior to the long dearth of winter is perhaps the most dramatic of examples that illustrates the important role that the male honey bee plays in addition to mating with the queen. For the humble drone is to the hive as the tail is to certain salamanders that can make their tails fall off when in the clutches of a predator, allowing them to scramble off to safety and regenerate a new tail over time. The sacrificial role embodied by the drone should not be overlooked in its importance in helping to increase a colony's resilience in the face of adversity, since the male honey bee may not only choose to end its life in sacrifice through mating, but may be fated from birth to be sacrificed by his sisters.

Observe the brood in a hive and you will notice that the drone brood tends to be located on the outer edges of the brood nest. In Langstroth-style hives this means that most of the drone brood will be found along the bottom or sides of the comb. In top bar hives, where the bees are allowed to build their comb without foundation, frames of drone brood will tend to be found between the honey and pollen frames and the rest of the brood area, in effect insulating the frames of worker brood from potential temperature fluctuations that may work their way into the brood area from the outside.

The placement of drone brood around the outer edges of the brood nest rather than more centrally is a strategic choice that the hive makes. This decision ensures that, should the hive experience a period of cold weather that causes the cluster to contract in order to conserve heat and maintain the brood nest temperature, it will be the developing drones rather than the worker brood that will be exposed, get chilled, and die. An additional form of sacrifice that the male bees may be forced to make during their initial development stages comes during times of low food availability. During times of nutritional scarcity, the unfertilized drone eggs and developing drone larvae tend to be the first resources cannibalized by the workers in an effort to conserve and recycle protein and nutritional resources within the colony when there is simply not enough food to maintain the existing colony.

As beekeepers, we tend to focus on the workers and queen, who are responsible for the production of comb, honey, and brood. When we overlook the drones, or treat them as an inconvenience and do not make adequate accommodation in the hive for the raising of male honey bees, we are not only denying the colony the ability to naturally express itself, we also may be limiting that colony's ability to respond effectively to thermal and nutritional stress. Without unfertilized male eggs and drone larvae to sacrifice, a hive undergoing such stress would more quickly use up the precious sperm that the queen stored during her mating flights, because they would be forced to cannabalize worker eggs and larvae instead, which the queen would eventually have to replace. Such sperm is a commodity that a hive can obtain only through the risky and uncertain process of queen supersedure.

Thus, a hive that has the option to sacrifice drone larvae and unfertilized eggs when a reduction in the brood nest is necessary is the hive most likely to reach its fullest potential during its lifetime.

Rather than solely focusing on the limited role of the drone as the provider of male DNA during mating, holistically oriented beekeepers will keep in mind the more common sacrificial role that the drone has within the colony. Drones that survive to maturity, rather than getting cannibalized or dying from chill brood, get a chance to sacrifice themselves in the process of mating. Those drones that survive their developmental days but do not manage to mate with a queen will still be alive when the colony is preparing for winter, and they will make the ultimate sacrifice on behalf of the colony—being ejected from the hive to die. When we realize that the root meaning of the word "sacrifice" means "to make sacred," we get a broader glimpse into the larger role that the male honey bee plays within the colony.

One way to honor the sacrificial role of the drone is to position frames that contain significant amounts of drone comb on the outer edges of the brood nest. Such frames would include drone comb used to attract and remove varroa mites from a colony, or damaged frames that have holes in the comb, since the repairs to this comb are likely to be fashioned as drone-sized cells rather than the smaller worker cells.

Like the chess player who recognizes the value of the pawn despite its limitations and takes full advantage of the various qualities and attributes of these pieces in order to win the game, beekeepers may do well to consider utilizing drones in ways that mimic what the bees do naturally. The more I consider the way a hive naturally sacrifices the relatively few drones in their population when necessary in order to help ensure the survival of the much more numerous workers and their queen, the more comfortable I become with the concept of culling some of the drone brood from a hive when necessary in order to suppress the population of varroa.

Heat

There's nothing like a steamy, sweltering day to help shift the honey flow into high gear. In fact, there are certain plants whose blossoms seem to produce a significant amount of nectar only when the hazy, hot, and humid days of summer kick in. According to John Lovell, author of *Honey Plants of North America* (1926), clover plants will yield the heaviest when temperatures are between 80° and 90°F (27° and 32°C). Yet while bees may benefit from increased nectar flows during hot weather, there are drawbacks to such warmth. For one, temperatures above approximately 97°F (36°C) may cause bee brood to overheat and die. And although exposure to temperatures around 119°F (48°C) for short periods will kill mites without harming adult bees, prolonged exposure to such heat will kill bees as well.

On hot days, many bees become occupied with maintaining the temperature of the brood nest within the hive at approximately 95°F (35°C) and controlling the humidity level. Watching the bees line up in the same direction and start fanning their wings to induce a draft to regulate the temperature and humidity of the brood nest is fascinating. After many years as a beekeeper, I still enjoy feeling the warm breeze on my skin as I marvel at the hive's cooperative and efficient air-conditioning procedures. For those who are comfortable around bees, it is a joy to put your nose down by the hive entrance and breathe in the sweet smell that rides the air currents generated by the fanning bees. A deep feeling of contentment runs through me whenever I inhale that fragrant scent that indicates all is well within a colony.

The more individual bees are involved with providing air circulation to the colony, the fewer bees are available to forage in the field. Because of this, numerous schemes have been hatched over the years to provide additional ventilation during hot weather, from vented hive covers to solar-powered fans that provide forced-air ventilation. Always one to prefer the simplest route whenever possible, I find that creating a ¾-inch-diameter

FIGURE 5-12. Cocking the outer cover is a quick, easy, and low-cost method of providing additional hive ventilation.

upper entrance hole in the hive body will do much to improve the movement of air through the hive. Just be sure that any additional entrances that are drilled into the equipment are located below or to the side of the handhold so that the entrance is not inadvertently blocked when moving the equipment, which would increase the risk of raising the ire of the guard bees. Additionally, moving the outer cover back ½ inch so the forward lip of the telescoping cover sits on top of the front of the inner cover seems to provide enough ventilation up through the inner cover for all but the most extreme temperatures.

Some beekeepers like to shift an entire super or hive body back about an inch to improve airflow and help a colony deal with hot, humid temperatures. I don't like to use this approach because it allows rain to get inside the hives and will invite robbing in all but the strongest colonies. If additional cooling is required, the apiculturist can simply provide the bees with the resources they need to handle the situation themselves, by making sure they have access to plenty of clean water. The workers will forage for water and bring it into the hive, where it reduces the temperature within the colony as it evaporates.

One of the most basic ways beekeepers are able to influence the temperatures inside the hive is through the color they paint the outside of their equipment. White and other light colors help reflect sunlight and keep the interior temperature of the hive cooler. Thus, an extremely low-tech way to take advantage of the lethal effect of heat on mites is to simply paint the hives a dark color and place them where they will receive lots of direct sunlight. Unless your apiary is located in a desert or semidesert region, the preoccupation of numerous bees with ventilating such a hive, and the subsequent loss of honey production, may well be worth the effort of inhibiting varroa through the increased temperature. Although it is unlikely that a hive painted black and sitting in the sun on a hot day will approach the lethal temperatures needed to directly kill mites, such a hive, when used in combination with a screened bottom board, may well remove a significant number of mites that lose their footing due to their weakened state of heat exhaustion.

The observation that mature bees can tolerate high temperatures better than varroa mites has led to the creation of techniques for artificially heating up the honey bees' environment in order to adversely affect the mites. One such technique has been developed and studied by Vermont beekeeper Jeff Cunningham at the end of the twentieth century as part of a Sustainable Agriculture Research and Education (SARE) grant project titled "Evaluating a Heat Therapeutic Control of the Honey Bee Mite *Varroa destructor*." Jeff built a heating apparatus based on one he observed while visiting beekeepers in the former Soviet Republic of Uzbekistan, and he presented the results of his testing at a meeting of the Vermont Beekeepers Association. The unit consists of a portable wooden cabinet, a heating unit, and a wire mesh cage that is filled with bees and placed within the upper portion of the cabinet. Plexiglas windows on the top and sides of the cabinet provide means of observation and ventilation. By observing an interior-mounted thermometer and the number of mites falling to the bottom of the apparatus while the interior of the heating unit is warmed up, an accurate and practical indication of the effective temperature range can be established.

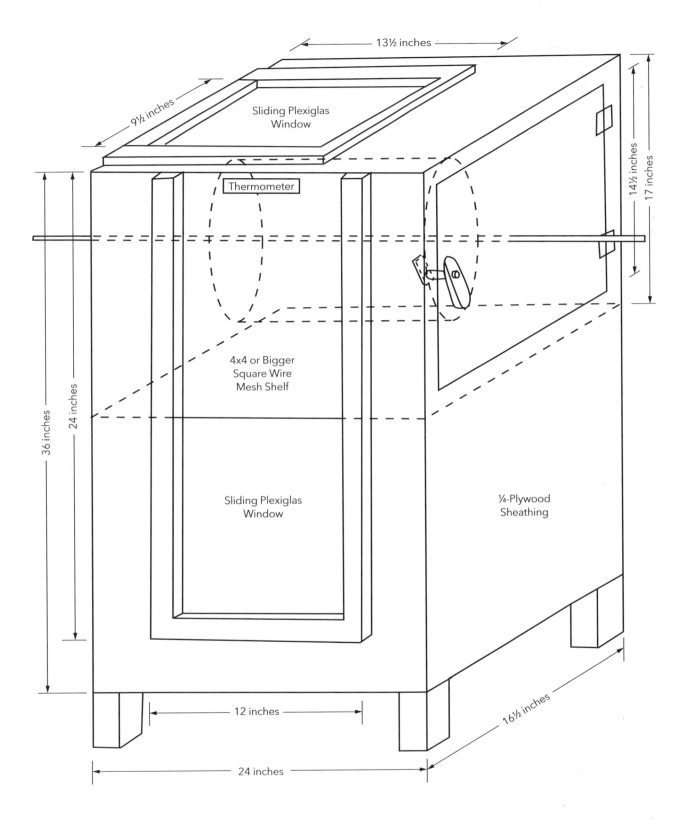

FIGURE 5-13. Heat treatment apparatus: dimensions of cabinet with cage in place.

Honey bees are shaken, brushed, or blown into a large funnel that facilitates the transfer of bees into the cage of the heating unit, much in the same way that packages of bees are prepared. The process is made easier by performing this task at night, when the bees are not as apt to fly. Jeff also found that cool weather made the job of transferring supers of bees into the heating unit easier. Also, spring is an easier time to treat with heat, because healthy hives tend to be at their lowest population right after winter.

Jeff's study indicated that a temperature of 116.6°F (47°C) for up to 15 minutes was ideal in causing the mites to drop off bees without causing any noticeable harm to the bees themselves. A key factor that affected the unit's effectiveness was the number of bees within the cage during treatment. Jeff found that the bees need to be able to space themselves out to allow ample room for the circulation of heated air. If the cage was overfilled with bees, the bees would form clumps within the heating chamber, and far fewer mites would fall than when treating loosely distributed honey bees. By the same token, the heated bees are easy to pour

back into their hive following treatment, but one should make sure to distribute the bees to allow for adequate cooling.

The heat source that Jeff utilizes is a 1,500-watt "utility" or "milk house" electric heater with a built-in fan, such as those found in most hardware and home supply stores. A generator of some type is necessary for applying heat treatments to apiaries that are not located close to homes or other electrical power sources.

Due to all the preparation involved, which includes caging the queen temporarily for safekeeping, and the relatively long treatment time needed for each hive, a significant investment of labor is required (approximately 30 minutes per hive) when using this heating method. The upside is that the financial investment is very low over the long run because the primary cost is the initial purchase of the materials to build and outfit the heating cabinet and wire cage. Ultimately, this approach is unlikely to be economical for large commercial operations, but it is an attractive option for hobbyists and sideline operations running up to

FIGURE 5-14. A small trailer is handy for transporting the heating chamber and associated equipment to and from the apiary. PHOTO BY CHARLES A. PARENT III.

FIGURE 5-15. After the hive's population of workers and drones is placed inside, the wire mesh cage is inserted into the heating apparatus to apply the precise mite-killing temperature for a specific period of time. PHOTO BY CHARLES A. PARENT III.

a few dozen hives. The biggest benefit of using heat treatments for varroa control is that the chances that mites will build up a tolerance to the treatment are extremely low, and there is absolutely no opportunity for contamination of honey or wax, making this option available for use at any time during the season.

Small-Cell Foundation

It has been proposed that small-cell foundation can be used to improve the ability of *Apis mellifera* to withstand the presence of varroa as well as other diseases and pests. It has been well established that the cell size of brood comb directly affects the size of the bee that develops within it, with larger cells producing bigger bees and smaller cells reducing the physical size of the emerging adult. A leading proponent of the small-cell approach is Dee Lusby of Arizona Rangeland Honey. Dee and her late husband, Ed, were reportedly the first commercial beekeepers to successfully overwinter bees on a regular basis without the use of high-impact varroa mite treatments of any kind. It is claimed that this success is a result of their converting to small-cell foundation. Nevertheless, it is important to note

that during the process of transitioning to small-cell foundation only about 10 percent of their colonies survived the initial reduction, and it was these hives that were used to rebuild their colony numbers; this may have served to weed out the vulnerable colonies in the operation.

According to Dee, most, if not all, of our current beekeeping woes can be traced back to the period between 1880 and 1939, when a concerted effort was made to increase the cell size of bee foundation to produce larger honey bees. The reasoning was that this would result in an increase in honey production, because bigger bees are capable of carrying larger loads of nectar and pollen and have a longer proboscis to help them reach deeper into blossoms. As a result, today's foundation is typically made to accommodate a 5.4-millimeter or larger cell size. The major problem with this artificial enlargement, according to Dee, is that, although the bigger cell size has increased the average body size of honey bees and the amount of honey they are able to produce, it has decreased the vitality of the honey bees' immune system, making them generally more susceptible to disease and parasitic mites—the full consequences of which we are experiencing today.

A skeptic would argue that the reason the Lusbys were able to forgo the use of chemicals and antibiotics is more likely due to the fact that their hives are located in Africanized honey bee territory, and either their bees have picked up some of the AHB's natural genetic resistance to mites, or their European honey bee queens have actually been replaced by their Africanized cousins. It is also likely that the initial large-scale losses that Arizona Rangeland Honey experienced during its conversion to natural-sized comb selected for the hives in the operation that were most resistant to the mite. Add to this the fact that the intense heat of the Arizona desert plays a role in creating a less hospitable environment for the healthy growth and reproduction of *Varroa destructor*, and you have a strong argument explaining the mite tolerance these bees display.

Despite these potential mitigating factors, there is evidence to support the theory of healthier bees as a result of smaller cell size. In the first place, the average cell size of foundation used in European honey bee colonies has indeed been artificially increased over the years. In the early days of foundation manufacture, cell size equaled approximately five cells to the inch, just shy of 5.1 millimeters per cell.[13] Amos Ives Root (1839–1923), the founder of the A. I. Root Company, hired a diemaker to manufacture the first foundation rollers in the United States. Although initially the foundation mill was built on the basis of 5 cells per inch, Mr. Root later decided to increase the cell size to 4.83 cells per inch (approximately 5.26 millimeters per cell).[14] Cell size continued to be enlarged, to the point where today's standard wax foundation typically has cells that are about 5.4 millimeters in size. Nevertheless, cell size will tend to vary somewhat, because foundation is prone to stretching during the manufacturing process or when being installed into frames.

The second piece of evidence to support the theoretical benefits of controlling varroa with small-cell foundation stems from the fact that *Varroa destructor* does not cause great damage to smaller species of bees that naturally live closer to the equator, such as varroa's natural host, the Eastern (or Indian) honey bee (*Apis cerana*), or the infamous Africanized honey bee (*A. mellifera scutellata*). The Indian and African honey bees will bite at mites and remove them from their bodies, which is one way they combat varroa—but they also have smaller worker-comb cell diameters and abbreviated developmental times for the workers and drones to grow from eggs into adults when compared to the European honey bee. This shortened pupation period limits the number of varroa that are able to be raised to maturity and helps keep their population in check. This suggests that *Apis mellifera* raised on a smaller-sized cell foundation may also have a shorter birthing cycle from egg to adult, which would provide a measure of natural mite control similar to that enjoyed by their cousins living nearer to the equator.

Getting back to a cell size that is closer to *Apis mellifera*'s natural state, as proposed by Dee Lusby, is not a simple proposition, if for no other reason than a queen bred on and accustomed to laying eggs in 5.4-millimeter cells is likely to be too large to comfortably fit her abdomen into 4.9-millimeter cells. Indeed, when building comb naturally without the benefit of foundation, bees raised in and on 5.4-millimeter cells seem to construct cells that measure about 5.4 millimeters or slightly smaller. According to Dee, worker bees raised on large-cell foundation encounter significant difficulties when trying to draw out foundation of a smaller cell diameter than that upon which they were raised. To deal with this complication, Dee recommends making the transition in two steps. First, reduce the cell size of the comb from 5.4 to 5.1 millimeters, and then, after a year, once the bees have stabilized on the new comb, reduce their cell size again, down to 4.9 millimeters. It is important to remember that the honey bee's current cell size is the result of incremental cell enlargement that began over 100 years ago. It is not a good idea to expect to reverse such an ingrained and established biological process overnight.

There is considerable controversy surrounding the small-cell phenomenon. Some beekeepers who have made the switch to small-cell comb swear that it's the best thing since sliced bread and that they no longer treat for mites, yet their bees continue to thrive. Other beekeepers who have tried it report that it didn't seem to help at all and their bees died due to mite infestation anyway. So far the scientific community has not been especially helpful in clearing up the debate either. There are some studies that seem to indicate that bees raised on natural-sized comb may indeed suffer less varroa infestation,[15] but other studies show that the level of mites in the colony are not impacted at all by comb size.[16] One problem with the research to date is that it has tended to focus on varroa levels within colonies raised on the differing cell sizes. The theory being explored is that in small cells, the mites do not have enough room to move about, which inhibits their reproduction. Study methodology has tended to monitor for mites that fall through a screened bottom board either naturally or after treatment with a chemical miticide. Such monitoring, however, is subject to false readings, chemical-resistant mites, mites reproducing in sealed cells, and other problems. I am not aware of any trials that have looked at the development time from egg to adult of small-cell bees compared to bees raised on conventional comb.

Some academics believe that the development time of the honey bee is controlled genetically. However, on his website, beekeeper Michael Bush states that he has observed a decrease in development time of about two days in worker bees raised on small-cell comb.[17] If such an observation can be duplicated and observed to not occur on today's standard comb, this would help verify that small-cell comb plays a role in reducing varroa populations in a hive, because we do know that the longer the incubation time of the developing brood, the more mites a female varroa is able to raise to maturity.

In an effort to resolve this controversy, I decided to experiment with my hives by regressing them down to small-cell comb and then stopping mite treatments to see what if any effect it would have on the bees' ability to tolerate varroa mites. The results I experienced were mixed. At least one of my small-cell hives died over winter with signs of varroa stress (dead mites in hive and small clusters of dead bees with small abdomens). Yet at the same time, there has been a pronounced difference in colony strength and health in the spring, with the small-sized bees being noticeably healthier with a larger population compared to hives containing combs with standard cell size. Unfortunately, I cannot determine whether the increased vitality of my small-cell hives is due to the cell size, or due to the fact that all their combs were recently replaced with fresh new comb (which studies indicate typically contain significantly fewer pathogens and toxic chemical contaminants). As a result, I cannot say with any degree of certainty that small-cell bees deal with varroa any better and are healthier than bees raised

on conventional comb. Despite these disappointing results, I am continuing with my effort to convert all my hives to small-cell comb simply because I like the idea of keeping bees that are the size nature intended them to be. This practice seems to be more in line with the concept of natural beekeeping.

I do wish that I had started my hives on smaller-sized foundation to begin with, because it is very time-consuming and requires a large financial investment to remove all the combs in a hive and shake the bees onto new foundation of a smaller cell size. Given that this process must be done twice over a two-year time period, from 5.4 mm comb onto 5.1 mm foundation the first year, and from 5.1 mm comb to 4.9 mm foundation the second year, the more hives you have to convert, the more time-consuming and expensive the process will be. If I were starting out with bees today and wanted naturally sized bees, I would try to buy packages of bees or nucleus colonies raised on small-cell comb from the growing number of bee suppliers that make them available. If small-cell bees were not available, I would buy a package of conventional bees and start them on 5.1 mm foundation, which would cut the process of regression in half and cost half as much. It used to be a challenge obtaining foundation with the appropriate cell size in order to make the transition to natural-sized comb. However, today some beekeeping supply companies offer small-cell foundation (see resources on page 272).

Fungi

Not being native to Vermont and the Champlain Valley, I have noted with a degree of consternation the pessimistic tendencies of most of the beekeepers here, which tends to match that of farmers in this neck of the woods. This attitude was eloquently expressed to me by beekeeper Bill Mraz, who noted shortly after the arrival of *Varroa destructor* that if we really wanted to control the mite, we should start trying to raise and propagate the species. It would then be just a matter of time, he postulated, before some type of disease or critter would come along and start killing off the mites we were trying to raise. There may be something to this pessimistic attitude shared by so many in Vermont agriculture, because it turns out that just such an organism has been uncovered by researchers at the ARS Beneficial Insects Research Unit (BIRU) at Weslaco, Texas, who have worked with a strain of the fungus *Metarhizium anisopliae*. This particular fungus has proven itself to be highly pathogenic to *Varroa destructor*, yet it is not toxic to humans and does not harm bees or interfere with the queen's reproduction. Although the fungus does not work immediately and takes three to four days after being introduced to a hive before peak mite mortality occurs, the acaricidal effects have been found to continue for at least forty-five days following the application of the spores. The fungus was observed to spread to other colonies through the natural drifting of workers and drones between hives, a situation that is likely to prove beneficial.

Although the fungus does not seem to penetrate the cappings on sealed brood cells, treatments are not limited to periods when no brood is being produced or when brood levels are low, given that its extended time of activity within the hive allows it to be effective over the course of several of the mites' reproductive cycles. According to the researchers, *Metarhizium anisopliae* was as effective as the miticide fluvalinate (the active ingredient in Apistan) when used on mites that were not resistant to fluvalinate.[18]

When it comes to high-impact mite treatments, the ideal is a substance that is deadly to varroa, is effective over long periods, is not toxic to bees or humans, and will not contaminate honey. *Metarhizium anisopliae* seems to fit this description well. In addition, the use of biological controls that do not carry the potential for the buildup of resistance by the mite harmonizes with the organic philosophy of honey bee husbandry. As a result, *Metarhizium anisopliae* is a potentially exciting tool that the organic beekeeper may have available in the future, at least until the day when a truly varroa-resistant honey bee becomes available to make such measures unnecessary. The major challenges

that must be overcome if *Metarhizium anisopliae* is to become a useful tool in honey bee management is to produce spores that have an adequate shelf life, as well as to develop an effective way of introducing the fungi into the hive so that it comes into contact with enough mites to make it a viable control option.[19]

Treatment-Free Beekeeping

Let's start with a disclaimer: "treatment-free" is not a totally accurate term when it comes to bees. In the context of beekeeping, "treatment-free" really refers to treating a hive in such a way as to suppress mite population levels without introducing any kind of foreign substances into the hive. *Chemical-free* would be a more accurate term, but I am sticking with treatment-free, because that's the term in common usage today.

Most of the queen breeders and bee suppliers who advertise their bees as treatment-free have obtained genetic stock from at least one of the three strains of honey bee proven to have an established level of mite tolerance: Russian, hygienic, or varroa-sensitive hygienic (VSH). Such suppliers typically make and sell lots of nucleus colonies each year. This means that they are making nucs, and splitting their hives, therefore interrupting the colony's brood cycle, while at the same time removing old combs from their hives and, more often than not, selling them to unsuspecting buyers. In fact, everyone I have met to date who is managing to keep bees fairly successfully without introducing some type of foreign substance into the hive to control mites is practicing a combination of three of more of the following cultural management practices:

- obtaining genetic stock from strains of bees with some level of proven mite resistance
- making regular splits and nucleus colonies
- removing old combs on a regular basis and replacing them with fresh combs
- using screened bottom boards
- trapping mites or culling sealed drone comb from the hive

These practices either reduce the mite levels in hives by a relatively small percentage or encourage an environment in which the bees are able to tolerate a higher population of mites than normal. In isolation, none of them are typically enough to reliably keep a hive alive year after year, but each technique when added together can help a hive deal with the mite population. Generally speaking, the greater the number of these practices that are combined into one's yearly management schedule, the greater the survival rate of the colonies without the need to implement a treatment of any kind during the year.

While it is too early for me to tell for sure, I suspect that the treatment-free approach will result in a higher percentage of yearly losses compared to operations that rely on soft-chemical treatments to reduce varroa pressures, and do so in a way that limits the amount of stress introduced into the hive as a byproduct of the treatment. Nevertheless, this is the path I am adopting in my apiaries. As it becomes increasingly clear that, in terms of sustainability, humankind must change the way it does almost everything, I have begun looking at my beekeeping operation with a more critical eye. Most organic farming operations are not sustainable as they are being managed today. They typically rely on inputs from off the farm, such as horse manure from the farm down the road. A truly sustainable vegetable farm, for example, might incorporate animals to produce the manure that will build soil fertility, which will feed the plants and keep them healthy so they can in turn close the circle and feed the people and animals on the farm. In my apiaries I endeavor to mimic the same model and figure out how to make or produce everything my bees need within my beekeeping operation instead of relying on some distant corporation to manufacture, package, and ship what is needed to me.

I conclude this discussion, though, by noting that the treatment-free approach requires advanced skills and increases the risk of failure. For beginners, the use of one of the following soft chemical varroa treatments is the recommended path.

Soft Chemical Treatments for Varroa

The use of soft chemicals is the kinder and gentler offshoot of the industrial business model that looks for the "magic bullet" that, when added to the hive management protocol, solves the problem of the mite. Soft chemicals are much less toxic than hard chemicals like coumaphos, yet they are still able to get the job done. More importantly, they can allow the beginner to keep mites under control using more natural means in a way that is highly effective and takes little skill and time. These mite-control solutions can play a critical role in helping neophytes develop confidence in their beekeeping skills so that they may transition to utilizing fewer treatments in the hive or eventually go treatment-free.

It is important to note that the use of chemicals within the hive, no matter how safe they are for humans or other life forms, is really just a temporary solution, a stepping-stone to help us make the difficult transition from bees that are completely vulnerable to those that can hold their own and even thrive in the face of varroa. Like their hard-chemical cousins, soft chemicals are inputs that come from outside the apiary, and thus we require the efforts of others to make them available. They also require additional labor and resources on the part of the apiculturist. This goes against the basic philosophy of self-reliance and the closed-loop system of farming that lies at the heart of the organic approach. Such reliance on chemicals can generally be considered unsustainable in the long run, unless the chemicals can be produced on the farm, or unless their use has the potential to eventually bring such health, balance, and vitality to the hive that eventually chemicals are no longer needed.

Soft chemicals are often used in organic agriculture because they fit with the mind-set of the conventional farmer turned organic—the idea that applying a liquid or powder will solve the problem through the miracle of modern chemistry. Many organic farmers have a conventional farming background, so this mind-set is common and understandable. Soft chemicals also allow those producers with no other alternative to move in the direction of a more sustainable approach and lessen their reliance on toxic chemotherapy management.

Grease Patties

When *Varroa destructor* first appeared in Vermont in the early 1990s, there were no established options for dealing with this new threat. As a result, I started looking for clues that might lead to successful treatment protocols that would help keep the bees alive. I found myself scanning articles in bee journals and eavesdropping on conversations at bee meetings, always on the lookout for any mention of what the beekeepers who had already lived with varroa for several years were doing to counter the mite's damage.

One of the first references to a potential treatment I came across in my search was the observation by some beekeepers that grease patties seemed to be helping hives weather the onslaught of varroa. Traditionally, grease patties have been used to introduce antibiotics into the hive for the treatment and prevention of American foulbrood (AFB). Unfortunately, introducing antibiotics into the hive in this manner has only accelerated the emergence of antibiotic-resistant strains of AFB.

What, then, might be causing grease patties to have a positive effect on bees living with varroa infestations? It has been theorized that the grease from the patties, when it is distributed throughout the hive and covers the bees with a light coating, adversely affects the mobility of the varroa in much the same way it affects the tracheal mite (more on the tracheal mite later in this chapter). Unfortunately, because varroa mites are so much larger than the microscopic tracheal mites, the ability of grease patties to reduce mobility has less effect. Nevertheless it is logical to assume that combining grease patties with essential oils (rather than antibiotics) could prove to be effective for varroa control, while at the same time helping to hinder the mobility of tracheal mites. Essential oils that are toxic to mites will increase the percentage of mites that lose their grip and fall to the bottom of

A GREASE PATTY RECIPE

Combine 2 parts saturated fat (Crisco vegetable oil) with 1 part white sugar and thoroughly mix. (Small amounts of honey or essential oil may also be added.) Form this mixture into uniform patties of about ¼ pound (4 ounces) apiece; store between sheets of wax paper so that they do not stick together. To use, place a single patty on the top bars of the brood chamber.

An easier application method when treating large numbers of hives is to simply take a spoon and place a dollop of sweet grease equaling approximately ¼ pound (4 ounces) on the top bars of each brood nest. Although this method is not as precise as using preformed and preweighed patties, it is less work and much faster.

the hive; therefore, the use of a screened bottom or a screened bottom board is indicated in conjunction with oil-laced grease patties.

Essential Oils

An essential oil is the oil of a plant or flower that contains the "essence" of the plant. It will typically smell like the plant or flower from which it is derived. Because it takes large amounts of plant material to extract small quantities of oil, essential oils are very potent. Herbalists have traditionally made use of these oils in their treatment of disease conditions. A number of essential oils kill bacteria, fungi, microbes, and other pathogens. As such, it is not too big a jump to think that oils derived from flowers and herbs that are used to treat human ailments could be used to treat diseases of the beehive as well.

From my experience as an organic grower, I am familiar with the use of dormant oil (oil with no noticeable scent or therapeutic qualities) as an effective treatment for apple tree mites. The oil kills the mites by clogging their breathing tubes and causing suffocation. In the mid to late 1990s, I began to come across the occasional bee journal article or letter to the editor reporting people's experiments for controlling varroa with essential oils derived from eucalyptus, thyme, and members of the mint family (Lamiaceae). In fact, some beekeepers were even claiming that bees that foraged on peppermint blossoms were benefiting from the minute amounts of peppermint oil they brought back into the hive in the nectar.

Initially I soaked a paper towel in a mixture of eucalyptus and thyme oil, wadded it up, and tossed it into my hives just above the brood chamber. As the oils evaporated and affected the mites, the bees would proceed to chew up the towel and lug the pieces out of the hive, helping to further spread the essential oils throughout the colony in the process. Thanks to the bees' efforts, this method of introducing essential oils to the hive had the added benefit of eliminating a second trip to the bee yard to remove the paper towel. The drawback to using paper towels was that the length of time the oils were present in the hive was inconsistent, with strong hives removing the oily towels much more rapidly than weaker colonies. In addition, the mixture of oils that I used varied in effectiveness, as I had no scientific research and testing available on which to base the oil mixture to obtain optimum results.

After reading an article in a bee journal about research being conducted with essential oils absorbed in a spongelike material used in floral arrangements, I obtained a block of Oasis Floral Foam from my local florist. This Oasis material has the ability to absorb a large amount of liquid and is similar to the material described in the research article. The florist sticks the flower stems into the sponge and adds water, so that the sponge acts as both an anchor for the bouquet and a reservoir for water to

keep the cut flowers fresh. As was discussed in the article, I cut wafers from the sponge block, soaked them in my essential oil mixture, and laid them on the frames in my hives just above the brood nest. The bees chewed these up as well, but they lasted significantly longer than the paper towels.

For several years I continued to play around with using paper towels, Oasis wafers, and grease patties, all laced with thyme and eucalyptus oil, in my yearly management cycle. Not only did this have some effect on both the varroa and the tracheal mites, but the patties made from grease and sugar would sometimes mean the difference between a hive starving over the winter or surviving into the spring. On a number of occasions, after popping open an inner cover of a hive while unpacking it in early spring, I would find a small cluster of greasy bees living off the patty as they anxiously awaited the return of nectar-bearing blossoms.

Colony winter survival rates improved with the use of essential oils. Unfortunately, although my efforts with essential oils seemed to help some, none of my efforts were successful enough in reducing mite numbers to consistently ensure the survival of the majority of colonies.

Then came published research indicating that thymol, a component of thyme oil found in some foods and considered safe for humans, is extremely deadly to varroa, but not to bees or people. I purchased some thymol and started to add it to my essential oil mixtures, which up until that point had been based on intuition and guesswork rather than anything scientific. Luckily, I didn't have to continue floundering along this way much longer.

Api Life VAR is a thymol-based acaricide that is manufactured in Italy and is likely the final product of that early research on essential oils that I read about. Now available in the United States, Api Life utilizes a mixture of approximately 74 percent thymol, 16 percent eucalyptus oil, 3.7 percent L-menthol, and 6.3 percent camphor, all soaked into a foam wafer that looks suspiciously similar to the Oasis material used by florists. Distributed in the United States, Api Life has been shown in controlled studies to reduce varroa levels by as much as 95 percent within the hive for a high-impact effect.

Each of the ingredients that make up Api Life is approved for use individually in organic agriculture. The primary active ingredient, thymol, is used in food products and is generally recognized as safe. As the essential oil mixture evaporates from the wafer, the vapors travel throughout the hive. Because the mite-killing compounds are delivered primarily in gaseous form, the opportunity for the comb to absorb residues is reduced, when compared to treatments that spread the active ingredient throughout the colony in solid or liquid form.

The rate at which essential oils evaporate into the air varies with temperature, and this has a great impact on an oil's effectiveness as a mite control mechanism. According to the package directions, Api Life is most effectively applied when the average daily temperature range is 60° to 70°F (16° to 21°C). The use of Api Life in temperatures around 90°F (32°C) or higher causes the queen to stop laying eggs temporarily, and it may increase the potential for queen supersedure or bee and brood mortality. A single treatment protocol calls for the breaking of a single wafer into several equal-sized pieces and placing the pieces on the top bars in the corners of the brood nest near the outer edges of the brood area. After 7 to 10 days, the wafer pieces are replaced with fresh ones, and a third dose is applied after another 7 to 10 days and left in for up to 12 days.

As with all foreign objects placed within the hive, the bees will naturally want to remove these strange essential-oil-soaked foam tablets. To prevent this, Api Life VAR may be enclosed in an envelope of wire screen or mesh. I find it helpful to staple down the corners of the screening to the top bars of the frames to keep the bees from gaining access to the acaricide within the wire envelope. This pesticide treatment should not be used when the honey flow is on and honey supers are on the hive.

Although the instructions that come with Api Life indicate that two treatments a year

FIGURE 5-16. Varroa-treatment products made with the active ingredient thymol (a component of thyme oil), although temperature-sensitive, are easy and relatively safe to use, making them a good choice for beginner beekeepers.

are recommended (once in spring and again in autumn), I have found that a single treatment protocol in autumn is adequate so long as it is combined with a variety of low-impact varroa controls—such as trapping, making splits, genetically resistant bees, and screened hive bottoms—throughout the rest of the season. Without the aid of these low-impact controls, a single autumn treatment of Api Life may not always suffice unless another highly effective treatment is applied to reduce mite levels in early spring. Unfortunately, in my experience, adding Api Life to nucleus colonies when they are freshly created in the spring seems to slow down their development. It almost seems as though the essential oil vapors cause the queen to slow down or stop laying eggs for a period of time. This occurs even though the average daytime temperature falls within the proper treatment range as described on the label. It may be that the smaller population of the newly formed colony is not as effective at keeping the hive adequately ventilated as a larger, well-established hive would be. Whatever the reason, I recommend the use of Api Life only as a fall treatment and suggest some other high-impact protocol when a spring treatment is desired.

Another essential-oil-based treatment consists of a thymol-containing gel sold under the trade name Apiguard. Like Api Life, the gel is introduced into the hive above the brood chamber. Research indicates that this product is effective against varroa in a variety of geographic locations and at different times of the year.[20] Apiguard is also reported to help control tracheal mites and chalkbrood and, like other essential oil treatments, is temperature-sensitive.

Food-Grade Mineral Oil

Although I have never ventured to try it myself, food-grade mineral oil (FGMO) has received a fair amount of attention as a treatment for varroa mites. Research indicates that the use of straight FGMO offers some benefit against mites, though not enough to successfully allow mineral oil to be the sole control method used against varroa.[21] This contradicts numerous testimonial reports of good results against varroa from beekeepers using FGMO around the United States. The basic treatment protocol incorporates an electric or propane insect fogger as the mineral-oil dispersing mechanism.

Used by itself, it seems that FGMO has limited effectiveness in the area of mite control. In much the same way as the grease patty works, a mineral oil fog spreads a light coating of oil throughout the hive, which provides an impediment to sure footing for the mites. It also might provide enough irritation to increase levels of grooming behavior among the bees, which may in turn dislodge a greater number of varroa from their hosts than would otherwise be the case. In this regard, it seems logical that any use of FGMO should be combined with a screened bottom or screened bottom board to capitalize on its potential benefits. It is possible that a fogging of mineral oil could also coat some of the mites with oil, clogging up their breathing tubes and suffocating them in the process. Indications are that FGMO can be used most effectively as a carrier for more potent acaricides, such as thymol, that have proven toxic to mites but safe for people.

FGMO EMULSION RECIPE

Here is a mineral oil/thymol/honey emulsion formulation that is reported to have successful results.[22]

Ingredients:
- 1 L (33.8 oz.) food-grade mineral oil
- 1 kg (2.2 lb.) beeswax
- 1 kg (2.2 lb.) honey
- 100 g (3.53 oz.) thymol
- 50 ml (1.69 oz.) ethyl alcohol (minimum 90%)
- 100 one-meter-long (39.37 in.) pieces of cotton cord (or ⁵⁄₃₂-in. upholsterers' welt cord)

Preparation:
1. Heat mineral oil and then add beeswax. Stir well until wax melts completely.
2. Remove from heat and mix in honey until honey is dissolved.
3. Dissolve thymol in alcohol.
4. Thoroughly stir diluted thymol into oil/wax/honey mixture.
5. Add cotton cordage pieces to mixture and allow them to become fully soaked.
6. Store in a sealed container in a cool place until ready to use.

The downside to the use of FGMO applied with an insect fogger is the threat of fire posed by the fogger, especially when used in dry conditions or during periods of drought. Due to its flash point, mineral oil has the potential to ignite and potentially injure bees should a blast of flame be emitted from the fogging mechanism. It was also noted by researchers that potential health hazards to operators, from both inhalation and exposure to the skin, are unknown.[23]

A much safer method for utilizing mineral oil in its capacity as a carrier for stronger mite-killing substances is to soak cotton cordage in a FGMO/thymol/honey emulsion and lay it on top of the brood nest. The addition of honey to the mineral oil mixture attracts bees, who promptly engage in chewing up and removing the emulsion-soaked cotton cord. Whether fogger-applied heated FGMO, or any of its by-products, may contaminate honey stored in combs or be absorbed by wax are questions that remain to be answered. In theory, the cool application of mineral oil using cotton cords should pose less of a concern with regard to any contamination issues.

The Sugar Solution

For the honey producer, it is somewhat ironic that a safe, effective way to control mites has been developed that makes use of sugar. Sugar has long been a major competitor of honey when it comes to the shopper's grocery dollar. This situation is a sore point for many commercial beekeepers given that the US beekeeping community receives no government subsidies for the production of honey and the all-important side benefits of crop pollination, whereas US sugar producers receive subsidies that are often in excess of the world's market price for sugar.

This new, nontoxic, natural alternative to chemical insecticides is based on sugar esters, which act to both suffocate the mite and break down the outer waxy coating that covers the varroa's body and normally keeps it from shriveling up from loss of moisture and dying from dehydration. As such, these esters must come in contact with the mites to kill them. These sugar esters are unique in that their lethal effects wash quickly over the target pests and yet they leave no harmful residues in their wake. Nontoxic sugars and fatty acids are all that are left after the tide of death recedes.

The commercial result of this breakthrough technology for beekeepers is a product called Sucrocide. Just follow the directions and add the appropriate amount of this biochemical miticide to water, fill up a basic garden-variety sprayer, hose down the bees, and watch the mites fall. Although the product is relatively safe to handle, care must be taken in its application, because it is possible to kill honey bees when higher-than-recommended dosages are used. Even after proper treatment with the correct concentration and quantity, the bees will look "washed up" and bedraggled, but with time they will recuperate and return to normal. The key to preventing a bee-killing overdose is to calibrate your sprayer so that the prescribed 1½ ounces of biopesticide is applied to each deep frame. A good way to calibrate has been outlined by the Washington State University Department of Entomology in the July 2004 *American Bee Journal*.[24] Their procedure calls for measuring 6 ounces of water into a disposable container and marking the level on the side. With a sprayer adjusted to deliver a fine mist, determine how many seconds it takes to fill the container up to the 6-ounce mark. For example, if it takes 16 seconds to fill the container, then the rate of discharge is 1½ ounces every 4 seconds. At this rate, each side of a full-depth frame should receive a 2-second blast of Sucrocide. The nozzle of the sprayer should be kept at a distance of about 9 to 12 inches from the frame during application. To ensure even coverage, make sure to direct the spray across the frame at a consistent and steady rate.

This treatment protocol results in soaked honey bees, and, like people, honey bees do not fare well when they become wet and chilled. Therefore, it is not recommended that Sucrocide be applied when temperatures are expected to drop below 55°F (13°C). Because the biopesticide must come in contact with the mites to be effective, treatments applied when the bees are clustering will inhibit effective application.

Sucrocide can be applied at a faster rate by simply spraying the appropriate amount down between the frames, rather than removing and spraying each frame individually as directed by the label. The Florida Department of Agriculture and

Consumer Services Division of Plant Industry has developed a commercially viable advancement to this approach. It is a spray wand that incorporates a manifold with nine nozzles attached to it that allows a liquid to be applied between the frames from the top of a hive in a single pass.[25] Although not quite as effective as spraying each frame individually, this fast method requires much less labor and works well enough that, with regular applications, adequate control has been reported. Care must be taken that the amount of brace comb and burr comb in hives is kept to a minimum so that it does not impinge on the efficacy of these topside applications. Like the use of confectioners' sugar in conjunction with screened bottom boards, Sucrocide treatments kill only the mites that are on the bees at the time of treatment, so several applications are required, spaced three to five days apart, in order to affect high population reductions.

Sucrocide is composed of sucrose octanoate esters, which are synthetic sugar esters and therefore require special regulatory approval to be utilized in certified organic production. However, the high level of efficacy and incredibly low toxicity of this biopesticide should allow it to be added to the National Organic Program's official list of allowed synthetic substances, thereby making it available for future use in organic honey production.

Organic Acids

In general, the use of acid to control varroa is ideal because it is corrosive but not toxic, and it does not bioaccumulate in honey or wax, thus avoiding contamination issues. Several acids, such as formic, acetic, oxalic, and lactic (which is utilized in human metabolism), are all found naturally in honey.[26] As a result, federal regulators exempt such acids from tolerance requirements for honey and beeswax. Because acids used for mite control act physically on the target organism, the chances of the mites building up resistance is remote. When applied appropriately to a hive, acids are beneficial against both varroa and tracheal mites. These characteristics make acid an ideal candidate for use in organic hive management.

Although not toxic, acids are extremely caustic and can burn holes in clothing or cause severe burns to exposed skin, making them dangerous for the applicator as well as the bees if not used carefully. Wearing protective clothing, such as gloves and eye protection, is strongly advised when working with acids. Care must also be taken by applicators to position themselves upwind from the acid source so fumes do not overcome them during treatments. There are four main types of acid that can be utilized as high-impact controls for varroa and tracheal mites.

FORMIC ACID

Formic acid is used fairly extensively in several countries throughout Europe, as well as in New Zealand and Canada, but it was not until almost 20 years following varroa's arrival that it received approval in the United States for use against mites. Not only will formic acid kill mites clinging to adult bees, but under certain circumstances, it also appears to pass through the cappings and kill varroa mites that are in sealed brood cells.[27] When formic acid is employed as a liquid that must be vaporized through natural evaporation, it is temperature-sensitive. The ideal temperature range for delivery is approximately 60° to 80°F (11° to 27°C).

Beekeepers in Canada, who have typically used formic acid liquid in concentrations of 85 percent by squirting it on the bottom board and letting it evaporate up into the hive, have reported some difficulties. Apparently, due to the close correlation between dosage and temperature when using the acid, such treatments may prove to be very hard on the queen, often leading to queen failure or supersedure. Most Canadians who utilize this approach now further dilute the 85 percent acid down to the 60 to 65 percent range so that the treatment will be less problematic for the bees.[28] Due to the chance that the hive may experience queen issues, acids generally are best applied in the spring, so that there is plenty of time for the hive to sucessfully supersede its queen or be given a replacement and still provision itself adequately for winter. One method that has been utilized with apparent success is to apply lemongrass oil to the hive at the same time as the organic acid treatment. Geraniol and citral, which are components of the essential oil that are also found in the bees' homing pheromone, help to keep the hive smelling normal to the bees and seems to reduce the tendency for the bees to want to replace their queen even when temperatures or the dosage of formic acid are too high.

An early effort to take advantage of the benefits of formic acid while minimizing the potential hazards of applying it was the release of a formic acid gel pack approved for use by the FDA against *Varroa destructor* and distributed by Betterbee, a company based in Greenwich, New York. I had the opportunity to use these gel packs for two seasons,

and they worked very well. Unfortunately, the use of the formic gel pack was short-lived. The product was pulled from the market after the first year when it was discovered that the packaging would fall apart as a result of the action of the acid.

Following the failure of the gel packs a new product, the Mite Away II pad, became available. After a number of years this product was removed from the US market and was replaced by the Mite Away Quick Strip (MAQS) in 2011. The directions call for treating with two strips placed across the top of the brood frames when average daytime temperatures are between 50° and 92°F (10° and 33°C). While the directions say that the treatment should last seven days, beekeepers can also leave them in the hive and let the bees chew them up and remove

FIGURE 5-17. Though temperature-sensitive, the Mite Away Quick Strip (MAQS) can be used even when honey supers are on the hive and, unlike other treatments, will kill mites that are reproducing in sealed brood cells.

them. Company testing indicates that the strip not only kills varroa mites, but unlike most other mite treatments it will kill up to 95 percent of the varroa reproducing and being raised in capped cells of brood. Since residue contamination is not an issue, the MAQS can also be used even during a honey flow when honey supers are on the hive, a feature that most other mite treatments cannot offer.

My experience with these strips showed that while they are very effective at killing mites, they can be hard on the bees during the first few days. I noticed undertaker bees had removed immature brood, apparently killed or injured from the treatment, from the hive. This minor damage to the brood seems to be most pronounced when daytime temperatures are over 80°F (26°C). MAQS can also cause the queen to stop laying temporarily. I also found that, like Api Life VAR, the strips are most quickly opened by running the sharp end of a hive tool along the seam of the packaging. This operation not only cuts open the package for easy access, but it also divides the two strips each package contains in half. For those who wish to reduce the acid's detrimental impact upon the brood, two of these half strips can be used in place of the usual two full-size strips, as long as one does not mind the reduction in mite control that will result.

Other formic delivery systems include the Mite-Gone evaporation system developed by Bill Ruzicka; the Nassenheider Evaporator; plastic bags containing Homasote (tentest board), developed by the Ontario Beekeepers' Association; as well as various evaporators made by do-it-yourselfers.[29]

OXALIC ACID

Oxalic acid can be utilized in much lower concentrations than formic acid, making it a safer alternative for the naturally inclined beekeeper. Aerosol applications of oxalic have been used successfully for many years throughout Europe and have been promoted in the United States by biodynamic enthusiast Gunther Hauk of the Spikenard Honeybee Sanctuary in Virginia. Treatment consists of lightly spraying the bees with a fine mist of a 2.1

percent dilution of oxalic. The treatment protocol calls for the application of ½ ounce per side of each brood frame. The correct concentration may be obtained by mixing an ounce of oxalic acid with a quart of water. Because the low concentration of acid will not penetrate into sealed brood cells, oxalic acid is best applied when there is a minimum of brood and the majority of varroa within the hive will be exposed to the treatment, such as during the winter months.[30]

Unlike treating with formic acid, each frame in the hive containing bees is traditionally removed and sprayed with a very light mist of oxalic that covers the bees without wetting them down. This is a very labor-intensive procedure and often requires the beekeeper to enter the brood nest at a time when the natural inclination is to leave the bees alone and not disturb the winter cluster. Gunther Hauk reports, however, that such treatments are not harmful to wintering bees and, when correctly performed, may result in 98 percent effectiveness, even in temperatures as low as about 40°F (4°C). An oxalic-acid evaporation device developed by the Swiss company Andermatt Biocontrol AG is an attractive alternative to those with concerns about the amount of labor involved in treatments and the effect of opening up the hive and breaking the cluster during winter applications.[31] US beekeepers can obtain an oxalic-acid vaporizer from Heilyser Technology of Sidney, British Columbia. (For more information, see resources on page 273.)

As an alternative to spraying each frame individually, one can dribble a mixture of a 3.5 percent concentration of oxalic acid in sugar syrup made up in a ratio of one part sugar to one part water. This can be obtained simply by mixing 35 grams of oxalic with 1 liter of sugar water. This solution is trickled down between the combs and onto the bees with the use of a syringe or garden sprayer. Daytime temperatures should be at least 38°F (3°C) during application, which is typically conducted once each season, in autumn. This dribbling method is much faster and easier than hand-spraying each frame individually and is well suited to larger commercial operations. The fact that no fine particles are formed by a spray mist that may be accidentally inhaled makes dribbling a much safer option for the person doing the application.[32]

LACTIC ACID

Long appreciated for its nutritional benefits when consumed in lacto-fermented foods such as sauerkraut, lactic acid is another acid that has been used by beekeepers with good results against mites. Application of lactic acid is similar to that of oxalic acid, except a concentration of 15 to 18 percent is used. Four or five treatments are applied throughout the year. When used during the winter when temperatures are over 50°F (10°C) and no brood is present, a 99 percent mite kill can be expected.[33]

ACETIC ACID

Acetic acid (vinegar) is also being used to control varroa and tracheal mites. As with some of the other organic acids, users report suppression of various hive diseases in addition to mite control. Strong vinegar, with a 20 to 25 percent dilution of acetic acid, is typically used, along with a vaporizing apparatus for application. One commercially available unit manufactured by Better Way Wax Melter Honey Processors incorporates a blower working in conjunction with the vaporizer to help distribute the acid fumes throughout the hive. Once again, it is important to note that extreme care must be taken to avoid getting acid on clothing and skin, or inhaling the fumes.

An Herbal Treatment

One varroa mite control product that has recently become available commercially is HopGuard. Based on the same herb used to make beer, this biopesticide is composed of hop beta acids made from food-grade material. It works on contact. While it is safe to handle, you do not want to get this

FIGURE 5-18. HopGuard, derived from the herb hops, is one of the only treatments that can be used when supers of honey that are to be harvested are on the hive. It is also one of the only commercially available Varroa mite treatments that is easily adaptable to top bar hives.

stuff in your eyes, and gloves are advisable simply because the product is messy to handle. Cardboard strips covered in a thick brown hops-derived liquid are placed between brood frames within the hive. Unlike most other treatments, it is safe to use even during a honey flow. Nevertheless, it is not a good idea to use the strips on frames that are directly above honey frames, since the hop goo may drip down onto the honey. This treatment is not temperature-dependent, so it can be used almost any time of year. Its use is most effective during times when the hive is broodless. Observations are that this form of commercial mite control is very gentle on the bees and queen and may even help with some hive diseases such as American foulbrood.

After about three days the HopGuard strips have a tendency to dry out, and this greatly reduces their effectiveness. One way to overcome this is to treat colonies during times of the year when the queen has stopped laying eggs so that no mites are sealed inside capped brood cells, and thus all the mites in the hive will be exposed to the acaricide. Another approach is to take advantage of the fact that HopGuard can be safely used up to three times a year and replace the dried-out strips twice at four- to five-day intervals. Since all worker and drone brood cells are typically capped on the ninth day after the egg has been laid, replacing the cardboard strips every fourth day will provide the hive with 12 full days of protection. This will expose all the mites in the hive to the miticide with the exception of the mites that were sealed into drone brood cells during the three days immediately prior to the first strip being inserted into the hive. By spacing out the three treatments by four days and applying treatments two and three on the fifth and tenth days respectively, all the mites should be exposed, although the exposure of the mites on the days immediately prior to when a new strip is applied will be low due to the drying of the previous strip. Since HopGuard is applied as a strip that is hung between the frames, it is the only commercially available treatment that can be easily and effectively used by beekeepers who use top bar hives.

• TRACHEAL MITES •

The honey bee tracheal mite (*Acarapis woodi*) was first identified in England in 1921, and its discovery spurred the creation of the Honeybee Act of 1922 by the US Congress. This legislation regulated the importation of new bee stock into the United States in an effort to prevent the introduction of tracheal mites, as well as other pests and diseases. It performed the job admirably well until 1984, when mites infesting the breathing tubes of honey bees were found for the first time in the United States along the Mexican border.

The US beekeeping industry, led by Roger Morse of Cornell University, focused in the late 1980s and early 1990s on breeding strains of bees that showed levels of resistance to tracheal mites. As a result of these types of efforts, *Acarapis woodi* is not considered a major threat by most beekeepers today. With luck, our tracheal mite experience may be a foreshadowing of what we can expect from *Varroa destructor*. As with its larger cousin, *Acarapis woodi* was considered a minor problem in its home country but caused large-scale losses to the American beekeeping industry during the first dozen years or so following its introduction on American shores.

Detecting *Acarapis woodi* infestation is complicated due to the microscopic size of the mite. Tracheal mites are visible only with the aid of a microscope or optic lens. As a result, a quick, easy, and widely utilized method of mite identification in the field has yet to be developed. One indication of widespread tracheal mite presence is numerous bees within a hive that have formed their wings into what appears to be the letter *K*. Honey bees have two pairs of wings, and each pair is hooked together when they fly. Bees can disconnect their wings for grooming purposes, but the telltale *K* orientation is usually found only when mites are in attendance. This symptom is thought to be associated with a virus that the mite carries and spreads among the bees.

Although not a definitive diagnosis, the observation of bees crawling around outside the entrance of an unmolested hive when temperatures are low or below freezing is another sign that a heavy infestation of *Acarapis woodi* is probably present. The bees act as if their tracheae are so clogged up with mites that they are having trouble breathing, and they go outside to have better access to fresh air. Bees suffering from this condition will often linger around the outside of the hive's entrance so long that the cold will work upon their joints, making them stiff and immobile, until they eventually lose their footing and flop over into the snow, where their numbers pile up throughout the course of the winter. Thus, the all-too-typical sign of tracheal mite infestation is a large number of dead bees lying on the ground in front of the hive early in the spring after the snow melts. Nevertheless, it is possible that other conditions may cause bees to behave similarly, so the only sure way to test for the presence of *Acarapis woodi* is to capture some live bees and inspect them under a microscope. Your state bee inspector may provide this service for free under certain conditions.

As with all pests and diseases of the hive, natural tolerance or resistance is the best form of control. Buckfast bees from England are widely held to possess a genetic tolerance to *Acarapis woodi*. Aside from breeding, beekeepers have long utilized the fumes given off by menthol, a component of peppermint oil, for the treatment of hives infested with tracheal mites. The only commercially available menthol treatment for tracheal mites is Mite-A-Thol. This product is made up of menthol crystals that cause the microscopic mite to die from dehydration. However, because the menthol fumes must reach inside the tracheae of the honey bee, Mite-A-Thol can be used effectively only when average daytime temperatures are warm enough to allow the menthol to evaporate (between 60° and 80°F/16° and 27°C).

The grease patty is a homegrown remedy for tracheal mite infestations (see "A Grease Patty" recipe earlier in this chapter on page 146). Because of

FIGURE 5-19. Workers spotted hanging around the hive entrance when temperatures are below 40°F (4°C) may be an indication that tracheal mites are present.

Acarapis woodi's microscopic size, the oils from a grease patty, when lightly covering the bodies of the bees, inhibit the movement of the mite from bee to bee, which out of necessity must occur outside of the honey bee's breathing tubes. As a result, grease patties work to help prevent the spread of mites from infected bees and limit the damage they may cause within a colony. Due to the ability of an oily surface to hinder the movement of the tracheal mite, it is likely that a fog of food-grade mineral oil sometimes used for varroa control could prove to be an effective agent in the control of *Acarapis woodi* as well.

Most beekeepers today do not go to the trouble to treat colonies for tracheal mites, and because many of the soft-chemical varroa mite treatments (thymol products and organic acids) also kill the tracheal mite, treating for *Acarapis woodi* has become largely redundant.

His labor is a Chant—
His idleness—a Tune—
Oh, for a Bee's experience
Of Clovers, and of Noon!

—EMILY DICKINSON

Insects make up about 80 percent of the animal population on Earth. Unfortunately, beekeepers have a tendency to place their primary focus in the apiary on the honey bee and remain unaware of the other insects in the vicinity until the hive is under attack. Then we rush in to exterminate the offenders without realizing that we might also be harming other beneficial bugs, including the bees themselves.

As organic beekeepers, we should attempt to learn as much as possible about the behavior and biology of the other insects that live in and around our apiaries. Frequently, a little knowledge about how certain insects function can provide us with the ammunition we need to prevent pest damage before it becomes necessary to use invasive and potentially toxic control methods.

• WAX MOTHS •

A strong aversion to maggots is not a good trait to have if you intend to keep honey bees. A couple of insects that will give you plenty of opportunity to practice getting over your maggot repugnance are the greater wax moth (GWM), scientifically known as *Galleria mellonella*, and the lesser wax moth (LWM), *Achroia grisella*. Prior to the arrival of the small hive beetle, the wax moth was America's greatest threat to unprotected honeycombs during periods of prolonged warm weather. The

FIGURE 6-1. Wax moths hang around beehives looking for the opportunity to lay their eggs in cracks and crevices where the bees can't get at them. When its wings are folded, the moth can be difficult to recognize.

adult wax moths will lay their eggs among the combs, or in cracks and crevices that the bees residing in the hive are unable to reach. Once they hatch, the wax moth larvae (also called wax worms) will feed on the combs and their contents—cast-off honey bee larva skins, pollen, and honey. As a result, they prefer older dark combs that contain high levels of impurities and larval castings rather than newly created combs that feature the beautiful golden-yellow coloring of freshly formed beeswax.

The moth larvae bear a resemblance to the maggots typically found on carrion, reaching a maximum body length of about 1 inch, though their

FIGURE 6-2. Wax moth with wings unfolded.

FIGURE 6-3. Mother Nature's cleanup crew, the wax worms, can make short work of combs that are uninhabited by a sufficient number of honey bees. PHOTO BY STEVE PARISE.

final size is largely dependent on the type and amount of food available. As they grow, the feeding larvae will chew their way through the combs, leaving silk-lined tunnels of webbing to mark their path. This webbing becomes littered with the wax moth larvae's feces and gets so thick that it shields them from any nest-cleaning worker bees that may try to sting the larvae or remove the larvae from the hive. Once the larvae mature, they spin a cocoon and pupate in a protected site within the hive. The cocoons are usually spun in oval depressions that the larvae chew into the surface of the wood, and to which they adhere themselves. The adult wax moth will then emerge from the cocoon to begin the cycle all over again. The speed of growth of the

larvae and pupae depend a lot on temperature, with the process taking much longer when cool temperatures prevail than during warm periods.

In Vermont, the danger of GWM and LWM damage begins primarily in late June and early July, after the moths have either migrated from the South or hitched a ride north with migratory beekeepers. The threat lasts well into September and even October, until the first hard frosts kill off the adults, eggs, and larvae. The wax moth can survive the cold northern winter only by living on combs in a heated building. In the South, though, the wax moth is a threat on a year-round basis. Combs that have become home to wax moth larvae can be totally ruined if not caught in time. If found early enough, most of the webbing may be torn off the comb and the frame placed into a strong hive that will remove the remaining webbing, as well as clean up, repair, and replace damaged comb if the destruction is not too severe.

It was while pulling a chunk of webbing off an infested frame to expose the maggotlike larvae squirming around in the center of the comb for the first time that I felt a great revulsion and—I must admit—a wave of nausea that rippled through my stomach from the unappetizing sight. Although I do not wish wax moth infestation on anyone's hives, I am happy to report that one's initial revulsion subsides with repeated exposure to these little critters.

To control the damaging effects of the wax moth without chemicals, the organically minded beekeeper can make use of two things that keep most everything healthy: sunlight and fresh air. Adult moths prefer to lay their eggs in dark crevices deep within the hive, and they are reluctant to lay in combs that are exposed to light and fresh air. Because of this, protection from the moths may be obtained by simply storing empty hive bodies and supers of combs on end with the tops and the bottoms of the frames exposed to the sunlight and ventilation.

By far the best form of control for the GWM and LWM are the bees themselves. A strong hive will do a fine job of removing any moths or larvae that are unfortunate enough to try to make a home in

FIGURE 6-4. During the pupation phase, wax moth larvae chew depressions in the wooden parts of the hive, in which they embed the silken cocoons that they spin around themselves.

their hive. It is the weak hive that is susceptible. A common mistake among novice beekeepers when they find a dead colony filled with moth larvae is to think that wax moths killed their hive. The wax moth is an opportunist, a scavenger to whom nature has delegated the task of cleaning up dead hives, thereby acting as a natural form of disease prevention to help keep other colonies in the neighborhood healthy. The only way wax moths can be found thriving in a hive occupied by bees is if the honey bee colony lacks sufficient numbers to adequately defend and patrol the entire cavity composing the hive's interior. For this reason, I believe it is a good practice to manipulate the size of the hive to match its occupants. I make it a practice to add empty supers to a hive only when a colony has almost filled the space it has. This way the area that the bees must keep free of invaders is minimized. The exception to this practice comes after the honey harvest, at which time I place the freshly extracted supers, still wet with honey, between the inner cover and the outer cover of the hives. The hives are typically very strong and populous at this time of year, and they are often overcrowded given that they now have to occupy a smaller space than that which existed prior to the removal of the excess honey supers. Such hives have more than enough bees to crawl up through the inner cover into the empty supers above, cleaning up leftover honey and prohibiting wax moths from gaining a foothold in the stack of empty combs.

FIGURE 6-5. In order to expose empty stored combs to light and provide ventilation so as to reduce wax moth damage, some bee-keepers build a shelter where frames can be hung up out of the weather and out of reach of mice. Another option is to stand supers on end by a window in an outbuilding.

The one major drawback to this approach, however, is that if there is a late honey flow, a strong colony will often start to store the freshly gathered nectar in the supers above the inner cover. These honey-filled combs are more likely to hold uncapped, unripe honey, as the cool evenings tend to prevent the bees from maintaining their honey-ripening duties around the clock. As a result, such honey, if harvested, must typically be used within a month or so—otherwise it may start to ferment and pick up a vinegary taste. This can sometimes be avoided if the bees combine this high-moisture honey with a significant amount of the remaining fully ripened honey that is left over on the wet frames that have gone through the extraction process and have been stored above the inner cover. Depending on how efficient your extracting process is, this leftover honey may be enough to offset the high moisture content of the unripe honey, preventing fermentation.

Because all my honey-extracting equipment has been cleaned up and put away by the time the first hard frosts arrive, supers that become filled with late-season honey are typically placed on a hive that contains borderline honey supplies to provide additional honey stores to help that hive survive winter. However, it is not a good idea to use a partially filled super for supplemental feeding at this time of year, because the bees are unlikely to have the chance to fill out the remaining frames with honey. The winter cluster may then eventually move into the area that contains the empty comb and starve.

If none of your hives can make use of an additional super of honey, late-season honey supers can be placed a couple hundred feet away from the apiary on an unseasonably warm day, allowing the bees to rob out the supers prior to your placing them inside for winter storage. Leaving the supers containing honey any closer to the apiary is likely to create a feeding frenzy that will encourage bees from neighboring hives to steal honey from one another. This can cause hives to become overly defensive, weaken strong colonies, or cause weak colonies to die out.

I have also found that when storing extracted supers on top of hives, it is preferable to place shallow supers directly above the inner cover and position deep supers higher up on top of the shallows. This is because queen bees seem to prefer to lay their eggs in the larger span of comb that is found within the frames of the hive bodies, and deep supers stored directly on top of inner covers following the harvest can sometimes end up with brood in them by the time they are ready to be put away for winter. In this situation the entire hive body, if filled up 75 percent or more with brood and honey, can be placed below the inner cover and left as part of the overwintering hive. Again, under no circumstances should empty or partially filled honey supers (less than 75 to 80 percent full) be left on top of a hive for the winter season. To do so would invite the cluster to move up into them and risk starvation by eating themselves into a corner late in the winter or early spring.

Another form of wax moth control can be found in the form of a special strain of the microscopic creature known as *Bacillus thuringiensis* (Bt). Available commercially in a product manufactured by Vita called B401 (Certan), Bt is 100 percent effective in controlling wax worms in empty supers of drawn comb. B401 must be used as a preventive measure, however; it will not be as effective after an infestation has taken place. A single application

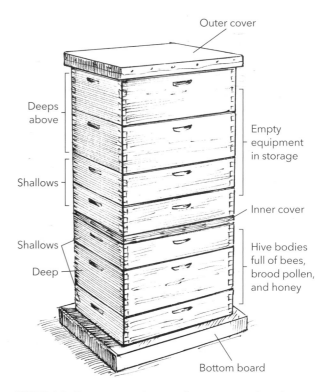

FIGURE 6-6. Keep wax moths out of your extracted equipment and allow the bees to clean up the leftover honey on the combs by storing them on the hive until the first hard frosts arrive.

of a diluted Bt concentration sprayed over the wax surface of each comb will provide protection for up to a year or more, until the equipment is placed back into use. Because *Bacillus thuringiensis* becomes toxic only when exposed to the unique digestive environment found within the wax worm's gut, it is a safe alternative to mothballs (paradichlorobenzene), which can contaminate honey and wax. There have also been reports of beekeepers repelling wax moths by placing tobacco leaves between supers that are placed in storage with an outer cover on top of the stack.

Although the US Environmental Protection Agency (EPA) has approved paradichlorobenzene for use to protect honeycombs from wax moth damage, as a synthetic chemical that has not been included in the National Organic Program's national list, it is not allowed in organic production. However, wax moths should not be a major problem for the conscientious beekeeper who takes

FIGURE 6-7. To extend the amount of time that unused supers can be stored outside without being damaged by wax moths, stack them so that light and air can circulate between the combs.

the time to perform regular inspections of hives and takes steps to protect empty supers that are not occupied by bees. Keeping all equipment filled with healthy bees is the best way to confront the wax moth threat. In the Champlain Valley of Vermont, this means that all supers of drawn comb should be in use and inhabited with bees by about Independence Day (early July). If conditions prevent frames of drawn comb from being used early enough during the course of the season, it is a good precaution to spray each frame with a Certan solution, fumigate them with carbon dioxide, store the combs where they will be exposed to fresh air and sunlight, or place all unused supers in a freezer for at least 24 hours prior to storing the equipment in such a way that the wax moths cannot gain access

to them. These are the only methods of control that are acceptable to organic certifiers. Should you find yourself faced with equipment full of baby wax moths despite your best efforts, I hear the worms make great fishing bait, and I have seen chickens go after them with gusto.

• SMALL HIVE BEETLES •

The small hive beetle (SHB), technically known as *Aethina tumida*, is indigenous to Africa. It was first discovered in the United States at the end of the twentieth century. *Aethina tumida* beetles are reddish to yellowish brown in color immediately upon emergence from their pupal stage. Once fully mature, the adult beetles turn dark brown to nearly black. They are about ¼ inch or so in length, and slightly smaller than a ladybug, although beetles of varying sizes have been reported and are probably the result of differing food sources and climates. *Aethina tumida* is distinguishable from most other similar beetles in that the midsection of its body containing the wings is sectioned off from both the head and the tail areas in such a way that the insect looks like it is composed of three distinct segments. Its small size, quickness, hardness, and covering of fine hair make the beetle difficult to pick up by hand.

SHBs are similar to wax moths in that they are opportunists that prosper by scavenging on weak or dead hives, as well as stored equipment that contains drawn comb. Thus, it is not wise to leave stacks of empty equipment sitting around. Sanitary conditions around the bee yard should be maintained by removing broken combs, burr comb scrapings, and so on, because they may act to attract beetles and support their reproductive cycle. This advice also applies to the honey house. Supers filled with honey should not sit too long before being extracted, and it's not a good idea to let cappings sit exposed for extended periods. Small hive beetles are also attracted by the honey bee alarm pheromone given off by hives under stress due to disease or during management techniques that cause alarm within the colony. Beetles have reportedly been able to detect the bee alarm pheromone from over 8 miles away.[1]

Beetle eggs are tiny—about 1.4 millimeters long and 0.26 millimeters wide—and similar in appearance to honey bee eggs. Adult beetles will lay irregular masses of eggs in cracks and crevices throughout a hive or stack of equipment. The eggs typically hatch within one to six days depending on the relative humidity and temperature. SHB eggs will not hatch below 50°F (10°C) and require humidity levels above 50 percent for larvae to successfully emerge.[2]

Once hatched, the SHB larvae feed on honey, pollen, eggs, and brood until they mature. They then crawl out of the hive and tunnel into the ground to pupate. Beetle larvae do not weave the same messy mass of webbing that is characteristic of wax moths, and upon first glance they resemble the common maggot even more than wax worms do. They also move faster than wax moth larvae. Luckily, the SHB larvae do not consume and destroy a lot of beeswax like wax worms do. The larvae tunnel through the combs in search of food and can heavily damage newly drawn-out and delicate comb; however, older, sturdier brood combs are better able to withstand heavy larval infestation without disintegrating. The greatest damage from the beetle comes from the destruction of the honey stored within the combs as the immature beetles eat their way through the hive and the converted nectar becomes contaminated by the larval feces and ferments in the comb. Infested honey bee colonies have been observed to abandon combs and even entire hives through the process known as absconding once they have become badly infested and fermentation has begun.

Temperature is a major factor affecting the spread of SHBs. The beetles are native to a tropical/subtropical climate in Africa, and the beetle is considered a significant pest in warm southern states such as Georgia and Florida. It remains to be seen how much of an issue SHBs will be in regions where cooler temperatures prevail. Like all insects, SHBs have a minimum temperature below which reproduction and development will not occur. In addition, the beetle's growth and metabolic rate increase as temperature increases.

FIGURE 6-8. Small hive beetles have been known to overwinter within the honey bee cluster in northern latitudes. PHOTO BY JEFFREY W. LOTZ, FLORIDA DEPARTMENT OF AGRICULTURE AND CONSUMER SERVICES.

SHBs' population density within a given geographic location is even more dependent on their success at breeding and overwintering, which in turn is tied to the number of degree-days that the temperature range is above the minimum needed for reproduction. Consistent reproductive success of the SHB requires temperatures above 68°F (20°C), the temperature at which approximately 50 percent of the developing beetle larvae will survive. At temperatures below 50°F (10°C), SHB larvae were found not to survive at all.[3] This is likely to be the major factor that has prevented SHBs from becoming a major issue thus far in northern regions.

Additional factors that play a role in healthy beetle growth and development are soil moisture levels and density. SHBs require soil moisture levels of 5 to 25 percent to pupate successfully.[4] This would suggest that areas in Western states that experience prolonged periods of dry weather, areas that see a lot of rain over lengthy periods, or areas where the soil type is such that moisture from local precipitation will accumulate over time will not support beetle reproduction, and therefore are unlikely to encounter significant long-term problems from SHBs. Choosing yard sites that receive direct sunshine will play a role in helping to keep moisture levels in the soil low. Soil compactness is also significant, simply because loose soils are much easier for SHB pupae to burrow into compared to dense hardpan or clay soils. It is likely that populations of beetles that survive from year to year and have the potential to grow in numbers are most likely to occur in locations that experience prolonged warm temperatures and have soils that are composed primarily of silt, which will hold some moisture and not become extremely hard, dry, or waterlogged.

In contrast to the wax moth, which must use stealth to enter the hive, lay its eggs, and then depart before being discovered and killed, *Aethina tumida* has the ability to move about inside the hive and even coexist to some degree with the bees within a colony. Honey bees of European stock have long been bred for gentleness, making them much less aggressive than their African relatives, and it is likely that this difference in behavior accounts for some of the contrast in severity of SHB damage between Africa and North America.

Nevertheless, the increased propagation of hygienically inclined honey bees may be helping to reverse this trend, as the aggressively hygienic bees mimic the imprisonment behavior that is a typical SHB defense expressed by the African Cape honey bee (*Apis mellifera capensis*). Guard bees will corner adult beetles in confinement areas built out of wax and propolis and prevent their escape. The imprisoned beetles are thus prevented from mating, laying eggs, and doing further feeding damage to the hive. SHBs will starve within about two weeks without food, and yet imprisoned beetles have been known to survive up to two months without access to food in combs.[5] This is because *Aethina tumida* has the ability to touch a worker bee's antennae in such a manner as to fool the bee into feeding the hungry beetle. SHBs and worker bees can often be observed scurrying around on the top bars of hives immediately after the inner cover is removed on a hive that is infested. The act of popping the inner cover frees the beetles that had been held captive between the cover and the top bars. Frequent hive manipulations in areas that support large SHB populations may lead to higher beetle activity and additional stress on the hive due to the repeated release of the beetles from captivity.

If not for the complication of beekeeper interference, European honey bees that exhibit a strong tendency to collect propolis might prove to be more adept at confinement behavior and better adapted to tolerate beetle infestation. In the event that beetle populations grow to high levels within a colony, European honey bees will again imitate African Cape honey bees through their tendency to abscond, leaving their old hive behind and finding a new home.

The first efforts to control SHBs of which I became aware were homemade traps that consisted of a small hole drilled into the bottom board and an attached jar beneath the hole filled with beer, vinegar, or mineral oil to attract the unwelcome hive guests. The beetles, in their search for a corner of the hive to hide in, would enter the jar through the hole in the bottom board and drown. Another device that has been partially successful in controlling the SHB is a hanging trap baited with fruit such as cantaloupe and similar in design to traps used for yellow jackets. The success of this approach is limited, however, due to the fact that comb filled with brood, honey, and pollen is much more attractive to beetles than fruit, which serves as their secondary feeding source. Unfortunately, honeycomb as bait within a trap has not proven much better than fruit at attracting adult beetles, because the comb bait in the trap is no more attractive to the beetle than the comb in nearby hives.

The first commercially available trap, and the one I still consider to be the best, that came onto the market specifically for use against the SHB was the West Beetle Trap. This trap consists of a tray filled with vegetable oil and covered with a grid that contains holes large enough for the beetles to pass through, but too small for the bees. The trap is placed on the bottom board of the hive; wooden spacers are needed to raise the hive body above it about 2 centimeters and provide clearance for the bees to enter and exit while the trap is in place. It is also critical that the base of the hive be leveled prior to setting the trap in place and adding the vegetable oil, to prevent the oil from leaking out and possibly killing the bees. Dadant & Sons of Hamilton, Illinois, recommend checking the trap about every two weeks and performing periodic cleanings of the device. A significant benefit of the trap is that repeated manipulations of the colony that disturb beetles in hiding or release them from imprisonment cause them to move around the hive and potentially find the trap sooner. A side benefit of the West Beetle Trap is that any varroa mites that lose their footing and fall from above are likely to land in the oil and perish along with the beetles.

Dr. Mike Hood of Clemson University in South Carolina has developed another commercially

FIGURE 6-9. The West Beetle Trap fits on the bottom board (with the help of some wooden shims placed between the bottom and the hive), holds more small hive beetles, and is easier to access when installed than other beetle traps on the market.

available trap that builds on the original jar-under-the-bottom-board idea. The Hood Small Hive Beetle Trap attaches to an empty frame that has not been fitted with foundation. The trap has separate compartments that are filled with apple cider vinegar or mineral oil and replaces one of the frames positioned next to the side of the hive body containing the brood chamber during the summer, or in an upper super during the winter. Beetles that enter the trap are attracted by the vinegar, drown in the oil, and are easily removed.

Aside from having to open up the hive in order to check and clean out the trap, the major disadvantage of the Hood Small Hive Beetle Trap is that by replacing a frame within the hive, it creates less room for the bees to store honey or raise brood. Also, you should be prepared to spend time cleaning the extra comb the bees will build in the open area of the frame surrounding the trap, as well as any propolis the bees may use to try to plug up the entrances to the beetle-killing device.

Additional beetle traps of various styles that have come on the market over the years include the Beetle Blaster, AJ's Beetle Eater, and the Beetle Jail, all of which tend to be small in size. Some are disposable, which is great for ease, low cost, and convenience, but not so good for the environment. I continue to prefer the West Beetle Trap, both because it can be maintained without having to open the hive and because it can hold a much larger number of beetles between cleanings than the other traps.

Another form of *Aethina tumida* control that some folks are recommending is removal of beetles by hand, either by crushing them with a hive tool, whacking them with a wooden spatula (aka the Beetle Swatter), or by sucking them up with a small vacuum. Although anecdotal reports from hobby beekeepers claim good results using such manual techniques, their efficiency is questionable, and the amount of time such control measures would take when relied upon exclusively makes these methods unrealistic for commercial beekeepers. Nevertheless, this inefficiency does not in any way diminish the strong sense of immediate gratification and accomplishment that comes from personally defending the bees by crushing small hive beetles with a hive tool as they scurry around looking for cover when a colony is first opened.

The use of pheromones has been suggested as a beetle control mechanism because such use of male beetle pheromones is effective in control of other pest beetle species. Also, male SHBs have been reported to enter honey bee colonies before the females, which indicates the possibility that similar hormones are involved in colony invasions.

Research on nematodes has indicated that at least a few species are lethal to the small hive beetle. The nematode species *Steinernema carpocapsae*, *Steinernema riobrave*, and *Heterorhabditis indica* all are reported to cause over 75 percent mortality in exposed beetles,[6] suggesting that these nematodes have the potential for use as a biological control of the SHB.

Another strategy that has received a lot of attention in the search for adequate SHB control is targeting the larvae when they enter their roving stage in search of adequate ground to pupate in. Some of the more natural management techniques that focus on the larvae include the use of moats, setting larva traps around the hives, sprinkling the ground around infested colonies with diatomaceous earth,

and allowing chickens to roam around the hives. Biological controls include the use of soil fungi (*Aspergillus*) and nematodes that will attack small hive beetle larvae while they are pupating in the soil. Beekeepers who frequently move their hives to new apiary locations have reported reduced beetle problems, presumably because they break the beetle's reproduction cycle and eliminate ready access to hives for beetles emerging from the soil following pupation. Unfortunately these approaches all seek to address infestation after the fact, and none of them prevents larval damage to the hive or equipment. It is much easier and more effective to deal with the adult beetles before they have the chance to lay eggs and raise larvae.

Luckily there has been success in renovating combs that have been slimed with larval feces and drip with fermented honey. Measures that seek to salvage equipment after infestation by SHB larvae include household products that kill beetle larvae and improve honey bee acceptance of affected combs. These products include regular bleach, white vinegar, dishwashing detergent, and vegetable oil. A dilution of 50 percent bleach is reportedly 100 percent effective in killing larvae within 4 hours, while a mixture of dish detergent at 1 percent strength kills 85 percent of the beetle larvae within 24 hours.[7]

Reports indicate that wintergreen essential oil will repel small hive beetles. As a result, folks who still use wintergreen to control tracheal mites may find that their problems with the beetle are reduced compared to their neighboring apiculturists.

I do not spend a lot of time and resources attempting to repel, trap, or kill small hive beetle adults or larvae. Nevertheless, I do admit to sometimes crushing a few with my hive tool as they try to scurry away immediately after being exposed when the inner cover is popped open on an infested hive. For the most part, my SHB control management has been limited to keeping hives strong and populous and simply not leaving any empty equipment lying around the bee yard once the weather turns consistently warm. I am also careful to extract harvested honey supers within several days of taking them off the hive, and so far this strategy has worked out well. I find that spotting the occasional hive beetle crawling around on frames of honey that the bees and I have worked so hard to produce and harvest provides more than enough motivation for me to extract the honey as soon as possible and banish any thoughts of procrastination.

One fall when I was in the process of packing up hives for winter, I came across a colony that had died out sometime after I had harvested its excess honey and applied an essential-oil mite treatment approximately three to four weeks earlier. Due to robbing, there was not a lot of honey left in the frames, and when I entered the brood area within the hive and pulled out a frame for inspection (to be sure the colony had not died from disease), I was confronted with comb that was swarming with maggotlike SHB larvae. This was my first encounter with beetle larvae. Many years of encountering wax moth larvae had served me well, as I felt no disgust or nausea at the sight. In fact, I found it fascinating that, when looking down between the frames from above, the combs looked perfectly normal, with no sign of damage or anything out of the ordinary. It wasn't until I had removed and tilted the frame and looked into the comb that the hundreds of developing beetle larvae became visible in their womb, deep within the cells along the midrib of the waxen structure. As the faint smell of fermented honey wafted up, I quickly placed the frame back into the hive body from which it came.

Now I felt panic setting in. I had never dealt with combs full of these creepy, crawly critters before. Luckily, I had not yet read the numerous warnings that I would later come across that indicate stacking supers containing beetle larvae onto healthy hives is a sure way to spread the infestation from one colony to another. Nor was I aware of the "authorities" who suggest that bees are reluctant to clean up combs filled with fermented honey and prefer to abandon such combs instead. I was not quite sure what to do, and I decided to treat the dead hive filled with SHB larvae as I would have

treated a hive infested with wax moth larvae so that the frames of comb would still be salvageable. I placed the boxes of larvae-filled frames on top of the inner covers of the strongest hives in the bee yard and hoped for the best.

Upon returning to the yard a week later, I had visions of maggots having overrun my strongest colonies, oozing out of every orifice of the hive as the former colonies were slowly consumed like the dead, bloated carcass of an animal that has not been discovered by the turkey vultures and other scavengers. Much to my relief, however, all signs of the beetle larvae had disappeared like magic, and although the frames of wax comb invariably had suffered some damage, on the whole they were still intact and appeared undamaged, as if my whole experience the previous week had been nothing more than a bad dream, with only the components of the dead hive sitting on top of the colonies providing proof of the reality of my experience. A quick check of the empty combs revealed no sign of SHB larvae or of any fermented honey, and all of the hives upon which the infested supers were placed overwintered well and were in fine shape come spring.

If the dire warnings not to stack or store infested equipment on top of strong colonies are to be believed, then I got lucky. Perhaps my efforts to incorporate hygienic queens into my apiaries and raise bees with hygienic behavior have been more successful than I realize. (I have never actually tested any of my hives specifically for the hygienic trait, relying instead on the test of winter survival.) Then again, maybe the cool September temperatures hampered the larvae's growth and vitality, making them more vulnerable to the defending bees. I imagine that, because most of the honey in the supers had been robbed out before the beetles moved in, the bees had an easier time restoring the slimed combs to usable condition than would otherwise have been the case. Whatever the factors involved, I am glad I chose to break up the dead hive and place only a single infested box on top of each strong colony. Stacking all the supers containing larvae on a single hive could have been detrimental, because such a large excess of combs might well have proven too much for the bees to properly defend and clean up.

As a result of the small hive beetle, the days of stacking hives high with empty supers early in the season and returning at the end of the season to harvest them are long gone. Even before the days of varroa, this was not a good hive management program anyway, because the bees would typically fill out the center frames of each super and ignore the combs on the outer edges of the colony. Of course, for those who didn't have the time to occasionally visit their colonies during the course of the season, this approach was a necessity. With the arrival of the SHB, however, hives should receive an empty super only when they have the majority of their current space filled. The rule of thumb I use is to try and time the addition of a new super to a colony to coincide with the bees starting to work on the outermost frames on either side of the uppermost box after filling and capping most everything else below and between. In this way, the bees have enough space for additional honey storage and to keep them from feeling overcrowded, but not so much extra room that they are unable to prevent parasites from taking up residence.

• SPIDERS •
AND OTHER INSECTS

Various spiders (technically called arachnids and having eight legs) and insects (which by definition all have six legs) may live in, on, or around beehives, and many will feed on bees, pollen, wax, honey, brood, hive debris found on the bottom board, or even the wooden hive parts themselves. Fortunately most pests are only local or occasional in nature, and most do not require special control measures.

Many insects such as cockroaches and earwigs can be found living in beehives but do no apparent harm. Nevertheless, these insects can be very offensive especially if they make their way into a jar of extracted honey. Some spiders and larger insects such as dragonflies and praying mantises will capture foraging honey bees, but they seldom cause significant losses to a colony. The same can be said for bullfrogs. Spiderwebs on or around hives are often littered with the remains of dead bees and can be easily taken down with a flick of a hive tool.

One insect that has raised concerns due to its ability to cause bees to become disoriented is the parasitic phorid fly (*Apocephalus borealis*). This parasite has been nicknamed the "zombie fly" for its ability to cause bees to walk around aimlessly, leaving their hives and flying off at night. The fly lays its eggs in the bee, and, as the larvae mature, they eat away at the inner organs of the bee's head, causing the host bee to exhibit the unusual behavior before it dies. While the fly is historically associated with a number of bumblebee species and paper wasps (*Vespula* spp.), it was observed paracitizing honey bees in California and South Dakota in 2009 and 2010 and in Washington state in 2012. Some researchers theorize that the unnatural crowding of enormous numbers of hives during events such as almond pollination in California may be helping to select for a more aggressive phorid fly that has developed the ability to reproduce on honey bees. While it provides tabloids and news outlets with a great sound bite, *Apocephalus borealis* has yet to be identified as a major factor in any widespread colony deaths, with only 10 to 20 percent of foragers in a particular hive involved in a worst-case scenario. The fly has previously been known to inhabit the states of New Mexico, Utah, Oregon, Idaho, Minnesota, and numerous states along the East Coast, from Maine down to Georgia.[8]

Various types of ants can often be found living in or around the hive. Most of these ants live off hive debris and are more of a nuisance than any kind of serious threat. Termites and carpenter ants that attack wooden hive parts can be a severe problem, however. To repel ants, use stems with leaves and flowers from the tansy plant (*Tanacetum vulgare*). This plant is also known as mugwort, bitter buttons, golden buttons, and cow bitter. By placing the tansy stems around the hive and under the inner cover, you should find that the ants are repelled by the plant and will leave the hive in short order.

Another organically appropriate deterrent to these pests is an oil barrier. This requires a hive stand positioned so that the legs of the stand sit in containers of oil; another option is to apply a grease coating to each leg of the stand to create a barrier to the pests. Such arrangements have also proven effective in protecting hives from deadly fire ants, which have been known to invade and destroy bee colonies in southern climes.

Hive stands have the additional benefit of keeping colonies up off the cold, wet ground, thus helping to keep hives warmer. The onset of wood rot is also delayed by reduced contact with the damp earth. Stands that raise the hive entrance above the level of grass and other vegetation help ensure that the honey bee's flight path is not interrupted in the event that the beekeeper gets behind in mowing around the beehives. Beekeepers with busy schedules may choose to place their hives on an old concrete foundation, or on top of a piece of old carpeting or landscape fabric to eliminate the need to mow. In general, keeping the grass mowed and the bee yard clean of debris will go far in reducing the likelihood that ants and other critters will take up residence.

CHAPTER 7

Four-Legged and Feathered Pests

My grandfather always said that living is like licking honey off a thorn.
—LOUIS ADAMIC

Many of the creatures in the animal world that wield offensive weapons—poison glands or powerful claws and teeth—are colored conspicuously to warn potential adversaries to leave them alone. The bee's striped colors are effective in advertising such a message, so much so that other insects such as the drone fly and bumblebee moth, both stingless creatures, manage to evade the attention of many predators because they possess coloration similar to that of bees. When looks are not enough to deter a would-be predator, the honey bee has the capacity to resort to a cooperative team effort in an attempt to discourage a marauder. Despite the intimidating defense a beehive can muster, however, a number of animals have developed defenses and strategies that allow them to feast on a hive's contents.

• SKUNKS AND RACCOONS • AND BEARS, OH MY!

Although not normally considered a major threat to the honey bee, skunks (*Mephitis mephitis*) can weaken strong hives and cause lethal damage to already weakened colonies. The skunk is renowned for scratching around the hive entrance, swatting the bees that come out to see what all the fuss is about, and then contentedly munching on them in their dazed and injured state. Skunks are active primarily at dusk and dawn from late autumn through early spring. Signs that skunks have been dining on your tab include scratch marks around hive entrances, matted grass and scratched-up earth in front of hives, skunk tracks, and small piles of feces in and around the bee yard that resemble cat droppings, but contain bee parts. However, the very first signal that skunks have made your bee yard a regular stop on their rounds is often the dozen or so bees that start buzzing around your head with a tone of angered annoyance as soon as you show up. Colonies that are molested by skunks, or anything else, on a regular basis will tend to be very jumpy and behave in a highly irritable manner.

Although much less prevalent, raccoons (*Procyon lotor*) have been known to cause damage in the apiary. Though known to feed on insects, the raccoon is more likely to reach up under stacks of empty supers stored in an apiary and rip apart the frames of comb for a sweet snack. The raccoon closely resembles the bear but is not large enough to knock over hives and rip apart supers filled with bees and honey. Damaged frames of drawn comb removed from stacks of empty equipment and scattered around the bee yard, with occupied hives left untouched, is a potential sign that raccoons have been at work.

One very effective way to discourage skunks and raccoons is to take some pieces of plywood and fill one side with nails or screws that stick out ¼ to ½ inch on the opposite side and are spaced about ½ inch apart. The plywood should be at least 18 inches by 12 inches in size. These boards can be placed in front of each hive to make it uncomfortable for any critter that wants to pull up a seat and snack on your bees or equipment. There have been instances where such boards have been pushed aside by a determined animal. To prevent this, it's a good idea to embed a few sticks or tent stakes in the ground on either side of the boards to help anchor them and to prevent their easy removal. Just remember to remove the boards or turn them over when working around the hives, or you run the risk of nails penetrating your shoes.

A similar approach that removes the inconvenience of having to move the skunk boards when working with the bees is the use of carpet grippers or tack strips, which are available at your local carpet store. These thin wooden strips have tacks sticking out of one side so that, when they are nailed to the floor along the walls of a room, wall-to-wall carpeting can be stretched over the tack strips and tucked into the edge between the strip and the wall to hold the edge of the carpeting in place. When cut to the width of a bottom board opening and nailed in place on the area of the bottom board that extends out in front of the hive, the carpet gripper becomes an effective skunk deterrent. When the carpet tack strip is nailed to the alighting area with the beveled edge facing away from the hive, the angle of the tacks will point outward and will be most effective in discouraging skunk feeding activities. In areas prone to frequent skunk activity, making these strips a standard part of your bottom board construction will go a long way toward easing your mind of concerns about skunks. However, if you plan on screening in the bottom board entrance in order to keep the bees inside while moving the hive, or expect to install a mouse guard in autumn, it is important to position the tack strip forward, away from the entrance, so there will be room for installation.

Another method that works well is to place some metal fencing material in front of the hive entrance. Sheep fencing rolled into cages for tomatoes works great when performing double-duty for this purpose. A loose coil of fencing creates enough of a barrier that critters can't easily forage at the hive entrance.

Some folks place their hives up on stands that raise the colony about 3 feet off the ground to prevent access to the hives by skunks. A hungry skunk that must stand on its hind legs and expose

FIGURE 7-1. Stake down your skunk boards, or the crafty critters are likely to move them out of their way.

its vulnerable underbelly to feed on a raised hive can quickly become discouraged. Keeping your hives on a stand is also likely to reduce the amount of bending that is necessary to work the hives, though a small ladder or step stool may be required during a good honey year.

No wild animal causes more fear and trepidation among beekeepers than the bear. Even the smallest of these in the United States, the black bear (*Ursus americanus*), can wreak havoc and major destruction within an apiary. Due to its size and strength, the bear is capable of the greatest destruction and economic damage because it not only destroys bees and combs, but can easily crush and tear apart frames, hive bodies, and supers.

As with other threats to the honey bee, prevention is the preferred method of controlling a marauding bear. Once a bear has gotten a taste of the gourmet delicacies packaged inside those cute

FIGURE 7-2. A skunk would have to be pretty handy with a hammer to remove a tack strip nailed to a bottom board. Note the mouse guard in place behind the tack strip.

little wooden boxes sitting within easy reach, about the only ways to stop it from coming back again and again until the whole bee yard is devastated are to shoot the bear, capture and remove it to another location far away, or move the bees. Clearly the latter two options represent the more humane and sustainable approach. However, the substantial amount of time and resources involved with such endeavors can make the inconvenience of erecting some type of bear deterrent around the bee yard seem well worth the effort. If bears have been sighted around your area, or you observe tracks, feces, or scarred trees that have been used as scratching posts, do not delay in providing some bear-proofing to your apiary. Otherwise the next signs of bear activity you are likely to come across are flattened beehives that have been torn apart and eaten.

The value of simple preventive measures cannot be overstated. First and foremost, as when dealing with skunks and raccoons, a clean bee yard will be less inviting to a curious bear. Leaving old supers or frames lying about the area will make your bees more noticeable and therefore a more likely target. Bears typically prefer wooded areas and are often found in mountainous regions; by simply locating bees away from prime bear habitat whenever possible, much damage can be avoided. If you must place bees in bear territory, bear-proofing measures are a must. One of the first things that should be done is to register the bee yard with your state bee inspector. Many states and Canadian provinces will provide some compensation to beekeepers who lose their bees to bears, as long as the bees have been properly registered.

FIGURE 7-3. Large stones or cinder blocks placed inside the coil of wire will prevent these barriers from being moved by a hungry skunk.

I am fortunate to have had only a few experiences with bears foraging in my bee yards, all in recent years, but I have heard enough horror stories to know that it is unlikely that there is any such thing as a 100 percent bear-proof fence once a bear has gotten a taste of the contents of a bee yard. Reports claim that a determined bear will tear down, walk through, climb over, or burrow under anything that stands between it and the golden ambrosia contained within the lemon-yellow wax combs of the hive. And an apiary does not have to be located in an isolated wooded area in order to become a target. Instances abound where bears have sacked hives kept in populated areas close to homes, even where dogs are in residence. Hence, the first order of defense requires that the bear be strongly discouraged from getting that first fateful taste of honey. In fact, it is best to put up a fence or take other measures the day your bees arrive. This is one time when it is best to not delay.

There are a number of options available for protecting one's bees from bears. A common choice is the electric fence. This is primarily because a portable electric fence is easier to install than a permanent fence, and it is easier to move in the event that the bees must be transported to another location. If the bees are expected to permanently reside in a single location, then a permanent fence can be used. When installing a permanent electric fence, rugged construction is essential due to the bear's powerful nature. Wooden or steel posts should be buried deep in the ground. It is also advisable to use extra-thick fencing wire and ceramic, rather than plastic, insulators. Place the

FIGURE 7-4. Bears are incredibly strong. Not only can a bear tear apart a hive, but the animal's digestive system seems to have no problem handling live bees, wax combs, bits of wood, wire, nails, and other frame parts that inevitably get chewed up and consumed while the bear feasts on a hive.

insulators on the inside corners of the posts so they will be less likely to break or be pulled out of the posts should an angry bear swat its paw at the wiring. Four or more strands spaced at approximately 1-foot intervals and a fully charged battery round out your bear protection. To help ensure adequate grounding, some folks recommend that every other wire be grounded.

If the bee yard is close enough to a power supply, the fence can be plugged in directly to an outlet. In isolated locations, solar panels are a great way to help ensure that the battery stays fully charged. Be sure to keep the grass and weeds well trimmed under the wires, or a short circuit may develop that will compromise the effectiveness of the fence. Some folks find that a strip of landscape fabric or carpeting about a foot wide works well to keep down the growth of vegetation under the fence. Place the strip down first and then pound the posts into the ground through the cloth or carpeting to hold it in place.

A side benefit of an electric fence is that, if the lower strand of wire is placed close enough to the ground, it may also be effective in deterring skunks and other critters. Another idea that Bill Mraz suggests is to use strands of barbed wire in place of the standard electric fence wire, so as to further deter an animal from pushing through the fence.

FIGURE 7-5. Temporary and portable, fiberglass electric fence posts like those commonly used for livestock are a low-cost option when trying to deter bears from invading your apiary. Note the backup straps around each hive.

I have heard folks recommend hanging a strip of bacon, aluminum foil coated with peanut butter, or some other bait from a hot wire so that a bear approaching the fence will touch the tempting morsel with its nose or tongue (both wet and sensitive organs) and get a good zap that should discourage further investigation. Nevertheless, it seems to me that hanging bait on the fence is likely to make the bee yard much more attractive to a curious bear than when no bait is used at all.

For additional security, it is a good idea to always have a backup system in place. Electric fences are not perfect—batteries can go dead, fences can get short-circuited, electronics can fail—so I like to

ensure that every hive is securely strapped together as a backup. This way, should a bear knock over a hive, the hive will not easily break open. By making the hive more difficult for a bear to break into, there is a chance that the animal will become discouraged and give up, leaving the hive alone and sparing you and your bees a tragedy.

Standard nylon straps will serve for strapping a hive, but they will weather and deteriorate over time, and there is the possibility that a bear's sharp claws will cut through a nylon strap. When attending Apimondia (the international beekeeping congress) in Australia in 2007, I was impressed with the single-piece metal straps that held together

FIGURE 7-6. Sturdy 4×4 posts help make an electric fence that is very bear resistant, but not easy to install or move after installation—making it a preferred choice for permanent apiary sites only.

every hive I saw there. Beekeeping in Australia is very migratory and apparently the majority of beekeepers there use the straps to keep their hives together as they move them from one location to another. I was thrilled when Mann Lake Ltd. in Minnesota began to sell these galvanized steel straps that will hold up to the weather and a bear's claws. While I admit it takes additional time to constantly unstrap and then restrap each hive during inspections and hive manipulations throughout the season, the straps make up for the inconvenience by not only providing backup protection against bears, but by eliminating the need to tie or weigh down covers to keep them from blowing off while overwintering hives.

Another low-tech approach involves creating a bed of sharp thorns that will deter a hungry bear in much the same way as recommended for skunks and raccoons. Pieces of plywood containing nails

or screws long enough to protrude between ½ and 1 inch and embedded on one side of the board at 2- to 3-inch intervals create a powerful, low-cost, low-maintenance deterrent when compared to a fence. The boards should be at least 4 feet square or larger and placed all around the hive so that a curious bear will not miss them when approaching from any direction. Care should also be taken to prevent the crafty bear from catching the edge of a board with its paw and turning it over or shoving it out of the way. Pushing the bear boards up flush against the hive or adjacent boards and staking them with wooden sticks or tent stakes around the edges can accomplish this. I have also heard that some beekeepers use carpet remnants in place of plywood. Although carpeting has the advantage of not warping like wood does, care must be taken to use carpeting thick enough to hold the nails or screws rigidly enough that they will not be easily pushed over.

FIGURE 7-7. These bees strapped for transport were not unstrapped at the time of delivery. By the time the beekeeper returned to remove the straps, a bear had knocked over all the hives. The hives did not break open, and the bear apparently got discouraged and left. All the beekeeper had to do was stand the hives back upright. PHOTO BY CHRIS HARP.

As with skunk boards, bear boards or carpeting will need to be moved or turned over when you are working around the hives. Thus, using large sheets of plywood that are not easily shoved out of the way or flipped over (by a bear) works well as long as they are not too difficult for you to move when entering the bee yard.

Other techniques that have been used successfully to deter bear damage include placing hives on a platform high above a bear's reach, enclosing an apiary with chain-link fencing, and placing hives inside buildings, such as on the second floor by an open window in an old barn that can be locked up. One must remember, however, that there is no such thing as 100 percent bear-proofing. Bears have been known to break into cabins, take down fences, flip over bear boards, and rip hive bodies apart with their bare claws. The way I see it, if you have taken care to set up numerous systems that back each other up and a bear still manages to break through your electric fence, get past your bear boards, and overcome the strapping on your hive, that bear deserves all the honey, pollen, and brood it can get!

FIGURE 7-8. Galvanized banding straps from Mann Lake Ltd. in Minnesota are super-rugged, hold up to weathering, and, most importantly, are not easily cut by a bear's sharp claws.

• MICE •

Mice may not be quite as much of a problem down South, where winters are not so long and cold, but in the northern reaches of the United States, a nice, relatively warm and dry beehive filled with wonderful things to eat proves a strong temptation for mice. If given the chance, mice will move into the bottom super or the hive body of a colony after temperatures are consistently below 50°F (10°C) and the bees have formed their winter cluster, leaving the majority of the hive's interior uninhabited and unprotected. The mouse can build its nest on the cluster's periphery with little chance of being molested until the temperature increases up above the 45° to 50°F (7° to 10°C) range.

This in itself wouldn't be so bad if the mice would refrain from chewing holes in the wax combs and damaging wooden frame parts while building a nest in the hive. Along with nest debris and holes in the combs come little mouse turds and urine. As long as the hive is strong and healthy, the mice will keep their distance and stay in the bottom portion of the hive, well below the cluster above, since the bees don't like having mice in their midst any more than most beekeepers do. Should the weather warm up to the point where the bees can break their cluster and venture down to investigate the area that the mice have annexed, the mice will simply leave until the temperature falls again and it is safe for them to return. However, should the hive find itself in a weakened condition or die out

FIGURE 7-9. Wax comb damaged by mice can be cleaned up and repaired by the bees if cleared of most debris and placed in a strong hive during a honey flow. Just be sure that the hive is level from side to side or the new comb will not be built straight.

completely, the hive you expect to be bustling with bees in the spring will be transformed instead into a little mouse condo.

Reducing the width of all hive entrances that are large enough to admit a mouse, so that bees can still pass but mice are excluded, will prevent such disappointing springtime discoveries. A piece of ½-inch hardware cloth that is 3 inches wide and cut to the width of the bottom entrance provides excellent mouse protection when bent at a 45-degree angle lengthwise and wedged into the space between the bottom board and the first bee box (see figure 7-2). Mouse excluders manufactured from pieces of wood with holes through which the bees can come and go are also available from bee supply companies. Unfortunately, they will also reduce airflow, negatively affecting ventilation within the hive during the crucial winter months.

The key to the efficacy of any mouse guard, however, is positioning the device in the entrance *prior* to the onset of cold weather. The tardy beekeeper who installs mouse guards late in the season after the first cold weather has arrived may find that, instead of keeping the mice out, the mouse guard has served to lock the mice inside, guaranteeing that the damage to the hive will be worse than it would have been without a guard in place at all. As a result, it is best to install mouse guards on warm, sunny days when the bees are flying strongly. This way any mouse that may have moved in during a recent cold snap will have deserted its nest due to the increased bee activity and will be waiting for the

FIGURE 7-10. This beekeeper used a mouse guard and couldn't understand why his hive was full of mice in the spring—until he found this hole chewed through the old rotten bottom board that had been on the hive.

bees to reform the cluster in the cool of the evening before returning to its anticipated winter abode.

You may benefit if there are rock walls, wooded areas, and other places that make good habitat for snakes near your apiaries, as many snakes will happily make a meal out of any mouse they come across that they can fit in their mouth. Nevertheless, in places like Florida, extra care must be taken because many of the snakes that live in the area are venomous and like to hide underneath the beehives. They can provide quite a surprise to anyone who happens to be moving hives around and carelessly reaches underneath a hive without first making sure the area is unoccupied.

If you are employing the service of snakes to help keep the mouse population in check around your bee yards, keep a sharp eye out when mowing the grass within the apiary. One day while mowing a bee yard, I spooked a garter snake that had been hiding amongst the hives. I watched as it ducked into the bottom entrance of a nearby beehive, only to exit less than a minute later looking like a porcupine, bristling with freshly laid stingers throbbing and pulsing all along the length of its back. The poor snake was not moving anywhere near as fast upon exiting as it had when it entered the hive, and it became more and more sluggish, dying a short time later.

Don't let all the action at the hive entrance distract you from what may be happening behind the front lines in this battle against the mice, either. Those empty honey supers, full of frames that the bees worked into drawn comb and that you have extracted, can be very attractive accommodations for a mouse. After all, there are no bees in these empty stored supers to worry about. Organic honey standards require a pest-free honey house and extracting area. Since prevention is the first line of defense, utilize barriers and good housekeeping and turn to controls if preventive measures are unsuccessful.

Prevention starts with a solid building. Inside, consider stacking supers kept in storage on top of one another, with the bottom super sitting on a flat surface, and use something mouse-proof like an outer cover to seal up the top of the pile. Be sure to keep the stack straight, with each super lined up directly on top of the one below. It takes only one skewed box in the pile to create an opening large enough for a mouse to squeeze through. If preventive measures fail or additional peace of mind is required, the assistance of some traps or the employment of a cat or two can be useful, though both these approaches tend to work slowly and may allow for some damage to occur before all the mice have been removed from the area. Although it may not seem to be in keeping with organic philosophy, the use of poison is often permissible in organic production as long as it can be demonstrated that the rodenticide will not affect the integrity of the final product in any way.

• THE BIRDS AND THE BEES •

For the most part, birds are a minor difficulty for beekeepers outside of Africa and Asia, which are home to avian species such as bee-eaters, the beeswax-eating birds called honeyguides, and at least two species of bird that will lead honey bee predators such as honey badgers or baboons to a hive and then dine on the spoils left behind after the larger mammal has exposed the nest and eaten its fill. In North America, heavy local predation of honey bees may occur by species of birds such as titmice, swifts, flycatchers, and some shrikes, which may present a problem for queen breeders and others working to rear their own mother bees.[1]

I have also observed on several occasions, while working in one particular bee yard located just a couple miles from the Canadian border, a flock of seagulls circling high above the apiary for long periods, and I have wondered whether they were dining on some high-flying bees. Other than the possibility of seagull slaughter or the capture of a queen in flight, the bird that catches the occasional bee on the wing in North America does not present any serious threat to beehives, with one notable exception: the woodpecker.

Although not a common occurrence, a woodpecker will occasionally peck holes in a hive to gain access to the tasty honey bees within. A woodpecker can seriously harm a weak hive, though the primary damage these birds cause is to the hive equipment itself. The one instance of this that I observed was damage that occurred after the fact when a woodpecker had opened up a large hole in and around the thin layer of wood located by the handhold on a hive body. The size of the hole was fairly large, which seems to indicate that the woodpecker spent a considerable amount of time feeding at the hive, either in one sitting or during repeated visits. The damage was discovered early in the spring and no further woodpecker activity was observed during the course of the season, which suggests that perhaps the woodpecker attacked the hive during the cold weather, when other insects were scarce and the number of honey bees that would emerge from the hive to investigate such a disturbance were few and slow-moving. Presumably, once the weather warmed up and the number of bees venturing forth increased to the point where the bees were able to land a number of stings, the woodpecker lost interest and moved on to easier pickings.

Controlling damage in such an instance is difficult at best, and although I know of nobody who has experience dealing with an active woodpecker attack, it is possible that introducing a barrier, such as a cardboard box slipped over the hive, may create enough of an inconvenience to deter such activity once a woodpecker is actively engaged in feeding on a hive. Alternatively, if a power source is available nearby, a motion detector connected to a floodlight or a radio may provide enough deterrence to discourage all but the most persistent bird. When necessary, woodpecker holes can be patched up by nailing or screwing a piece of sheet metal over the inside of the opening.

Botany Bees

Full many a tomato plant
Would never blush nor bear,
Without the bee to gallivant
And shift some pollen there.

She travels in the honey line,
But sets the vines aglow;
Which shows the finest things we do
Are not the things we know.

I do not care for honey much,
And yet I prize the bee;
The fair tomatoes that I love,
She makes 'em blush for me.

—CHICAGO DAILY NEWS

Challenges to a honey bee colony's existence are not limited to just other animals. Additional threats can appear in the form of vegetation and human activity that can sap the vitality of the hive and lead to its demise. Organic beekeepers should be aware of the types of plants within foraging range of their apiaries and the human activities that will occur nearby. Such threats may go beyond endangering the health of the hive: they may also damage the organic integrity of the products that will be harvested from the apiary. Therefore, the apiculturist must maintain a level of awareness that extends far beyond the world of the bee yard and reaches out into the surrounding region and human community.

• MOLD AND MILDEW •

Mold and mildew are much more reviled by humans than by the honey bee, and they can often be found on the combs in dead hives or on the outer fringes of colonies that are weak, especially in early spring. Such a discovery is unsightly and will often cause concern among beekeepers new to the craft who are worried about the spread of disease. There is no need for concern, however, because mold and mildew are not contagious and will not cause a hive to die out. Mold and mildew grow within the hive only when the humidity is too high. In such cases it may be found on wooden interior parts of the hive, wax combs, or dead bees or brood still in the comb. It is important to remember that, like the wax moth and small hive beetle, mold moves into the hive following the decline of the bees and is not the cause of the colony's woes. However, mold may make it difficult to identify the cause of a hive's demise, because it may mask the signs of diseased brood and brood cells.

Despite their distasteful appearance and odor, hive bodies, supers, and frames of drawn comb that contain mold or mildew do not need to be discarded, nor do they need to be cleaned other than to remove burr comb upon the insides of the supers or on the top, bottom, or side bars of the frames. It is also helpful to dislodge clumps of dead bees that may still be clinging loosely to the face of the comb in a hive that has died out before returning the now-empty equipment into service. The mold will not damage the wax, and moldy frames placed within a colony will be cleaned and polished up by the bees in no time. In fact, a strong hive can rehabilitate moldy old combs and put them back into service so fast it will make your head spin.

• POISON PLANTS •

By and large the plant kingdom offers few threats to the honey bee. The greatest may well be the threat of having a large branch from a nearby tree fall on top of a hive during a storm. Plants do produce a myriad of toxic compounds, however, and these can spill over into the nectar and make their way into honey on occasion. In 1936, Burnside and Vansell identified a dozen instances where honey bee mortality was reported from sources of yellow jessamine (*Gelsemium sempervirens*), California buckeye (*Aesculus californica*), locoweed (*Astragalus* ssp.), California false hellebore (*Veratrum californicum*), and leatherwood, also known as titi (*Cyrilla racemiflora*).[1] Reports of poisonous honey obtained from rhododendrons date back to antiquity, and they appear more recently in E. F. Phillips's book on beekeeping published in 1928, which also names mountain laurel (*Kalmia latifolia*) as a source of toxic honey. Nevertheless, a large number of honey bees are kept in the Appalachian Mountains, where many of these plants are found, and there are no consistent reports of serious problems. According to Phillips, "If all the honey from these sources were poisonous there would be an epidemic of poisoning annually in this region."[2] This is not the case, though, and it is likely that the dilution of toxic nectar with nontoxic sources is a factor that reduces poisoning episodes in both bees and humans. Thus, it is possible to reduce the effect of toxic nectar on a hive by feeding it honey syrup, sugar syrup, or bee tea in an effort to reduce the concentration of poisonous honey stored in the combs. It is also reported that, in the case of rhododendron, yellow jessamine, and mountain laurel honey, only the unripe, uncapped honey is poisonous to humans; the capped honey that has been fully ripened no longer presents a danger.[3]

Most dangers from plants are inconsequential to the honey bee. It is really the beekeeper who runs the greatest risk, not from poison honey, but from plants that are poisonous to human touch and may grow in or near bee yards. In the eastern United States and Canada, the poisonous plants that the apiculturist is likely to encounter are poison ivy (*Rhus toxicodendron*) and its cousin, poison sumac (*R. vernix*)—with poison oak (*R. diversiloba*) of primary concern in the western regions of these countries. Slightly less common, but no less dangerous, are giant hogweed (*Heracleum mantegazzianum*) and wild parsnip, also known as poison parsnip (*Pastinaca sativa*). It is important for the apiculturist to be able to identify these plants and to locate bee yards away from areas where such vegetation is found. In the event that poisonous plants such as these are unavoidable or are discovered within the confines of established yards, there are a few nontoxic, organically approved options available to combat them. They include digging them up by the roots (with the aid of gloves and protective clothing), which is practical only if the number of plants is small; smothering them with a material such as mulch or black plastic, which works well in areas where growth is concentrated within a relatively small and accessible area; or spraying with an organically approved weed killer that won't contaminate the groundwater or soil.

• CHEMICAL POLLUTION •

The most well-publicized danger to the hive stemming from human activity is chemical poisoning. This typically occurs from spray drift or honey bee exposure to poisons on blossoms following an application of pesticides—this despite the fact that all pesticide applicators are required by law to adhere to label directions, which are designed to reduce the chance of honey bee death. Chemical poisoning may manifest in the form of dead adult bees piled up in front of the hive entrance, often with their proboscises, or tongues, sticking

PLANTS TO AVOID

FIGURE 8-1. Poison parsnip is most easily identified in its second year of growth when it grows 2 to 5 feet tall and blooms, with flat-topped, umbrellalike, yellowish flower clusters.

FIGURE 8-2. Poison ivy typically grows close to the ground and features three leaves grouped on a small stem that is attached to a larger main stem. Flowers tend to be yellow or green, and the plant will bear white fruit (berries).

FIGURE 8-3. Poison sumac is normally found in boggy areas and can grow up to 15 feet tall. Stems will hold from 7 to 13 smooth-edged leaves and feature glossy, pale yellow to white berrylike fruits.

FIGURE 8-4. Poison oak can grow as a low shrub or long vine with oak-shaped leaves that develop typically in groups of three; it produces clusters of yellow berries.

FIGURE 8-5. Giant hogweed can grow up to 15 feet tall and features white flowers that resemble overgrown umbrellalike blooms similar to those of Queen Anne's lace.

out. High exposure may reduce the population of a colony to the point where there are not enough workers to care for the brood and keep the brood combs covered and warm. Pesticide spraying that must be carried out when bees are in the vicinity are best conducted at night. If spraying must occur during daylight hours, all colonies should be moved out of the area to a minimum distance of about 2 miles away. Whenever honey bee colonies are killed or injured by pesticides, the incident should be documented and reported.

For many, speaking up about the pesticide poisoning of hives creates an uncomfortable social situation. If a beekeeper is using miticides—given the established sublethal effects of approved miticides on bees both when used alone and in combination with other chemicals—complaining about someone else's pesticide use can seem like the pot calling the kettle black. This in itself is enough to prevent many poisoning incidents from being reported.

Even those who don't use chemical mite controls may find themselves in a situation where the pesticide applicator may be working for, or even be, their friend or neighbor. While one may choose to approach such a neighbor or friend to discuss the matter and try to reach a mutual understanding, there is little that can be done in the way of convincing a person to change their approach if they are uncooperative. This is especially true if label application instructions have been followed and the force of law is on their side. Combined, these factors alone prevent many pesticide kills from being reported. This is an unfortunate situation because the lack of reports of pesticide poisoning of honey bee hives creates the impression among regulatory agencies that there is no problem.

For most of us, state agencies and regulators are more accessible than their federal counterparts in Washington, DC. A call to your state bee inspector can often be helpful in obtaining support when investigating a pesticide incident in an apiary. In recent years many states have been reducing or eliminating their bee inspection programs due to budget cuts. However, numerous state-run

A HOMEMADE HERBICIDE

Here's a natural, nontoxic homemade weed killer recipe that can provide a measure of control for poison ivy. Mix 1 gallon 20 percent vinegar with ½ cup orange oil and 1 tablespoon natural Castile liquid soap. Spray this mixture onto the leaves of the plant. Poison ivy so treated will typically wilt and then resprout, requiring additional applications. This recipe works best during periods of dry weather, when rain will not wash the herbicide off the leaves immediately following application or nourish the roots of the plant and encourage rapid new growth.

Cautions: This contact herbicide will kill most any plant upon which it is sprayed, not just poison ivy. Be sure to thoroughly clean your hands and all tools, clothing, goggles, and protective equipment following application to remove noxious oils or acetic acid contamination. Also please note that 20 percent vinegar is much stronger than the 5 percent acetic acid content of the table vinegar we splash on our salads. It will cause skin burns upon contact, and it may cause severe lung irritation if the fumes are inhaled. Eye contact can result in severe burns and permanent corneal damage, so please take appropriate precautions to protect yourself, children, and animals.

Author's Note: A couple years after this book was initially published, the Environmental Protection Agency (EPA) cracked down on companies that were selling 20 percent vinegar and forced many of them to stop because they were selling an unapproved herbicide. As a result, you may have a hard time procuring this type of vinegar today.

apiary inspection and support programs are still in place throughout the country. According to the National Honey Bee Advisory Board (NHBAB), some state investigators strongly discourage the filing of incident reports or act negligently when investigating timely reports. State pesticide officers are reportedly discouraging or refusing to test samples for pesticide residues, citing the high costs associated with testing. The NHBAB also has indicated that many states take no enforcement action in response to confirmed bee die-offs from pesticides, claiming that label directions are vague and unenforceable, and that such incidents are often not reported to the EPA Ecological Incident Information System (EIIS) database since such reporting is not required. Depending on the state you live in, you may have to have samples tested and report the results yourself if you want anything to happen at all.

The National Pesticide Information Center (NPIC), a cooperative agreement between Oregon State University and the EPA, has developed a Web-based portal to gather information on ecological pesticide incidents that adversely affect nontarget organisms such as bees. To file a report, visit the NPIC's Ecological Pesticide Incident Reporting page (see resources on page 274), which has been set up to collect information directly from beekeepers, as well as government organizations, academia, wildlife rehabilitation centers, and conservation societies. The report questionnaire is quite thorough and includes questions not only about when and where the incident occurred and what the weather was like at the time of the incident, but also about the pesticide name, product registration number, active ingredients, and formulation. Needless to say, this does not make the site very user-friendly, since these are required fields and it is impossible to finish filing the report unless all requested information is filled in. For those who want to file a report without having all the pertinent information on hand, you can call the NPIC directly (800-858-7378) and file your report with one of their staff.

You can also report the incident directly to the EPA by sending them an e-mail. When contacting the EPA, try to provide as much detail as possible. Information should include when and where the incident took place, the nature of the effects, the number of colonies impacted, and the environmental conditions associated with the loss. Any information you may have about the surrounding area and what the bees may have been foraging on will be helpful. It is a good idea to also include whether state government representatives were notified and whether an investigation was conducted.

Your ability to fully file a report is greatly increased if you witness a pesticide-related bee kill event firsthand. Nevertheless, reporting a pesticide incident that you witness—or even one you don't witness but do observe its devastating aftermath—will carry greater weight if you can provide additional proof in support of your explanation of the event. Obtaining additional evidence of some sort is critical to the reporting process; having an independent lab test samples from affected hives is extremely valuable. The NPIC provides a link on their website (under "Testing") to the American Association for Laboratory Accreditation (A2LA), which can help you find a facility near you to conduct the testing required to detect suspected chemical contamination of bees or other hive components such as wax, pollen, or honey. Unfortunately, proper testing, especially of numerous samples, can be very costly.

Maryann Frazier at Penn State University regularly accepts honey, nectar, pollen, brood, adult bees, and wax samples for pesticide testing. The lab charges to test for 172 pesticide-related chemical residues that may contaminate a hive. In the past there have been limited funds available from a Project Apis m. (PAm) grant-funded cost-sharing program that will cover half the fee for having samples tested through Penn State. The beekeeper pays the other half of the cost. By participating in the program, your test results become part of a confidential database that is being used to track the chemical exposure of bees throughout the United States and its relation to bee health. Alternatively, samples can be sent to the USDA for testing (see resources on page 274).

When sending in samples for testing it is important to collect a large enough sample and to preserve the integrity of the sample by using an opaque container that will not shatter easily. Normally a minimum of 2 ounces of the material collected for testing is required, whether it is bees, comb, honey, wax, or pollen. Since a pound of bees consists of around 3,000 individuals, a bee sample should contain at least 400 bees in order to be sure a minimum of 2 ounces are sampled. Once the sample is gathered, immediate freezing will help preserve the sample and slow down the natural degradation of any chemical residues that the sample may contain. If possible, ship the sample in an insulated package with an ice pack as well. Since each lab may have their own guidelines for handing and shipping samples, contact the testing facility you intend to use prior to collecting the sample(s).

It is a good idea to document the event with photographs and include a witness holding a copy of the local daily newspaper in the photo in order to establish the date of the incident. If you can videotape the collection of the samples and the sealing of the packages or containers holding the samples, all the better. Your account of the incident is strengthened if you can include photos when e-mailing reports of the event to various agencies and organizations. It is also a good idea to report the event to your state and local beekeeping associations so they can help spread the word.

To help prevent honey bee deaths from pesticide poisoning in instances where pesticides must be used, consider making the following suggestions to local farmers:

- Apply pesticides only when needed. Without regular pest monitoring activity it is difficult to determine whether or not the pest population has reached the level where pesticides are called for.

- Always use the lowest effective rate of the recommended pesticide.
- Use the pesticide that will control the target pest and is the least toxic to bees. If all recommended pesticides are equally hazardous to bees, use the one that breaks down the fastest.
- Use pesticide sprays or granules instead of dusts or powders.
- When applying pesticides near beehives, it is preferable to use ground equipment rather than aerial application.
- Pesticides and chemically treated seeds are best applied or planted late in the afternoon or at night when foragers are not as active.
- Avoid pesticide drifting onto other plants that may be used as forage by bees.
- Notify area beekeepers several days in advance of a pesticide application so they can take steps to protect their hives. Notification, however, is not a release from responsibility.

Whereas exposure to agricultural chemicals tends to have an immediate effect on a colony, the release of everyday poisons that pollute the air, ground, and water has the more degenerative effect of shortening the life span and weakening the overall immunity of the bees, making them more vulnerable to diseases and viruses, and making it more difficult for forager bees to locate forage through their sense of smell.[4] Due to air pollution in some areas, floral scents that bees could detect at over 800 meters away in the 1800s are now detectable at less than 200 meters from plants.

In this day and age, almost every species of plant and animal on Earth is suffering some form of stress from the release of pollutants into the environment. This pollution is having not only a direct effect on soil, water, and air quality, but also an indirect effect through the shifting of atmospheric weather patterns and the abnormal warming or cooling of various regions of the Earth. The honey bee is affected along with the rest of nature.

• BIOLOGICAL POLLUTION •

A new form of agriculture-related pollution which honey bees have had to contend with over the last 15 to 20 years or so is the release of genetically modified organisms (GMOs) into the environment. Unlike traditional breeding methods that produce hybrids from similar plants and animals, genetic engineering mixes the genes from dissimilar organisms, even going so far as to mix animal, virus, and plant genes all into a single organism. Such a crossing of species would never take place naturally, and this new technology has resulted in environmental, philosophical, and moral concerns being raised that were not publicly debated on a wide scale in the United States prior to the release of GMOs into the environment and the food chain. Many years from now, I suspect that students of history will look back with astonishment and disbelief at the idea that humankind once had the arrogance to play around with the very foundation of all life on Earth: DNA (deoxyribonucleic acid), the carrier of genetic code. Our inability to use the technology of genetic engineering appropriately to protect non-GMO crops from contamination by GMOs threatens the very viability of organic agriculture. This conflict pits farmer against farmer—whereas the multinational agribusinesses reap the profits and have, so far, been able to avoid any responsibility for the damages inflicted by their patented technologies on non-GMO crops, as well as on livestock and people who consume GMO foods.

Despite lack of evidence to support the claims, the engineering of the *Bacillus thuringiensis* (Bt) toxin into crops such as corn, soy, canola, alfalfa, sugar beets, and cotton is touted by the biotech industry as an example of the beneficial use of genetic engineering. Bt has long been a mainstay of the organic farming community due to the ability of this microorganism—once consumed by the target pest and exposed to the unique environment within the digestive tract—to produce spores and crystals that cause paralysis and death. Although traditional Bt products do not kill pests as quickly as most chemical pesticides, there are no harmful side effects. Because the toxin is not produced until pests eat a leaf upon which the microbe rests, it is safe and nontoxic to animals, humans, and beneficial insects, and it may be used right up until harvest. Unfortunately, geneticists have installed the gene for the Bt toxin into every cell of plants such as corn and soybeans, rather than the microorganism itself. As a result, not only is the European corn borer exposed to a lethal dose of the toxin, for example, but so is everything else that feeds on the genetically modified corn plants, from farm animals to insects and even humans, who must now adjust to consuming this poison along with their daily meals—as the toxin is integrated into the structure of the food and can't be washed off like a traditional pesticide. In addition, the constant exposure to Bt has quickly led to the development of resistance among target species. This is a situation that never would have occurred if Bt had continued to be used as it has been by the organic farming community for decades prior to the release of Bt-impregnated GMOs.

Bees rely on plants to provide them with nectar, resins, and pollen. Pollen, being the male fertilizing seed of flowering plants and a concentrated source of genetic material, brings with it the greatest potential for GMO contamination within the hive. It is through the collection and consumption of pollen that *Apis mellifera* may suffer direct detrimental effects from genetically modified (GM) plants, depending on the type and amount of GM material it is exposed to.[5] In areas where honey bees and genetically modified plants are in close proximity, colonies are likely to consume GM material and store GM pollen within the hive. Although little research has been done to date, some studies have been conducted on honey bees that were fed solutions of proteins expressed by GMOs. Depending on their exposure, bees had trouble learning to distinguish between the smells of flowers and exhibited a shorter-than-normal life span.[6] Such learning difficulties could also potentially make it difficult for foragers to successfully navigate back to their hive.

Other research findings indicate, despite industry's initial claims to the contrary, that genes from GMOs can survive in the intestinal tracts of humans and animals such as honey bees. German scientist Hans-Hinrich Kaatz from the Institut fur Bienenkunde (Institute for Bee Research) at the University of Jena demonstrated that not only did the patented genes survive in young bees that had been fed pollen from glufosinate-tolerant canola plants, but the DNA from these organisms could jump the species barrier. Kaatz's research showed that the patented gene had transferred into the bacteria and yeast fungi inside the bees' intestines.[7] Such belated discoveries regarding GMOs are especially troubling because, once these newly manufactured genetic traits are released into the environment, they migrate, reproduce, and mutate, which means there is no way to clean up this form of biological pollution. Due to their efficiency as pollinators, the unsuspecting bees play a major role in the spread of genetically modified material.

Unfortunately, much of the research surrounding GMOs is filled with secrecy and controversy. This is because most of the GMO studies have been conducted by biotechnology companies that manufacture and profit from these technologies, and their research is considered proprietary information and not subject to public scrutiny. Should an independent lab want to study the impact of GMOs on their own, they must first get permission from the biotech firm that holds the intellectual property rights for the genetically modified material. Needless to say, companies like Monsanto don't give permission for independent research unless they believe it is in their interest to do so, which of course it rarely ever is.

In the United States, many of the government regulators charged with overseeing the approval of newly developed genetic engineering technologies are former employees of the very companies they are charged with regulating. This creates serious conflict-of-interest issues and is indicative of the charges some have leveled of lax oversight by government regulators. For example, when we look at the studies that have been done on the effects of genetically modified plants on honey bees, adult bees have been the focus of the majority of the research, yet the EPA examined only short-term honey bee larval toxicity before approving Bt-cotton plants for the US market.

Just as with chemical pesticides, the lack of any meaningful regulation means that knowing what types of crops are being grown in the area surrounding your apiary is the only way to avoid finding yourself with hives and honey bee products that are contaminated by the products of genetic engineering. It is for this reason that most organic regulations require an apiary site map that includes the surrounding area and notes all potential threats to organic integrity.

• THEFT AND VANDALISM •

Unfortunately, the risk of theft and vandalism is a real danger in today's society. One beautiful early spring day, while checking hives to be sure they had enough honey left to get them through the season, I pulled into a secluded bee yard to discover that a hunter had used a hive for target practice while passing through the area during the previous fall's hunting season. Positioning a bee yard within sight of a home rather than in an isolated, out-of-the-way place that is out of view is perhaps the best way the apiculturist can discourage criminal or mischievous behavior. Additionally, branding or otherwise marking your hive bodies, supers, and frames can aid greatly in establishing ownership of hives and equipment that have been stolen. For greater piece of mind, high-tech options offered by companies such as Bee Alert Technology of Missoula, Montana, include such gadgets as radio frequency transmitters that, when embedded in a hive, can identify a specific hive from distances of up to 1,500 feet away with special equipment. Other security systems are able to sense when a hive is being moved and will automatically place a call to a prespecified phone number to alert the owner to any potential hive-rustling situations.

A novel approach to deterring the theft of frames of honey was administered by beekeeper David Fretz of Rutland, Vermont. After experiencing the disappearance of frames from his hives, David took a lumber crayon and wrote in bold letters on the top of each colony "CAUTION: Cyanide Experiment—Poison Honey." David reports that thefts from his apiary stopped abruptly after that.

Bees that have an extremely calm and gentle temperament may be a pleasure to work with, but it is the testy hive that is most likely to act on its own behalf to thwart attempts at thievery. Entering a bee yard and finding a hive with its outer cover removed and the inner cover askew is a sure sign that the bees have stood their ground and fought off a would-be thief. I must admit to feeling a small sense of pleasure in this kind of natural justice when envisioning the humorous scene that must have occurred to cause the intruder, perhaps bent on stealing a frame or two of honey, to forgo covering up the evidence of the raid and make such a hasty retreat.

As indicated in Chapter 3, urban beekeepers have found that rooftop hives can be camouflaged and made to look like an air conditioner or other part of the city landscape. The novel shape of the top bar hive can also assist city beekeepers in reducing the hive's visibility and making it less of a target for vandals and thieves.

• GLOBALISM •

As many foreign countries are encouraged to accumulate more debt than they can ever hope to repay, in the process becoming hopelessly indebted and thus subject to undue influence on the part of foreign governments and financial institutions, world trade has opened up global markets and made foreign goods more accessible to businesses. The beekeeping industry has benefited greatly from the increased genetic variety of honey bees that is now available to help develop resistant stock. One area of concern, however, is the lack of effective regulatory control surrounding the increase in global trade. It's bad enough that the market economy undercuts all social values in deference to business growth and profits, but, in addition, the wholesale exchange of goods between countries has directly contributed to the difficult situation that US beekeepers find themselves in today. This is due to the increase in the number of organisms that are benign in their native lands but when deliberately imported or unintentionally transported, either legally or illegally, to new locales, they become a problem. Tracheal mites, varroa mites, small hive beetles, and, to an extent, Africanized honey bees are European honey bee pests that all either were imported or migrated into the United States between 1984 and 1997. Government regulators are appointed the daunting task of adequately controlling the movement of invasive and destructive exotic species through international trade channels, and they clearly cannot be solely relied upon.

Despite the discovery of 14 million-year-old fossilized native American honey bee remains in Nevada's Stewart Valley Basin,[8] the European honey bee is not native to North America. One of the earliest importations of bees to North America was in 1622.[9] The fact that bees have had to adapt themselves to ecosystems in North America that were originally foreign to them may make them more vulnerable to certain diseases and pests within the United States and Canada.

• BEEKEEPER ERROR •

Apiculturists themselves may pose a threat to honey bee colonies. Beekeeper error—whether from lack of knowledge and inexperience or simply due to an accident caused by not paying close enough attention or being in a rush—is probably the single largest cause of colony demise, greater than all other human activities combined. From not leaving enough honey on the hive to enable the bees to get through the winter to not treating a hive appropriately for diseases or mites or accidentally injuring the queen while inspecting the brood nest, the beekeeper can sabotage a hive in an enormous number of ways.

The apiculturist should make every effort to learn as much as possible from a variety of sources in an effort to reduce the possibility of mistakes. Aside from reading books on the subject of beekeeping, time spent talking with those who have greater experience is often time well spent. Local and state beekeeping organizations play a crucial role in providing opportunities for inexperienced beekeepers to ask questions of those who have been there before them. There's a lot of truth to the old saw that five different beekeepers will often have six different ideas on how best to handle a given situation. This is where your evaluation and intuitive skills come in handy, to help sift through the often conflicting information and choose the path that seems best suited to you and your particular situation. Many bee associations offer mentorship programs where a novice is paired up with a more experienced apiculturist. Additionally, courses and workshops on beekeeping are offered throughout the country and are often a great way to obtain hands-on experience.

It is important for beginners not to allow the fear of harming the bees to keep them from working with the hives. As experienced beekeepers know, simply taking your time, paying attention, and staying focused are the best defenses against accidents and oversights. It is also important to keep up with the latest developments in beekeeping by attending bee association meetings and regularly reading bee journals where the newest research findings, industry trends, and commercial products are promoted.

Unfortunately, human error has a way of foiling the best-laid plans. By staying open to all potential sources of information and planning ahead to leave room for the unexpected and the occasional snafu, you can remove the fear of failure and recover faster from an incident. Adopting the attitude that "there are no mistakes, only learning experiences" is as helpful when working with bees as it is in navigating the trials and tribulations of everyday life.

If every one would do his best in watching for disease, existence would be trebly blest for all the honeybees. If every bee crank would inspect his brood-combs twice a year and with his weather eye detect the first germ to appear; then swat the thing right in the neck and knock it galley west, we'd run them off the map, by heck, if each would do his best. Bacillus larvae would not eat the baby bees alive. We'd drive them out, so help us Pete, from each and every hive. No sunken cappings would we find; no toothpick roping test; no gluepot smell, the luscious kind, if each would do his best.

—BILL MELLVIR

Honey bee diseases are caused by living organisms, including fungi, bacteria, and viruses. Luckily, none of the diseases of the honey bee are known to have a physiological effect on people who handle

FIGURE 9-1. Healthy hives will contain brood nests with developing bees clustered closely together, all of a similar age group, and without many empty cells scattered among the brood. PHOTO BY STEVE PARISE.

the bees or equipment; eat the honey, pollen, or comb; or handle or use the beeswax or beeswax products from an infected colony. Bees are naturally able to resist most disease organisms when they are vigorous and well nourished. The bees also make use of propolis, the most potent naturally occurring antimicrobial there is, by shellacking the interior surfaces of the hive with a thin layer of the material to help maintain a sterile environment within the hive. Researchers have found that not only are colonies healthier when their hive is encased in a cocoon of propolis, but that bees will actually seek out propolis when infected with certain pathogens in an effort to medicate themselves and bring about a cure.[1] However, despite the bees' best efforts, some microorganisms have managed to evolve to the point where they are able to establish themselves and thrive within the hive. The organic beekeeper can greatly assist the bees in keeping disease organisms in check with practices that encourage the honey bee's natural hygienic tendencies, reduce stress by ensuring an adequate supply of high-quality nourishment, prevent the introduction of contaminated equipment, and include regular hive inspections to catch issues early on before they develop into larger problems. Although there are a large number of diseases that can befall *Apis mellifera*, we will concern ourselves here only with those commonly encountered in the Northeast, since these are the ones I am most familiar with.

• AMERICAN FOULBROOD •

Before the advent of parasitic mites in North America, the most serious of all diseases, from US beekeepers' perspective, was American foulbrood

(AFB). Although varroa has overshadowed it in recent years, foulbrood is still the most important microbial disease that beekeepers must concern themselves with. As its name implies, the bacterium *Paenibacillus larvae* (formerly known as *Bacillus larvae*), which causes AFB, infects bee larvae in their early stages, eventually killing them, and in advanced cases the infected brood give off a foul odor that is very noticeable and vaguely resembles the smell of rotten glue or dead fish. Early in its progression, the disease will turn the pupae into a brownish slime resembling the color of coffee with milk. Over time, the pupa melts away like an ice cream cone on a hot summer day as the bacterium consumes it, causing what's left of the baby bee to ooze down into the lowest part of the cell. The remains of the brood eventually dry out in the bottom of the cell and create a black scale that adheres so tightly to the cell wall that it is difficult for the bees to clean up the mess, especially since the hive's population will be reduced from the action of the disease.

Once a diseased hive has progressed to this stage, the colony is typically so weak that neighboring hives have robbed honey out of the sick hive and spread the foulbrood spores. Compared to healthy brood, whose worker cappings will be slightly raised in shape and light brown in color, the cappings covering the cells containing infants infected with foulbrood will often be dark brown to almost black in color, will have a greasy appearance and a sunken shape, and may be pockmarked with pin-sized holes. Given that brood infected with active *Paenibacillus larvae* spores fail to hatch, diseased cells can often be found adjacent to healthy brood or surrounded by cells that have been abandoned by healthy hatchlings. As a result, frames of brood in a colony suffering from AFB often take on a spotty pattern, rather than the more solid pattern typical of healthy hives where the queen has laid eggs of similar age all adjacent to one another.

AFB can also be recognized in the field by testing the stringiness of the dead pupae, which are moist and slimy in the early stages after death.

When a small twig, piece of straw, or toothpick is inserted into the pupal mass, it will tend to stick to the probe and stretch ¼ to ½ inch or so as the probe is removed, much like the sticky, ropelike viscosity of mucus. Dead pupae unaffected by foulbrood disease will not stretch or become threadlike after being poked. American foulbrood is easily spread, and it can destroy an entire apiary if not attended to quickly. Thus it is vitally important that every beekeeper become familiar with and be able to recognize the symptoms of AFB. (See a summary of symptoms in the "Beehive Autopsy Results" table in chapter 3 on page 76.)

Another field test that can be used to detect the presence of AFB is a powdered milk test (Holst Milk Test). This test detects the presence of an enzyme that the foulbrood bacteria release in order to melt down the larval bodies in the final stage of infection. The enzyme will also break down milk protein and will cause a weak solution of powdered milk to turn clear within minutes. One teaspoon of powdered milk is mixed thoroughly in 100 milliliters of water (a little less than ½ cup). The watered-down milk is then poured into two clear glass vials or jars. As large a sample of brood as possible from a suspected AFB colony is collected with a toothpick and placed into one of the prepared vials. Nothing is done with the second vial of milk. Both containers are then placed in a

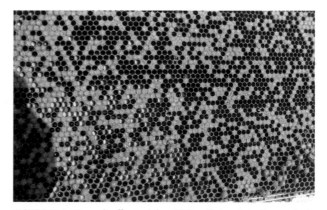

FIGURE 9-2. Brood affected by AFB will tend to have dark, sunken cappings (often with pinholes in them) and may imitate a shotgun pattern, with numerous empty cells scattered among the capped cells. PHOTO BY STEVE PARISE.

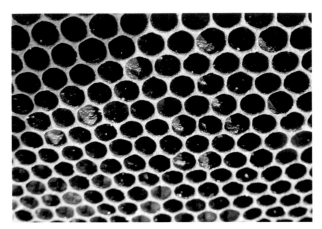

FIGURE 9-3. Black scales like these on the lower side of the cell are indicative of American foulbrood in its advanced stage. PHOTO BY STEVE PARISE.

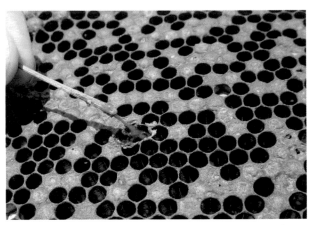

FIGURE 9-4. This test for foulbrood can be conducted only on recently deceased larvae that have not yet dried out and formed a black scale on the lower side of the cell. PHOTO BY STEVE PARISE.

warm place (such as the breast pocket of your shirt underneath your bee suit) for at least one hour. If the sample contains AFB the vial with the sample will become clear compared to the control vial.

Paenibacillus larvae is difficult to control because in its resting stage the bacterium forms spores that are reported to live as long as 70 years or longer, until favorable conditions allow it to bloom.[2] AFB is most often spread from colony to colony by bees from a neighboring colony robbing out a dead or weak infected hive and carrying the disease back to their own nest, or by drifters that visit from hives infected with the disease. Beekeepers may also unwittingly spread AFB by interchanging frames between sick and healthy colonies, by not cleaning up diseased equipment from a colony that has died from AFB before it gets robbed out, or by feeding colonies AFB-contaminated honey taken from diseased hives. Mixing the honey from a questionable source with water to create a syrup and then boiling the mixture for 20 to 30 minutes should kill any spores that may be present and make the feed safe, though not ideal, for use (see "Feeding" in chapter 3 on page 55).

For many years in the United States, the only approved treatment for AFB was to either burn the hive or use the antibiotic oxytetracycline, known commercially as Terramycin. The drug is applied either as the active ingredient in a grease patty or mixed with confectioners' sugar and sprinkled near the brood nest within the hive. However, the patties were not always consumed in a consistent and rapid manner, which would create a situation in which an incomplete dose of the antibiotic would be administered to the hive. The use of such patties has been blamed for the revelations of Terramycin-resistant forms of AFB now infecting hives. As a result, the FDA approved a new and more powerful antibiotic, Tylan (tylosin), for use against foulbrood disease. This new antibiotic is extremely stable, and rather than being used prophylactically, as oxytetracycline was, it should be applied only when signs of foulbrood are clearly visible, and it should be applied as a powder shaken into the hive to reduce the chances that *Paenibacillus larvae* will develop resistance to Tylan as well. (Antibiotics when applied in powder form should never be administered directly into the brood area, or they may kill the developing larvae and pupae.) Generic forms of Tylan are Pharmasin (Tylovet) and Tylomed-WS, both of which are also FDA-approved antibiotic treatments for foulbrood.

Most recently, the FDA has approved another antibiotic for bees: lincomycin hydrochloride (Lincomix). Unlike Tylan, Lincomix is approved for prophylactic use on beehives. To date, *Paenibacillus larvae* bacteria do not seem to have developed resistance to this antibiotic.

No matter how careful one is, however, antibiotic treatments (when they work at all) simply mask the presence of the disease, preventing viable spores from growing and reproducing. Such treatments do not remove the spores from the hive. As a result, hives treated with antibiotics tend to quickly show signs of reinfection with the active form of AFB once antibiotic use is discontinued. In addition, special care must be taken to ensure that honey being harvested does not become contaminated with antibiotics. The regular consumption of minute doses of antibiotics in the human diet has been linked to the development of antibiotic-resistant microbes that cause untreatable diseases.

To permanently remove American foulbrood from a colony, all AFB spores must be eliminated from the hive. With a strong dose of determination and an investment of time and resources, this can be accomplished within a single season by removing every bit of drawn comb and beeswax, pollen, brood, and honey from an infected colony. These are all the hive products that will harbor spores. Such an endeavor is best conducted early in the season, so that the hive has the maximum amount of time to build new comb and store adequate supplies of honey and pollen for winter. Also, by taking away a colony's drawn comb early on in its yearly cycle, before egg laying reaches its peak, a minimum amount of brood is sacrificed for the greater good.

The actual process of eliminating AFB from a hive consists of shaking each frame of bees from the hive into a new hive body filled with foundation. If you can spot the queen, it is best to transfer her into her new home by grabbing her wing and placing her on the top bars of the foundation frames so she can crawl down between the frames. (To reduce the chance of an injury that could affect her ability to lay eggs, a queen should never be held by any part of her body other than her wing.) Included in the new hive body of foundation should be one empty frame of drawn comb. This frame of comb is left so that the bees have a place to deposit honey they may have engorged themselves with during the

process of smoking and shaking them into their new home. The very next day, the frame of drawn comb containing these honey deposits should be removed and replaced with a frame of foundation. The colony should then be fed uninfected honey or sugar syrup so that the hive has something to get started on and won't starve should it immediately experience a week of cold, wet weather and not have the opportunity to forage.

By following this procedure, all the beeswax, pollen, brood, and honey that may possibly contain AFB spores are removed from the hive. All of the frames and comb from the infected colonies should be burned, and all hive bodies, supers, inner covers, and bottom boards should be scraped clean of burr comb, which should also be burned. The woodenware itself can be scorched with a propane torch or other device that will heat the surface of the equipment to a high enough temperature to ensure that any remaining AFB spores are destroyed.

When I first started beekeeping, I would use a dusting of Terramycin prophylactically in the spring about a month prior to the first major honey flow and again in the fall following the harvest. Over time, while making the effort to use antibiotics less often, I would treat a hive only after it showed symptoms of disease. Unfortunately, because the spores had become embedded in my combs, AFB would consistently pop up in my colonies when left untreated. I knew I needed to change my approach if I was going to get myself off the antibiotic treadmill and keep bees organically. I became motivated to experiment with an operation that I first read about in the book *Elimination of American Foulbrood without the Use of Drugs* by Mark Goodwin and Cliff Van Eaton, published by the National Beekeepers Association of New Zealand.[3] This book gave me the courage to believe that elimination of foulbrood without chemicals was indeed possible and could be successfully accomplished with my hives.

I removed all the drawn combs from my colonies at the end of April, right before the daytime temperatures typically become consistently warm (above 50°F/10°C) and the dandelions begin to bloom, providing the first major honey flow for the season in my area. I gave all the hives ½-gallon of sugar syrup to get them started until the dandelion flow kicked into high gear. I was concerned about whether the bees would be able to gather enough honey and build enough comb to supply themselves for the winter, and I was grateful that, during this particular season, we had a very good honey year. Not only did all the hives draw out their foundation and pack away plenty of honey stores for the winter, but enough surplus honey was produced that I was able to harvest a respectable crop that year. That was the year my honey crop was exceptionally light in color, a condition I attribute in part to the fact that all the honey had been stored by the bees in clean, new, lemon-yellow combs. That was over 10 years ago, and I have been blessed with AFB-free bees ever since, without having to once resort to antibiotic use. I am now extremely careful never to allow any questionable frames or equipment to become mixed in among my hives, and I am vigilant in working to prevent someone else's apiary nearby from spreading the disease to my colonies. My primary defense is the hive autopsy (see the "Beehive Autopsy Results" table in chapter 3 on page 76), because one of the most common sources of disease infection, and reinfection, are the neglected carcasses of dead hives from backyard (hobby) beekeepers close by.

The keys to maintaining foulbrood-free colonies are regular hive inspections and careful examination of every colony that dies for signs of disease. Once your hives are disease-free, regularly removing a couple of the old, darkened combs from the colonies each year and replacing them with frames of foundation will serve to reduce the opportunities for disease spores to build up within a colony's drawn comb reserves. This regular rotation of old combs out of the hive will also provide a steady source of beeswax to be rendered. By the same token, one should make it a practice to avoid frames of comb when purchasing equipment. Unless you know that hard chemicals and antibiotics have not been used on the colony and you have a high level of confidence in the source, accepting used frames is tantamount to taking on someone else's problems.

Should I ever discover another outbreak of foulbrood, I will immediately shake the surviving bees onto foundation and once again destroy all the hive's combs and honey. If this option is not immediately available, though, I will move the diseased hive to an isolated quarantine yard until it is possible to do so. It is important to eliminate all combs that exhibit AFB symptoms from your bee yard as soon as possible to avoid having other colonies nearby become infected by the disease spores.

• CHALKBROOD •

Caused by the fungus *Ascosphaera apis*, chalkbrood (CB) disease attacks both worker and drone larvae and pupae. The cottony, threadlike mycelium of this fungus will entomb the developing brood and form a white mass that will harden and mummify the larvae, causing them to resemble small pieces of chalk—hence the name. These mummies turn black as the fungus matures and approaches its reproductive stage. As a result, both white and black mummies can be found in hives with advanced cases of the disease. The chalklike mummies will tend to collect on the bottom board or by the entrance as house bees clean them out of their birthing-cells-turned-coffins. Chalkbrood appears to be a stress-related disease that will weaken the hive but rarely kill it outright. It's somewhat analogous to the common cold. As with a cold, which can be overcome with quality nourishment like chicken soup, CB will typically go away on its own following a good strong flow of honey. Similar to people, a beehive's natural immunity is strengthened when the bees are well nourished with high-quality food. Alternatively, a honey flow can be simulated by feeding the bees, which will tend to clear up chalkbrood as well.

Incidences of chalkbrood tend to correlate with high moisture and humidity levels within a

FIGURE 9-5. The appearance of dead worker and drone larvae as hard chalky lumps is caused by fungi and is associated with colony stress, colder than normal temperatures in the brood nest, high moisture levels in the hive, and hives that are shaded rather than in full sun.

colony. This suggests that increased hive ventilation may help to prevent this disease. I also find it helpful to keep hives out of the shade and in full sun in order to help reduce moisture levels in and around the colonies. As is the case for many maladies of the honey bee, propagating strains of bees that are resistant to chalkbrood and keeping colonies vigorous and strong are the best ways to prevent infection.

• NOSEMA •

Nosema disease is caused by microscopic organisms (*Nosema apis* and *N. ceranae*) that destroy the lining of the honey bee's gut, affecting the bee's nutrition and shortening its life span. Although the only sure way to diagnose nosema is to examine the bee's gut under a microscope, some signs of nosema infection may be observed in the bee yard, including fecal matter within the hive or excessive droppings by the hive entrance. However,

dysentery may also cause these signs and be confused with nosema. This occurred in my area of Addison County, Vermont, one year during an abnormally dry summer when the bees collected a lot of honeydew that produced a dark honey with a taste that resembled that of brown sugar. Although not a problem in itself, the indigestible matter in the richly colored honey that accounted for its dark complexion caused severe cases of dysentery within the hives that particular winter, which was a hard one for the bees, with long cold periods that kept them cooped up with few opportunities to go on cleansing flights.

It is said that removing the bee's head and pulling out its guts for inspection can provide a fairly accurate diagnosis of nosema. The color of a healthy bee's gut will be pinkish or brownish red, whereas a diseased gut will have a dull gray or whitish color. Historically, *Nosema apis* was the main organism of concern primarily during the winter months, when the bees are confined for long

FIGURE 9-6. Large patches of honey bee fecal matter on or in the hive is one indication that the colony is infected with the microsporidian *Nosema apis*. Hives infected by *Nosema ceranne* however, do not exhibit such symptoms and are unlikely to take food from feeders.

periods without the opportunity to relieve themselves of the indigestible matter that builds up in their digestive systems. Recently, CCD investigations have revealed that *Nosema apis* has been mostly displaced by *Nosema ceranae*, which is widely considered to be more virulent because it can weaken and kill hives even during the warm summer season. While the conventional approach and only federally approved control is the chemotherapy antibiotic treatment Fumagilin-B, an extract from the fungus *Aspergillus fumigatus,* the organic beekeeper would do well to start with resistant stock and work to keep the immunity level of the hives strong by ensuring proper nutrition, frequent comb replacement, ventilation, et cetera. A good wintering site is very important, especially in

northern climates. Hives should be exposed to full sunlight and entrances should face south to encourage bee flights. One way to help reduce instances of nosema is to avoid heavy fall feeding of syrup, because this has been correlated with a higher incidence of the disease. (Although this may simply be the result of nosema spores being passed around the hive during the processing of sugar syrup into honey rather than a consequence of sugar syrup itself being fed to the hive.)

Since I began keeping bees in 1992, I have never used Fumagilin-B or anything else to specifically treat nosema, and neither do most of the beekeepers I know. For some reason, nosema disease, which apparently can be a serious problem in some areas of the United States, has not been much of an issue in my neck of the woods. Whether most of the susceptible hives around me have died off, leaving only nosema-resistant colonies remaining, or the threat to colonies from this disease has been overblown, or I have just been lucky, I cannot say. My experience certainly indicates that, even without approved medications for nosema, the naturally inclined beekeeper can certainly raise strong, healthy bees most of the time.

Should the selection of resistant stock, a favorable apiary site, and good management practices fail to keep nosema at bay, there are organically acceptable options available. First of all, even though nosema spores have been known to live for 6 to 12 months in infected equipment, both species of nosema have proven to be susceptible to cold, with temperatures of 39°F (4°C) for 6 months reducing the viability of both species by over 60 percent.[4] Therefore, beekeepers in northern climates who store their unused combs in an unheated space during winter will enjoy the benefit of reduced nosema spore survival on their combs. At the other end of the scale, *Nosema ceranae* has proven tolerant of high temperatures, while temperatures of 140°F (60°C) are enough to kill *Nosema apis* within 15 minutes.[5]

As a result, frames of comb containing nosema spores can be cleaned up and the frames reused by placing the combs in a solar wax melter that will melt the combs down in the heat of the sun. Models range from a home-built box with an old window on top of it, to solar wax melters purchased from beekeeping supply companies. Research has also shown that exposure to a minimum of two hours of direct sunlight will break the cell membranes of the nosema spores, effectively killing them.[6]

Organic acids are reported to have a detrimental effect on nosema within a colony. Whether formic or oxalic acid, these compounds can kill off nosema spores in the hive, providing an added benefit to beekeepers who use acid treatments to control varroa.[7] The same may be said for beekeepers who use thymol products for mite control. Certain essential oils (such as thymol or emulsified lemongrass and spearmint oils) have the ability to increase the survivability of hives infected with nosema.[8] Along these same lines, the herbal preparation Nozevit, a nontoxic treatment made from oak bark, has shown some success in reducing nosema in some studies, but not in others.[9]

• VIRUSES •

More than 18 different viral organisms are known to infect beehives. This is the highest number of viruses to infect any insect that we are aware of, and many of these viruses are not well understood. They include black queen cell virus (BQCV); chronic bee paralysis virus (CBPV), aka hairless black syndrome; cloudy wing virus (CWV); deformed wing virus (DWV); insect iridescent virus (IIV); Israeli acute paralysis virus (IAPV); Kashmir bee virus (KBV); and sacbrood virus (SBV). To date, no vaccines or medications are officially approved for the treatment of honey bee viruses. These viruses are difficult to deal with because you can't see them and can only confirm their presence through expensive laboratory testing; or, in some cases, the presence of viruses may become suspected when related symptoms specific to a certain virus become obvious in a colony. Such observable symptoms include immature queens that die and turn pale yellow while the queen cell walls turn dark brown or black after the cell is sealed (these are symptoms of BQCV), or newly emerged bees with cloudy wings (CWV symptom) or misshapen wings (DWV symptom).

Once the symptoms of sacbrood virus become widespread enough for the beekeeper to observe, the disease is often too severe for the adult worker population to handle. Symptoms may include partially uncapped cells scattered about the combs, or capped cells that remain sealed after others on the comb have emerged (similar to American foulbrood). The shape of the infected larva often resembles a slipper inside the cell. The larva fails to pupate and turns from pearly white to pale yellow, to light brown, and finally dark brown; the head of the larva is the first part of the body to darken. The skin becomes tough and the body of the larva becomes watery, so that the pupa appears as a sack of fluid. In its later stages, the dark brown individual becomes a wrinkled, brittle scale that is easily removed from the cell (unlike AFB).

CBPV, or hairless black syndrome, can be identified by adult bees that exhibit an abnormal trembling motion of the body and wings. The bees are incapable of flight and can be found crawling up stems of grass near the hive. The abdomens of these infected adult bees may be bloated and their wings partially spread and dislocated resembling the letter K. The bees may also appear shiny and greasy due to their lack of hair. Such bees are often harassed by guard bees at the entrance to the hive; this behavior can be confused with signs of robbing.

The best defenses against viruses are cultural controls and disease-resistant bees. Requeening an infected colony will often help eliminate symptoms. The regular replacement of old comb within the hive can also help prevent viral organisms from building up in the beeswax. Since varroa mites are known to spread certain viruses while feeding, keeping mite populations under control can help prevent the outbreak and spread of many viral diseases. While yet unapproved, the only treatment that has shown some level of effectiveness in helping bees survive against viruses in the field is the drenching of infected colonies with emulsified essential oils of lemongrass and spearmint mixed with sugar syrup (see "Essential Oil Treatments" later in this chapter on page 202).

• COLONY COLLAPSE • DISORDER (CCD)

As the original version of this book was going through its final editing phases and being prepared for the printer (winter 2006–2007), word started traveling around beekeeping circles about the alarming and sudden loss of tens of thousands of managed bee colonies. This phenomenon is characterized by a rapid collapse in which the vast majority of the adult bees disappear from the hive.

Hives that seem to be fine one day are, two to three weeks later, found dead and relatively empty. If any living bees are found in the hive, it is usually only the queen and a handful of what appears to be relatively young bees that remain. These bees are not healthy. The immune systems of the remaining bees are severely stressed and the bees are suffering a raft of viral and fungal infections.

Once hives die out completely, no significant numbers of dead adult bees are found in or around the collapsed colonies, as would normally be the case. To add to the mystery, these hives typically have capped brood and plenty of honey and pollen stored away in the combs. Strangely enough, these honey stores remain initially unmolested by neighboring bees that would normally rob out such bounty with little delay. Even more interesting is a noticeable interval of up to two weeks or so before the usual parasites (wax moths and small hive beetles) move in and feed off the carcasses of the formerly bustling hives. The presence of the remaining capped brood seems to eliminate absconding as an explanation for the missing bees, because bees are not known to abandon a hive until all the capped brood are hatched.

Since its inception, CCD is often blamed for any colony that dies out with relatively few bees left in the hive. This prognosis is most common among beginner beekeepers. However, though many hives are indeed dying from CCD, many more hives are dying from other causes, some of which may mimic the symptoms of CCD. Prior to the 1980s, beekeepers would lose an average of about 3 to 5 percent of their hives in a bad year. After varroa arrived in the late 1980s, the average yearly/ winter loss rose to between 15 and 20 percent. Since the winter of 2006–2007, the average losses experienced by the beekeeping industry have risen to between 30 and 40 percent, with the exception of the past year (2011–2012), which saw an extremely mild winter throughout most of the United States and winter losses of only about 22 percent.

The telltale sign that CCD, and not some other issue, is indeed affecting a hive is an insufficient number of bees left in the hive to be able to cover and care for the developing larvae and pupae in the brood nest. Since the amount of brood, and the size of the brood nest, that can be raised and cared for is in direct proportion to the number of bees in the colony, too few bees to care for the brood is evidence that in the recent past many more adult bees existed in the hive, even though the bees have since disappeared.

No area in the United States seems to be immune to the ravages of CCD, with only the twelve states of Alaska, Hawaii, Nevada, Arizona, New Mexico, Nebraska, Kansas, Louisiana, Alabama, West Virginia, Vermont, and Maine free of any confirmed reports of colony collapse as of December 2009.[10] Despite being well known and widespread for the past five years, this significant blow to the beekeeping world is still poorly understood and threatens both honey production and the pollination industry in both the United States and in many other parts of the world. All of this is causing some people to openly wonder whether the US beekeeping industry can survive. Some states, such as Pennsylvania, now have less than half the number of managed colonies they had just 25 years ago.[11]

Researchers are beside themselves trying to figure out what exactly is going on. Dead bees and samples of comb, brood, honey, and pollen from CCD colonies and surviving hives from infected apiaries are all being collected, inspected, injected, and dissected in an effort to figure out the cause. Surveys have been carried out among both affected and non-affected beekeepers in an effort to evaluate the various management practices in use in case the way hives are being cared for is part of the issue.

So far the common thread that has revealed itself as a result of these investigations is the role of stress in bringing about an environment conducive to the onset of CCD symptoms. Common among the affected beekeepers interviewed is that all their operations suffered "some form of extraordinary stress" at least nine weeks before the initial hive deaths were observed.[12] The nature of this stress

varies widely: overcrowded apiaries, forced crop pollination on forage with minimal nutritional value, and a serious dearth of nectar and pollen, all three of which in turn lead to nutritional deficiencies. Additional stresses include drought or contaminated water supplies and the damaging effects of varroa mites. The one common factor is that all these conditions adversely affect the immune system of the bees, leaving them all the more susceptible to viruses and other disease organisms.

Added to all this information is the fact that every operation affected had used antibiotics in an effort to keep the bees healthy, and the vast majority had used a chemical miticide at some point.[13] The temporary support that such chemical and drug use offers the hive, combined with the sublethal buildup within the combs of toxic components of various compounds and the synergistic effects of more than one chemical or drug combining together, further compromises the natural immunity of honey bee colonies. Initial investigations of the numerous samples collected from these dead colonies have revealed a variety of pathogens within the hives and within the bees themselves. As a result, CCD cannot be attributed to any single disease organism, and because no specific cause has been identified as the source of the colony deaths, the new epidemic has been called a *disorder* rather than a disease.

From the natural beekeeping perspective, CCD sounds suspiciously like the results of the industrialization of beekeeping coming home to roost. The constant strain on hives that have been repeatedly pushed to their limits is bound to take its toll sooner or later. It is interesting to note that scientific and anecdotal evidence so far suggests that beekeepers practicing natural or organic apicultural techniques are less likely to experience CCD-like symptoms within their hives.

As I suspected and indicated back in 2007, at least part of what we are observing here is the result of the fouling of the interior of the hive's environment from the toxic pollutants and pesticides that are brought back to the hive by foraging bees, combined with the buildup of chemical agents used to control parasitic mites and other hive pests. The use, and abuse, of synthetic chemotherapy mite controls is helping to create conditions wherein the combs are no longer healthy places for bees to make a home. The key clue that points to this possibility is the reluctance of other bees and parasites to move in and scavenge over the remains of these dead colonies. Animals intuitively know which foods are good for them and which are not.

This situation reminds me of the anecdotal reports from farmers who notice that their livestock, when given the choice between conventionally grown grain and grain grown from genetically modified (GM) seed stock, will shun the genetically modified feed in favor of the conventional food every time. Such animals are observed to eat GM grain only when they are given no other choice. By leaving a stick of butter and a stick of margarine next to each other outside, we can observe a similar phenomenon. Birds, insects, and other creatures will devour the butter, which is full of vitamins, minerals, protein, and fats, and will ignore the margarine, which is composed of hydrogenated oils and trans-fatty acids and is generally so lacking in nutrition that it takes artificial coloring, multimillion-dollar advertising efforts, and a multi-pronged disinformation campaign to fool the consuming public into eating the stuff and thinking it's good for them. Could it be that the bees are telling us, in a similarly unambiguous way, that the methods we are using to manipulate hives full of bees, brood, honey, and bee bread, although well intentioned, are actually condemning the bees to death? By pushing our hives to be economic engines for honey production and pollination services, are we forgetting to observe the basic necessities the bees require in order to fully live out their lives in a healthy and productive manner? Granted, major losses of bees have been experienced by beekeepers in the past. However, none of history's previous massive honey bee dieoffs are reported to have been accompanied by the same set of symptoms we are seeing now.

I have since come to refer to the primary causes of CCD as the five P's: progeny (inbreeding), poor nutrition, pests, pesticides, and pathogens. As I initially suspected, and will explain more fully in the rest of this chapter, it is our industrial agricultural model, which is more about making money than producing healthy and wholesome food, that seems to be the source of much of the problem. Huge monoculture farms and feedlot beekeeping operations can create nutritional stress, which along with inbreeding and pesticide contamination can lead to increased pest pressures and ultimately death from diseases.

Essential Oil Treatments

Many beekeepers have turned to essential oils (EOs) to control the ravages of colony collapse disorder. The reliance on EOs by beekeepers experiencing CCD symptoms in their hives first came to my attention when I received a phone call from Michael Meyer of Springfield, Missouri. Michael manages about 700 hives and lost about 70 percent of them over the winter of 2008–2009 due to CCD-like symptoms. In an act of desperation, Michael mixed up a batch of Honey-B-Healthy (H-B-H), a commercially produced mixture of emulsified lemongrass and spearmint oils, with sugar syrup at four times the recommended feeding concentration (4 teaspoons per quart of sugar syrup instead of a single teaspoon). Since the bees would not eat such a strong mixture, he simply dumped about a cup in each hive, wetting the bees and the combs. This forced the bees to take it up as they licked themselves and their combs clean. Michael observed that all work in the hive stopped for about a half hour as the bees dealt with the mess, but otherwise it didn't seem to hurt the bees and their state of health promptly turned around. Michael indicated that others were having similar results.

After speaking with Michael, I started to make some phone calls of my own. I spoke with a number of other migratory beekeepers who kept anywhere from about 700 to over 70,000 hives, shuttling them around North America: from

Florida and Texas to Washington state; from Michigan to North and South Dakota, and even working up into Canada. Since there is no scientifically developed standard for treating hives with this essential oil blend, treatment protocols varied. One beekeeper mixed 1 gallon of H-B-H with 55 gallons of sugar syrup and gave each hive 8 ounces. Another beekeeper mixed a pint of the emulsified essential oils into a 5-gallon bucket of syrup and gave each hive 10 to 12 ounces, while a third drenched the bees three times with about a cup, each treatment spaced out four to five days apart. No matter what formula they used, they all reported that the treatments seemed to make the bees healthier and helped the colonies deal with varroa, viruses, nosema, pesticide exposure, chalkbrood, and queen issues that can arise when treating hives with organic acids, in addition to preventing mold growth in the sugar syrup.

Upon further questioning, I found out that a number of these beekeepers were also dealers of Honey-B-Healthy, and the fact that they were reporting such good results with essential oils while also selling it made me initially rather skeptical. After all, these anecdotal reports are all well and good, but where is the science-based evidence for the efficacy of using essential oils to kill viruses, molds, and fungi, to help prevent queen rejection, control nosema, and aid in reducing varroa issues in the hive?

Well, it turns out I was able to find enough studies to suggest that these varied reports are more than just old wives' tales (or old beekeepers' tales, as the case may be). Being highly concentrated, plant essential oils have proven antimicrobial, antifungal, and antiviral properties. Lemongrass oil in particular has been shown to have significant antifungal activity[14] and strong antiviral properties.[15]

Lemongrass oil also contains two honey bee pheromone components: *geraniol,* a major component of the Nasonov pheromone; and *citral,* a minor component of the Nasonov pheromone[16] (the pheromone with a scent attractive to bees and released by a worker bee's Nasonov gland). This would help explain successful reports of using a couple drops of lemongrass oil on a cotton ball as a lure in baited swarm traps. As it turns out, geraniol is one of the Nasonov pheromone components that have been shown to cause varroa mites to become confused and disoriented.[17] This could help account for the reports of H-B-H having a detrimental effect on varroa levels in hives.

Then I contacted David Wick of BVS, Inc. located in Florence, Montana. David's company screens and detects all types of viruses using an Integrated Virus Detection System (IVDS). The IVDS is an expensive detection device engineered by the US Army that can detect virus particles by their distinct size. Originally built for virus screening of humans, it is proving to have tremendous value in furthering our knowledge of colony collapse disorder, since one of the consistent factors in CCD is the presence of a plethora of honey bee viruses and other pathogens. As David reports, honey bee viruses can live on the surface of combs. Some of the viruses are quite hardy and can survive on combs for a long time, waiting for the right moment to replicate themselves. According to David, the viruses can even live entombed within the beeswax and not just on the surface of the comb.

So far David's work has resulted in observational data that sampled the virus loads of 20 hives over a period of 10 months. The number of the viruses in the hives skyrocketed after the hives were used for pollination of commercial almond groves and then dropped dramatically following a treatment that utilized LaFore's Essential Oil Patties, a product manufactured by Jeff LaFore of Milton-Freewater, Oregon, that contains a mixture of nine different EOs. At this point the data is only observational, and David does not know for sure if the EOs may be affecting the bees, the virus vectors (such as varroa), or just the viruses exposed on the combs.

David obtained grants to run experimental trials on essential oils and, as a part of these trials, he collected data specifically on the Honey-B-Healthy and LaFore essential oil products. He is still evaluating and analyzing the data at the time of this writing. I told David about beekeepers' reports of reviving hives that were crashing with CCD-like symptoms through the use of H-B-H drenches. David noted that such reports are consistent with what he has observed, and he hopes to have hard data to back up his observations soon.[18]

I also had a chat with Dr. Frank Eischen of the USDA Agricultural Research Service out of Weslaco, Texas. Dr. Eischen has used H-B-H in some trials and has found it to have some value against *Nosema ceranae*, though it was not as effective as Fumagilin-B. During the trials he conducted H-B-H was used only at the regular feeding strength, and he indicated an interest in evaluating the product at the higher drench concentration levels in the future.

At this point it would seem that EOs are what is keeping a large part of our industry healthy in the face of CCD. It should be no surprise to readers that as one who has championed the use of natural approaches over toxic chemicals and drugs, I consider these recent developments welcome news. However, we must remember that, by using EOs, we are dealing only with the symptoms of our problems and not the cause. Even if further testing of EOs proves that, when poured or sprayed onto bees, they are effective in killing viruses, fungal infections, molds, and other disease causing organisms while simultaneously confounding varroa, relying on these products is not ultimately the solution. While it's a move in the right direction, eventually we need to resolve the underlying problem by figuring out how best to deal with all that is stressing the bees' immune systems to the point where viruses and diseases threaten to gain the upper hand. Meanwhile, essential oils seem to be buying the industry time.

Stress and CCD

It appears that the combined impact of numerous stresses to the immune system, acting independently and synergistically with each other are the key factors bringing on the symptoms of CCD in hives located in America and around the world. These stressors can be divided into two basic categories: stress factors that we as beekeepers have a good deal of direct control over, and stress factors that we have minor, or indirect, control over.

Causes of Colony Stress over Which Beekeepers Have a Lot of Direct Control

ARTIFICIAL FEEDING

Honey bees are designed to eat just two things: honey and pollen. Honey is produced from the nectar of flowering plants. Bees add beneficial bacteria and enzymes to nectar in order to prevent it from spoiling while it is being processed into honey and also to break down the sucrose in it (into primarily glucose and fructose). Pollen collected by bees is made more digestible through a process of fermentation that utilizes the action of beneficial bacteria in the honey bee's digestive system to turn the pollen into bee bread. Unfortunately, when beekeepers feed their bees, they usually give them less-expensive substitutes such as syrups, made from sugar and corn syrup, or protein patties, which are often made with soy flour or brewer's yeast. Such artificial dietary regimens used only occasionally are fine for halting the immediate stress of starvation but are not suitable for use on a regular basis if healthy bees with strong immune systems are the goal. These substitutes do not contain the same balance of micronutrients, vitamins, minerals, and enzymes. The beneficial bacteria that live in the hive, and that are critical for helping the colony convert nectar into honey and pollen into bee bread, can not grow on artificial protein patties like they do on pollen.

Beekeepers can't be fully blamed for the train of thought that leads them to raise bees on an artificial diet. Thanks to the influence of advertising, we often feed our families and ourselves with items made from products and ingredients that bear little-to-no resemblance to real food, so it seems perfectly natural for us to treat the bees similarly.

However, as mentioned in chapter 3, I view artificial feeding of bees in much the same way as I view human consumption of fast foods or junk foods. Feeding hives an artificial diet to get them through a period of starvation stress is one thing, but to feed bees in this way on a regular basis fails to provide them with an ideal diet that leads to robust health and vitality. When bees undergo dietary stress, they become more susceptible to disease. Research has shown that colonies fed protein patties rather than pollen collected by foragers failed within six to eight weeks, and that, unlike natural pollen, protein patties are not stored and used as a critical winter and early spring protein source.[19]

PESTS

While varroa mites are the primary challenge for most honey bees these days, numerous other pests can also impact a colony and stress its immune system. Mites and other pests must be kept under control or the health of the hive will suffer. In this day and age, where the natural world is so out of balance, it appears that an extremely limited number of hives are able to survive long without assistance in areas that are no longer pristine. Thus, it is up to beekeepers to do what they can to reduce colony stress caused by hive pests, ideally using methods such as those found in this book, that reduce or eliminate additional stresses brought on

FIGURE 9-7. Dead bees with shrunken abdomens and sealed brood with pinholes in the cappings are potentially signs of parasitic mite syndrome (PMS).

by the treatments themselves. One of the most valuable things a beekeeper can do to deal effectively with any pest is to learn as much as possible about it. Once you are armed with information about biology, habitat, diet, instincts, et cetera, you will be able to figure out how best to deal with any issues that pest may create.

DISEASE

Since the advent of CCD, a number of new honey bee diseases and viruses have been discovered. Upon further investigation, however, it turns out that many of these "new" diseases are actually not new at all, but have been around for many years— we just didn't realize they were there because we were not looking closely enough. This chapter covers many of the most common diseases that can cause colony stress and immune dysfunction. There are other diseases with which I have had little contact, including sacbrood, European foulbrood, and stonebrood, that are also potential issues for a hive. It is unlikely that even the most disease-resistant bees will be resistant to all the potential bee pathogens that a hive may be exposed to. Familiarity with the symptoms and treatment/ prevention options for diseases, combined with regular hive inspections, give beekeepers a chance of keeping on top of disease problems.

INBREEDING

There are some three or four dozen major queen breeders and bee suppliers in the United States, primarily located in the southern United States and California. Even though there are also dozens of smaller breeders, most of the bees that are raised and sold commercially are not all that widely dissimilar genetically.[20] Add to this the fact that some research has indicated that fewer than 500 mother queen bees are used to breed around 900 thousand daughter queens each year for commercial sale in the United States, and the opportunity for inbreeding stress to occur is significant.[21] This is especially true for beekeepers who order most or all of their queens from only one supplier, and do so year after

year. The chances are strong that all those queens will be very closely related. Beekeepers who wish to minimize the potential for limited genetic diversity within their apiaries should vary the sources where queens are purchased as much as possible.

ANTIBIOTIC USE

The use of antibiotic drugs such at Terramycin (oxytetracycline), Tylan (tylosin), or Fumagilin-B to control various diseases in honey bee colonies is problematic because antibiotics are indiscriminate and will wipe out much of the good bacteria in the hive as well as the harmful bacteria. Research has only recently begun to establish that the use of antibiotics against hive diseases such as American foulbrood and nosema not only has the potential to weaken a hive's resistance to other diseases,[22] but increases the sensitivity of honey bees to harm from pesticide exposure.[23]

CHEMICAL PESTICIDE USE

As mentioned earlier, the use of toxic chemicals to control mites and other hive pests is a significant source of stress to a colony's immune system. Luckily there are numerous natural and organic alternatives to toxic chemical controls such as Apistan and CheckMite+. Even though some of these alternative approaches may have sublethal effects of their own upon a honey bee's health, their impact appears to be less than that of the hard chemicals.[24] See "Soft Chemical Treatments for Varroa" in chapter 5 on page 145 for more information on these alternatives.

Causes of Colony Stress over Which Beekeepers Have Only Minor, Indirect, or No Control

GEOMAGNETIC STORMS AND FLUCTUATIONS

It has been established that honey bees are influenced by magnetic fluctuations.[25] The crystalline substance known as magnetite (Fe_3O_4) in the bee's abdomen acts like a compass needle, helping bees to orient themselves in magnetic fields. Researchers have observed a remarkable correlation between increases in disturbances to the Earth's magnetosphere caused by the sun's solar eruptions and honey bee losses. Three separate time periods were looked at: the autumn/winter period of August through March, 1995–1996 and 1996–1997; the six autumn/winter periods between 2000 and 2006; and August through March of 2006–2009. During all three periods, as geomagnetic storm activity increased there was a corresponding increase in colony losses in the United States.[26] Since bees have the ability to use magnetic reception for orientation purposes during foraging flights, the potential for disorientation during solar storm activity is present. Researchers point out that honey bees becoming lost during foraging corresponds with the

primary CCD symptom of the disappearance of workers from the hive.

ELECTROMAGNETIC AND MICROWAVE RADIATION

It is well established that living organisms are affected by electromagnetic radiation fields (EMFs) and microwave radiation, such as that emitted by a cell phone.[27] As discussed above, honey bees can detect low-frequency magnetic fields and potentially make use of Earth-generated fields during navigation. A 2011 study found that active mobile telephones placed directly inside a hive induced worker piping signals, which are known to either announce the swarming process of a colony or signal a disturbed beehive.[28] Other research has

FIGURE 9-8. Dandelions are the first major honey bee nectar source to bloom in spring throughout much of America's Northeast.

shown that when a mobile phone is placed inside a hive, workers fail to return to the hive after only 10 days.[29] The same study also found that the energy emitted by a cell phone affected egg laying by the queen; the study recommended recognizing EMFs as a pollutant. It is important to note, however, that while previous studies have found that hives placed under high-voltage lines become impaired, these cell phone studies are not representative of real-life EMF exposure by honey bees to cell phone emissions and therefore fail to make a direct link between cell phones and CCD.

CHEMICAL POLLUTION AND THE ANTHROPOLOGICALLY INDUCED CLIMATE CRISIS

Pollution of our air, water, and land has long been recognized as a problem, and yet our air continues to become dirtier, our water filthier, and our land more contaminated. Research has shown that air pollution can interfere with a foraging honey bee's ability to smell sources of nectar, and that polluted water collected by bees can have a toxic effect.

The idea that the collective actions of humanity can increase the level of carbon dioxide in the atmosphere by about 40 percent and yet humanity does not expect this change to have a dramatic impact upon the planet is beyond me. As mentioned in chapter 3, the bloom time of many plants is changing, and this is affecting pollinators of all kinds. Recent updates to the USDA's plant hardiness zone map are a further indication of the Earth's rapidly shifting climate patterns. Here in Middlebury, Vermont, winters have become less severe in terms of cold and the area's plant hardiness zone has been reclassified from zone 4a to zone 4b.

INDUSTRIALIZATION AND CORPORATE GLOBALIZATION OF AGRICULTURE

The industrialization of society's economic activity, which may be beneficial in some industries, is proving to be a failure as it is applied to agriculture, an inherently biological activity. Industrial agriculture received a big boost during the Green Revolution that began in the 1940s and resulted in a 30-year span of research and development, combined with technology transfer initiatives designed to increase crop production. Unfortunately, this effort, while economically enriching a small group of businesses and individuals, will never be able to fully feed a world containing seven billion people and counting. This technology-driven form of agriculture, which relies on monoculture and large-scale mechanization, has proved itself too wasteful and inefficient. It is one of the primary causes of the growing degradation of our world's water quality, destruction of the natural fertility of the soil, increased worldwide soil erosion, loss of the planet's biodiversity, and global climate instability, both through the destruction of forest land for agricultural use and the production of greenhouse gases.[30]

Contrary to the numerous government and business reports that praise the efficiency of the modern agricultural system, the transition from internal-input to high-external-input agriculture has resulted not in an increase, but rather a decrease in productivity of over sixtyfold in the last 50 years.[31] The majority of people are not aware of this, in part because most studies and reports that are released on the subject focus on the amount of human labor required to produce our crops. Little, if any attention is paid to the collective costs society pays for these gains in labor efficiency that ultimately benefit only the few. However, more and more reports are being published that take into account a complete picture of our modern industrial agriculture system. These reports clearly indicate that the only way to feed our growing human population is by relying not on more industrial agriculture, but on smaller, human-scale, sustainable agricultural efforts that embrace practices like those performed by many organic, biodynamic, permaculture, and subsistence farmers throughout the world.[32]

Our current agricultural system is so broken and bankrupt that, without the subsidies it receives from the world's governments, it would collapse astonishingly fast. These subsidies are both direct agricultural subsidies and indirect subsidies that artificially deflate the cost of diesel fuel, farm chemicals, and artificial fertilizers. Backing up the subsidies are a labyrinth of laws designed to remove the common person's access to land, clean water, clean air, and seeds and put them into the hands of corporate agribusinesses. International treaties and laws codify the ability to sell crops below the cost of production, pushing small-scale farmers off their land, while at the same time preventing individuals from processing crops and butchering livestock without "approved" facilities, ostensibly as a public health consideration. Seed saving and planting has also been outlawed in many circumstances under the guise of protecting intellectual property rights.

A good example that illustrates the folly of industrial-scale farming is the yearly pollination of the California almond crop. This is the largest single commercial pollination event in the world. There is one area of California happens to have the perfect soil conditions and temperatures for growing almonds. It is a desert region, so there is not enough water to grow lots of almond trees there, but large amounts of water are diverted from the Colorado River for irrigation. As a result, about 750,000 acres in California produce over 70 percent of the world's almond crop. The price of almonds is on the rise, and thus another 100,000 acres of trees have been planted that have yet to mature and begin bearing nuts. This brings the total area planted in almond trees to around 850,000 acres, an area about the size of Rhode Island.

Now, to ensure maximum nut set, and meet crop insurance standards for pollination, growers like to have a minimum of two hives on each acre during bloom time. With the current estimate of 750,000 acres in production, about 1.5 million hives are needed in California during the months of February and March when the almond trees flower. Close to 1.7 million hives will be needed in the near future when the new plantings of trees reach maturity. Unfortunately, according to official USDA figures, there are only about 2.6 million honey bee hives in the entire United States. Since only a

small percentage of the hives are kept in California year-round, hundreds of thousands of colonies of bees have to be brought in to meet the pollination needs of almond growers, often from other states over a thousand miles away, and sometimes even from other countries.

Given the design of our capitalistic society, with its focus on profit, we don't often get what is best for society, but we do get what people can make money doing. To make money moving bees, the job must be done en masse, which is very stressful on the colonies. Hives are kept on pallets, usually in groups of four to eight, depending on how they are being managed, so they can be picked up quickly and easily by a forklift and loaded onto tractor-trailer trucks. A single 18-wheeler can hold 480 or so hives, depending on how they are loaded. Since it is not good beekeeping public relations to leave behind a trail of angry bees as you drive down the highway, a large net is thrown over the back of the truck to keep the bees contained. During the process of getting loaded and being jostled and bumped as the trucks make their way down the road, many of the bees come out of their hives to see what all the commotion is about. Of course, with hundreds of hives all on the same truck, the chances that these bees are going to find their way back into their *own* hive is slim to none. Thus, the bees get entangled in the net and buffeted by the wind and the rain. During the day the sun shines down on the hives and can cause them to overheat. To prevent overheating, water is often sprayed over the hives to cool the hives by evaporation. At night, temperatures often drop dramatically. This leads to wide temperature swings inside the hives; the bees try to adapt, but such conditions can interrupt the queen's egg laying. The process of long-distance travel is so stressful on the bees that it is not uncommon for 10 percent of the hives to die in transit. Beekeepers have learned to expect such losses, however, and they factor this into their business plan.

Once the trucks arrive in California, the hives have to be unloaded so they can be inspected and the dead hives weeded out. Hives that have lost too many bees and are too weak to meet the pollination contract specifications are equalized with frames of bees and brood taken from stronger hives. To accomplish this, the hives are all unloaded into what are called "holding yards"—large fields where up to tens of thousands of beehives are placed temporarily while the dead ones are sorted out. Remember now, this is February, the end of winter in California, and there are no plants in bloom at this time of year. As a result the hives that are brought in must be fed until the almond trees start to bloom. As stated earlier, beekeepers don't normally feed their bees what they are meant to eat—honey and pollen—and the colonies are instead given less expensive substitutes.

Finally, once all the hives are inspected, equalized, fed, and ready to go and the almond trees start to bloom, the colonies are moved into the almond orchards and the bees get to forage on and eat real nectar and pollen. However, since nothing is blooming at that time except the almond trees, the diet that the bees must live on is extremely limited. Bees are like us: they need a varied diet in order to obtain the proper nutrients needed to be healthy. Limiting their diet to a single crop is additional dietary stress. Also, because so many hives from different parts of the country are all brought into this one area at the same time, there is the increased opportunity for diseases and pests to spread among the hives.

Given the ordeal that these poor bees are put through, it is a wonder that they have lasted this long and we haven't seen large-scale problems sooner. Beekeepers are aware of the stresses that they are placing on their bees, and they really don't want to treat their colonies this way, but they have no choice. This is the industrialized system we have collectively created over decades to produce the majority of our food. Huge monocultures, combined with liberal herbicide use, have greatly diminished food sources for native pollinators. Combined with the destruction of available habitat for nesting sites, the fragmentation of pollinator

migratory routes, and the wholesale application of insecticides that indiscriminately kill the good bugs along with the target species, the loss of natural food sources has decreased native pollinators in most areas to the point that honey bees must be brought in during bloom or the crops won't get pollinated and the food system that Western civilization relies upon will collapse. Headlines such as BEEKEEPERS STARVE NATION would appear in our nation's newspapers should beekeepers refuse to expose their bees to such stressful situations. There really is no other option for America's beekeepers.

Given the wide variety of factors that contribute to colony stress and that are implicated in the phenomenon known as CCD, a growing number of researchers, academics, and beekeepers are pointing to a combination of variables that are behind the recent spate of colony collapses. Similar to Acquired Immune Deficiency Syndrome (AIDS), which has infected people all over the world, bees are being stressed by so many things that their immune systems are getting overwhelmed and are simply collapsing. Ultimately, I think chemical contaminants may play a larger role in the increase in honey bee die-offs seen in recent years than has been widely acknowledged up until now. Looking ahead, while we may see a decline in reported cases of CCD, primarily due to efforts to increase the overall health of our honey bee populations, it's my opinion that we are not likely to see CCD go away completely anytime soon.

I tend to believe, like many today, in the current thinking regarding CCD: that it is a result of multiple causes. However, one small detail about this theory has been bothering me for some time. Nagging from the back of my mind is the one characteristic of CCD that tends to be overlooked when discussing the cause of colony collapse. This characteristic typically appears at the bottom of the list of CCD symptoms: the notable delay in robbing by other bees and the slower-than-normal invasion of the collapsed hive by common pests such as wax moths, small hive beetles (SHBs), wasps, and hornets. As far as I can tell, this symptom differentiates the current mass die-off from previous large-scale bee losses of the past.

While I admit it is theoretically possible that honey bee diseases may have an impact on hive scavengers, and research is ongoing, to date no honey bee disease has been found that also affects other insect species, with the exception of other closely related species such as other bees and wasps. Even a clearly recognizable disease such as American foulbrood with its notorious foul odor is not enough to deter wax moths or beetles. Now I know there are those who will disagree with me, but I find it doubtful that we would be seeing this symptomatic delay in scavenging in such a variety of species if a honey bee disease were the driving force behind CCD. Don't get me wrong: I am not saying that diseases are not involved in the phenomenon known as CCD. The bees are dying, and they are often dying from diseases as far as anyone can tell. The real question is why are the bee's immune systems becoming so vulnerable to diseases all of a sudden?

One potential candidate for the cause of CCD that is currently being evaluated, and could potentially explain the symptoms of delayed scavenging, is toxic chemical contamination, which is primarily caused by industrial agriculture. Unlike the other suspects, chemicals have already proven themselves capable of repelling bees, as anyone who has used a fume board to harvest honey can testify. As a result, it is much more likely that chemicals are also the factor that repels wax moths and SHBs and are the underlying cause of the current worldwide honey bee die-off.

Research at Penn State University in 2007 identified 46 different pesticides and their breakdown products in samples of bee pollen, with as many as 17 pesticides in a single sample.[33] Such toxic compounds found in the hive are potentially from numerous sources, including chemicals from industrial agriculture, toxins emitted from the manufacture and use of consumer goods (including automobiles), industrial pollution, genetically modified organisms, and chemicals that beekeepers

use in and around the hive. Governments around the world have set up various agencies to regulate such chemicals. Even the bees seem to recognize when pollen is loaded with high levels of pesticides, and they have been documented trying to seal off the contaminated pollen from the rest of the hive by entombing the cells filled with such pollen with a layer of wax and propolis.[34] Unfortunately, efforts to protect the public and environment from harmful chemical contamination have been a dismal failure. However, it really isn't the fault of our regulatory agencies that their efforts to protect human and environmental health from potentially harmful chemicals are not even coming close to working.

First of all, government and regulatory agencies have limited funding and staffing and therefore tend to rely on the corporations that manufacture the chemicals being considered for registration to carry out the trials and provide the data that is used to evaluate the compound's safety. Talk about your conflict of interest! Corporations have a strong incentive to massage and tweak test results and design studies that will result in data favorable to their product, and they do so.[35]

When chemicals are evaluated for toxicity, they are studied in isolation. Little thought is given to the chemical's breakdown products, which can prove to be more toxic and longer-lasting than the original chemical itself, such as in the case of imidacloprid olefin, a substance produced as the systemic neonicotinoid insecticide called imidacloprid degrades. Once in use and released into the environment, chemicals and their breakdown products will combine with other chemicals already in the environment to form new compounds. The synergistic effects of some of these combinations have proven themselves to be hundreds of times more toxic than either compound on its own.

In addition, recent research into endocrine-disrupting chemicals (the kind often used as pesticides) reveals that the timing of exposure combines with the amount of exposure to produce a chemical's effect.[36] A certain dose of a chemical might be very toxic to an organism in its developmental stage, while not having any detrimental effects on the organism once it has matured, or vice versa. To make matters worse, sometimes a lower dose of a chemical proves to be more damaging than higher doses. While totally counterintuitive, such new understandings of chemical toxicity have proven wrong Paracelsus's roughly 450-year-old saying, "Poison is in everything, and no thing is without poison. The dosage makes it either a poison or a remedy." Today we know that often the timing is what often makes the poison,[37] and that sometimes less is actually worse.[38]

Add to this the many studies that now show that a cocktail of "insignificant" doses of several chemicals, each acting on its own, can combine to have significant results. In other words, exposure to very low concentrations of several chemicals at the same time can cause biological effects that none of the chemicals would have on their own.[39] Thus, when living organism is exposed to a mixture of chemicals, every component contributes to the overall effect, no matter how minute their concentration.

All of this makes the task of toxicity testing so complicated that, realistically, no chemical will ever be thoroughly tested for safety, either for humans or bees, before being manufactured and marketed. To do so we first would need to know which biological tissues or functions the chemical affects, in what ways, at what potencies, and whether vulnerable populations will be exposed to other chemicals that affect the same tissues or biological functions. Next would be tests of groups of chemicals in combinations at low and high doses, and several doses in between. We would then have to determine whether the creature being studied (mouse, human, honey bee, or whatever) is impacted by this combination of chemicals at one particular stage of life or another. In humans, for example, during gestation in the womb, exposure to certain chemical drugs during one particular week can produce effects not seen when exposure occurs during a different week.

However, none of this testing takes into account the potential synergistic effects of the multiple compounds that already exist in the environment. For example, suppose one wanted to test the synergistic actions of just 500 toxic chemicals in unique combinations of five chemicals each. A little mathematics indicates that over four trillion possible groups of chemicals would need testing. Even at the wildly optimistic test rate of a million combinations each year, it would take over four thousand years to finish the task. Consider that hundreds of new chemicals are introduced into commercial channels each year, and the enormity of trying to regulate these toxins becomes increasingly apparent.

At this point you might be asking yourself, how on Earth did this situation come about? How could we have managed to allow the impacts of our various economic activities to add up to a world so damaged that the Earth's natural capacity for self-renewal is being exceeded and permanent degradation has become evident? And, more importantly, where might we begin in our efforts to fix this mess?

One reason for the current chemical regulatory morass is that regulated chemical industries have influence over the formation of the regulations by which they are to be governed. The revolving door of chemical corporation executives becoming EPA officials, and then returning to their corporate jobs once a new administration takes over, is legendary. In addition, our legal and regulatory systems were never designed to limit the accumulation of small impacts. Instead, laws typically rely on cost-benefit analysis to justify individual impacts, a practice that has become obsolete as it destroys the planet as a place suitable for honey bee (and human) habitation.

The importance of addressing the colony collapse issue lies in the fact that it is not just honey bees that are disappearing. About a year after CCD hit, I was listening to the radio and a program came on that featured a professor who studies bats in New York state. He was talking about a sudden die-off of bats that he was witnessing. The dead bats could not be found in or around their winter caves as would normally be expected. Even more intriguing was that fact that many of the bats that were found still alive were infected with a white nose fungus, a new disease that researchers had not seen before.

While listening to this, I started thinking to myself, "This sounds a lot like CCD." Thanks to the Internet, I was able to track down this professor and call him to discuss his findings. During our conversation he mentioned that a colleague of his that studies moths was seeing a similar dramatic decline in population numbers. It turns out that practically all the world's pollinators are in decline: bees, moths, butterflies, birds, bats, et cetera. In fact, folks who study this stuff are discovering that it is not just pollinators that are dying in unprecedented numbers, but today, as I write these words, there is a rate of global species extinction affecting plants and animals that rivals the time of the dinosaurs and threatens the ability of human society to continue.[40] It has been estimated that as many as 70 species a day are disappearing from the face of the Earth and that the current rate of species extinction is 100 times or more that of the natural or average extinction levels.

When most people hear about this situation, they feel a sense of temporary sadness that usually passes quickly. This is because they tend not to think that species extinction affects them directly. But the fact is, it does. It is diverse species that form ecosystems, and it is the ecosystems that are responsible for such activities as purifying the air in the atmosphere, filtering the waters of the Earth, and recycling the mountains of garbage that we produce on a daily basis. Just as the hive is a superorganism, made up of many individuals, the Gaia hypothesis of James Lovelock[41] suggests that the Earth's biosphere can be viewed as a kind of superorganism where each plant and animal plays a role in helping to keep the atmosphere and global temperature stable. Imbalances created within the biosphere affect the whole planet.

What is astonishing about the fact that bees are dying out at this time is that bees are survivors. Fossilized evidence of the oldest known honey bees has been carbon-dated to approximately 100 million years ago. This is about 35 million years before the Cretaceous–Paleogene extinction event when, it is generally believed, a giant asteroid collided with the Earth. This is thought to have occurred about 65 million years ago and led directly to the extinction of most dinosaurs at the close of the Mesozoic era. Since bees survived this last major global extinction event, what is so different about the cause of their current demise?

In his 1980 book *Overshoot*, William Catton, Jr. states, "Infinitesimal actions, if they are numerous and cumulative, can become enormously consequential." This statement refers to the problem of cumulative impacts, where actions that seem harmless or tolerable at the individual level can degrade the planet's life-support systems if thousands or millions of people do them. One person fertilizing a lawn near the Chesapeake Bay, for example, makes no significant impact, but when thousands of people do the same all along the bay and along the numerous tributaries that feed into the bay, the water becomes severely degraded and blue crab populations decline dramatically.

When it comes to chemicals, the current regulatory approach to controlling pollution does not deal with global pollution. The main focus has instead been on the maximally exposed individual.[42] In the United States, we conduct risk assessments (used when conducting cost-benefit analyses) to evaluate the risk to a hypothetical "maximally exposed" individual. If the threat to that individual (or honey bee) is found to fall within acceptable limits, then regulation does not occur, and these so-called acceptable amounts of contamination are allowed to be released forever after. Then another risk assessment and cost-benefit analysis gives the go-ahead to another acceptable release or use of a different toxic substance or harmful activity. Then another, and another, so on and so forth. What we have not considered until recently is the total impact of all these "acceptable risks." Our society has assumed that it could tolerate unlimited small amounts of harm as a byproduct of economic growth. It is only when a particular activity demonstrably fails to provide a benefit to society that most of our property and environmental laws permit interference with economic activity. This must change. Not only do we need to start preventing companies from causing massive damage to our water, air, and soils, through the small but cumulative impacts of seemingly insignificant amounts of harm, we need to start forcing companies to bear the costs of cleaning up the messes and repairing the damage their products and activities produce, if not actually stopping them from continuing to damage the world in the first place.

This may seem like a commonsense proposal, but it is much different from the way decisions are being made today! Not only are our chemical and pollution regulatory control systems broken and failing miserably, but virtually all the systems that Western civilization has created and relies upon are proving themselves to be failures incapable of taking us into the future in the way we say we want to move forward. I have already outlined some of the ways in which our agricultural system is broken, but much the same can be said for our economic system, our health care system, our system of government, our energy policies, our judicial system, our education system. You name it, nothing is working the way we want and expect it to work. It's enough to cause one to ask, "Where are we going and what's with this hand basket?"

A big part of the problem is that the majority of us have lost touch with the physical reality of living. We have allowed the systems we as a society have created to interpose themselves between us and the real world. As author and activist Derrick Jensen has pointed out, we have created a society in which most of us have learned to identify with businesses, corporations, and governments as the foundations upon which our ability to survive and obtain vital resources such as food and water depends. When we identify with a system of corporations and governments as being necessary for our existence, we will naturally fight to the death for the right of those corporate and government structures to exist,

because we believe that our lives and well-being depend on them. If, however, we recognize the truth—that our lives really depend on our water coming from healthy rivers, streams, and lakes, and our food coming from healthy, fertile soils and vibrant, living oceans—then we would instead be defending the forests, oceans, rivers, mountains, and meadows with our lives.

The insanity of the currently prevailing worldview is that corporations and governments aren't even real things. They are simply ideas, concepts around which real people, materials, goods, and services are organized. Businesses, corporations, and governments exist only on paper. Yet we often act as if we must preserve the existence of business and governmental structures at all costs, even when such preservation is harmful and destructive to what is real. One of the great things about beekeeping is that it can act as a gateway that allows us to reconnect with the reality of the natural world and help us to break through the illusory web of thoughts and ideas that we create for ourselves. Once we start keeping bees, we are likely to be more mindful of all the other insects around us and our bee yards. Then we start to pay more attention to the plants in the area, both wild and cultivated: what kinds of plants they are, when they bloom, and so forth. This naturally leads to a greater awareness of the weather and how it is impacting the plants and the bees, and so on.

If I have not lost you and you have managed to continue to read this far, you may be wondering why I am going on and on about things that seem to have little or nothing to do with bees and beekeeping. The reason is that, just as a colony of bees has a huge influence upon the surrounding environment, the environment surrounding a hive has a huge influence on the colony of honey bees. If we really want our actions to help the bees be healthy and thrive, we can't just focus on the conditions inside the hive—we also have to pay attention to the conditions outside of the hive. If the world our bees inhabit is in a state of collapse, it matters little how chemically pure the wax combs in our hives are, how disease-free our bees are, or what the nutritional quality of the feed we provide our colonies with is.

Now, I realize that this is a heavy rap, not for the faint of heart. So I have to apologize to you, dear reader, because things are about to get a little heavier. You see, while we can blame the various systems we have created in our culture for being flawed and not taking us where we say we want to go, there is an underlying cause that precedes all the systems we rely upon. The root cause of all the environmental destruction, governmental and economic corruption, et cetera is sitting here at the keyboard typing these words, and sitting in your seat reading these lines: each one of us is a part of these failed systems. We all participate in them to one extent or another. Even if we don't make our living by working directly for one or more of these systems, we participate in them and support them by our actions, with our dollars, our votes, and numerous other ways. I found that as long as I viewed the current situation as someone else's fault, someone else's problem, then I was off the hook. I didn't have to change the way I lived or how I did things, or deal seriously with any of these issues, since it wasn't *my* problem. Once I recognized that I was part of the problem and accepted this fact, I found that I then had the responsibility and became empowered to do something about it.

The importance for human society to approach chemical regulation differently becomes apparent when we accept that humans, through the power of cumulative impacts, have become a force of global proportions and are degrading the biosphere upon which the honey bee (and ultimately all other species) is entirely dependent. This implies that the public, along with the media, the courts, public and corporate decision makers, and school children, among others, must all be informed that the world is new—it has changed because humans have become a global force and are now degrading the planet in ways that threaten the survival of much of the life that currently exists on Earth. This is a new awareness for most of us and requires a new understanding of our history; new thinking, new goals, new habits and attitudes, new stories and societal structures. Much of what we learned in high school, college, and graduate school is obsolete and stands in our way.

For example, a big part of the reason we collectively have been unable to adequately address any of our modern-day issues is because all the issues are related. We can't really fix any of our problems by focusing on them in isolation: we have to work on all the issues at the same time. For example, we are never going to have a chance of fixing the US economy until we stop our endless wars and radically change the way the health care system is designed and paid for in America. Our health care system will never be properly working until we fix the broken food system that allows so many toxic chemicals and artificial ingredients to be allowed in food production and processing. We won't be able to fix our agricultural system until we correct the deficiencies in our government's regulatory systems that allow the regular use of toxic chemicals without proper safety standards. Our regulatory systems will never get fixed until we rein in corporate power and corruption. And we won't be able to stop our constant militarism until we establish an economic system that relies on decentralization, the equitable and efficient use of renewable energy sources, and relocalized supply systems, rather than the current centralized technologies and vulnerable supply lines that need to be protected at enormous expense and risk to our civil liberties.

In taking on these efforts and more, we must give full recognition to the remarkable human capacity for self-deception and denial. We humans do not accommodate change readily; in fact, we tend to resist it. We seem wired (especially as we age) to deceive ourselves and deny reality; we look for scapegoats to blame and punish. Given the enormity of the environmental and social problems before us, it is too intellectually and emotionally difficult for many individuals to wrap their minds around the issues and accept our current situation as fact, because to do so would not only mean some level of personal responsibility, but also an admission that our entire culture and way of life, which has gotten us to this point, is fundamentally flawed and incapable of taking us forward into the future in the direction we say we want to go. This crucial reality must be thoughtfully acknowledged and successfully navigated if we are going to implement the changes that are needed.

Proper chemical regulation, environmental stewardship, and economic and social rehabilitation are not primarily technical problems but a human problem, one that encompasses the economy, political power structure, and societal rules and laws that are all based upon faulty assumptions, among other things. Most of these issues will need to be overcome before we can expect CCD—or any of the rest of the world's problems—to be resolved permanently.

Given that we don't understand all the impacts on our world of the chemicals we use on a daily basis, it is hard to imagine that these chemicals are not playing a significant role in the current rates of species extinction. This is a role that may be on par with the disappearance of our forests, the acidification and warming of our oceans, and the increase in carbon dioxide concentration in our atmosphere. Our future, which is intimately tied to the Earth's future, depends on what we choose to do on personal, regional, national, and international levels as a society. One place we can start today is by reducing or eliminating the use of chemicals in our hives. Then we can expand on this to reduce or eliminate chemical use and exposure as much as possible from the rest of our lives. This means, among other things, avoiding chemicals, as well as chemically laden products, which are often derived from petroleum, and choosing foods grown with organic, biodynamic, or permaculture methods. If we don't take effective action to come to grips with these entrenched problems soon, we may find that CCD is just a harbinger of things to come.

That being said, I would like to paraphrase Charles Dickens and say, "CCD has created the best of times and the worst of times." It is the best of times because the massive die-off of honey bees has generated unprecedented media attention. This has resulted in our political leaders and the general public finally beginning to understand the awesome importance of the honey bee in supporting our system of food production. It is the worst of times because if these reported colony deaths continue to escalate, the adverse effect from the yearly loss of hundreds of thousands of managed colonies and their pollination activities will ripple throughout the agricultural community. When growers are unable to secure adequate pollination for their crops, production and quality will drop precipitously, and food shortages—along with skyrocketing prices—are sure to follow.

Beekeepers have never had much political clout. However, this newfound awareness of the plight of beekeepers has created a situation where politicians have finally been forced to take notice and divert more resources to addressing the challenges of the honey bee and the beekeeping industry. Best of all is the renaissance in beekeeping that has occurred over the past five years. More people than ever before have been getting involved in beekeeping, and many of them are women and young people, two demographics that have been relatively scarce in beekeeping circles in recent history. Beekeeping associations have seen a huge jump in membership numbers, and beekeeping classes are often filled to capacity. Some people are getting involved because they are not seeing honey bees around any more and they simply want bees to return to their areas and pollinate their gardens. Others are motivated by their concern and their desire to do something to help the bees survive. This is a good sign, because the difficult challenges outlined in this chapter and many more are not going to be solved by others, but must be addressed by each and every one of us in all the ways we are able. Just as our cumulative impacts have led us to the edge of a global catastrophe, the cumulative impact of our seemingly insignificant acts that reduce harm and promote healing can turn the tide and bring balance back to our current situation. Things are not going to turn around until we turn away from a culture of death and domination toward a culture that is life sustaining. To see so many people willing to start making changes in their lives on behalf of the honey bee is enough to give a beekeeper a sense of hope for the future.

CHAPTER 10
The Honey Harvest

Give me a day
That is balmy and shimmering;
Give me a field
Where the clover blooms sway;
O let me linger
Where nectar is gathering,
Deep in the blossoms of flowering May.

Give me a shelter
Where sunlight is filtering,
Tempering music
Of humming and whirr,
Then I will gather
A treasure so glistening,
Fragrant as incense, and precious as myrrh.

—FLORENCE HOLT DAVISON

Of all creatures great and small, the wondrous little honey bee is the only one I am aware of—aside from humans—that consistently takes more from nature than it needs. (Even squirrels, for instance, *need* to stash away more nuts than they require for winter because they forget where many of them are buried.) As long as there is nectar available, good flying weather, and room in the hive to store and ripen the nectar into honey, the worker bees will continue to gather nectar, even after they have put away a far greater supply than is required to see the colony through the winter months. This lack of moderation puts the hive in a vulnerable position. What happens to all that extra honey? Your friendly neighborhood beekeeper comes along and takes the excess honey that the colony worked so hard to gather, process, and store. As a result the bees can teach us to check our greed and not take more than we truly need from the world around us, lest someone is tempted to plunder our excess in similar fashion.

Although apiculturists have little control over the weather and the ability of nearby flowers to produce nectar, we can easily modify the amount of space within a hive for honey storage. To obtain a large honey harvest, the beekeeper takes advantage of the hoarding instinct of the honey bee by adding honey supers to the hive at the appropriate time so the bees always have room above the brood area for honey storage. The bees will start to fill the brood nest with honey if additional supers are not added to the colony once its combs have all been filled. This is simply because, once a hive is full, the only room left is created when a fully matured honey bee chews its way out of its birthing cell. When a hive infringes on the brood nest for honey storage it is said to be "honey bound," a condition that will typically trigger a colony's swarming instinct. The timely addition of honey supers on a hive helps prevent the hive and queen from becoming honey bound and dissuades workers from building an abundance of burr comb in their efforts to create more storage space. Adding another honey super also maximizes the hive's honey-gathering potential by helping to reduce the overcrowded conditions that will cause a hive to cut back on its foraging activities and decide to swarm.

"Many hands make light work" is a maxim personified by the colony that keeps its workforce at full strength. The opportunity to add empty supers to a hive before the bees have completely filled the supers they have is just one of the many reasons why conscientious beekeepers conduct regular inspections of their hives. This regular inspection, hand in hand with an understanding of just how prolific honey bees can be, ensures that supering is done appropriately and in a timely manner, thus reducing the bees' swarming tendency.

Notice that I talk about *reducing* the tendency to swarm, not eliminating it. The swarm is the hive's natural means of reproduction, thereby increasing the number of colonies in the world. It is a healthy and necessary process for the bees, despite what a swarm may mean in economic terms for the beekeeper, and so it is not wise for beekeepers to do everything in their power to prevent swarming. In fact, if your hives are swarming, it is an indication that you are taking pretty good care of them and your bees are healthy and thriving; struggling, weak colonies do not have the resources to issue swarms. Because of our tendency to see everything in purely human terms, the literature all too often refers to swarming in a negative light. From the bee's perspective, however, the swarm is a wonderful thing. Like the birth of a baby, I would invite you to consider the swarm as the miracle of life that it is and celebrate what the world has gained from such an occurrence rather than dreading it and dwelling on what might be perceived as a personal loss.

When supering hives, it is generally preferable to utilize drawn comb and use foundation only when the honey flow is heavy. It is also important to note that strong hives will have an easier time drawing out foundation in a timely manner than weak hives. This is not to imply that one should not place drawn combs on a strong hive during a heavy honey flow, or that foundation should never be placed on a weaker colony. It's just that the bees are reluctant to draw out foundation when the honey is only trickling in. If foundation needs to be drawn, waiting until a good honey flow is in process, or creating an artificial nectar flow by feeding, is the best policy; and the more bees available to do the work, the faster it will get done. In fact, workers are more likely to chew away the edges of the foundation when little or no honey is being made, and they will often deposit the small amount of nectar that is gathered in burr comb rather than begin

drawing out foundation. In the absence of a strong nectar flow, colonies have also been known to backfill the brood nest with what little nectar they are able to collect, becoming honey-bound and then swarming rather than building comb on the foundation. Many beekeepers will feed bees sugar syrup or bee tea when foundation is on the hive in order to ensure that comb gets drawn out; however, because I don't feed my bees unless they are starving, this is not something I have experience with. Instead, I place foundation I want drawn out in my hives just prior to when the first major honey flow of the year is expected. Alternatively, the capturing of a swarm provides a perfect time to get new foundation drawn into comb, as a swarm is fully primed to produce wax and build comb.

The hive body typically consists of ten frames of comb, in order to provide the queen with the maximum amount of egg-laying room, but honey supers should contain only eight frames. By using eight frames in a honey super that is made to hold ten frames (or six or seven frames in a super designed for eight frames), the amount of space for the bees to store honey is maximized. At first glance, this might seem counterintuitive. After all, wouldn't nine or ten frames of comb hold more honey than eight in a ten-frame box? Unlike frames

FIGURE 10-1. Before: Using only eight previously extracted frames evenly spaced in a ten-frame super provides the maximum amount of honey storage without sacrificing neat combs built within each frame.

FIGURE 10-2. After: The bees draw out the wax comb to whatever width they need to create the proper bee space between frames. The frames must be of drawn-out comb and they must be evenly spaced for this to work.

of brood comb, which tend to be consistently built to a certain width, the bees will modify the thickness of the combs they intend to store honey in and build them to fill the available space. Thus, because there are fewer pieces of wood and metal, and two less bee spaces taking up space within a ten-frame honey super filled with eight frames, as compared to the same super filled with ten frames, there is more room for the bees to build honey comb and store honey. Because of the importance of maintaining the hive's bee space, eight is the minimum number of frames that should be used in equipment built for ten frames.

Given that the modern hive body and super are designed to hold ten frames, it is important to evenly space the frames of comb within the cavity of the box when using only eight or nine frames in order to preserve the bee space, or seven frames in boxes designed for eight frames. (You may even be able to get away with six frames; I can't say for sure because I don't use eight-frame equipment and have never tried it.) Commercial frame spacers can be purchased, but my preference is to use a thumb or finger as my spacing tool to position the frames uniformly. To accomplish this, first space out eight or nine frames equally in a ten-frame super or hive body. Then see which of your thumbs or fingers fits between each frame. You may find that you will need to hold your thumb or finger at a certain angle for it to gently touch

FIGURE 10-3. Why spend money on a frame spacer when you have a finger or thumb that will accomplish the same job? PHOTO BY ALICE ECKLES.

the frame on either side of the space. By inserting the same digit in the same manner between each frame when spacing future supers, adequately spaced frames can be achieved in short order every time. The choice is yours. You can either purchase premade frame spacers or simply use the tools you have on hand—literally! You can then enjoy the added benefit of never losing track of where your frame-spacing tools are, always having them "handy" when you need them.

When it comes to harvesting honey, the organic philosophy and most organic standards dictate that only surplus honey be harvested, and that care must be taken to preserve the integrity of the harvest. The practice of taking almost all of a colony's honey and then feeding the bees sugar or corn syrup in return, although logical from a macroeconomic viewpoint, is neither sustainable nor in the best interest of the bees, because it takes the honey bees' time, energy, and resources to turn nectar, sugar water, bee tea, or corn syrup into honey. The fact that most corn syrup and sugar from sugar beets will contain at least some genetically modified material, and that corn syrup is also contaminated with small amounts of systemic neonicotinoid pesticides further complicates this issue. As a result, ecologically minded beekeepers must leave an adequate supply of honey on the hives to provide for the bees' winter needs, and this amount will differ from region to region.

FIGURE 10-4. An escape board, with its small openings at each corner of the triangle, is a low-impact tool to aid in removing bees from honey supers.

Once the decision has been made to harvest a honey crop, the organic beekeeper has three basic options available for removing the bees from the honey supers: manual shaking and brushing; blowing; or using some type of bee escape. Shaking and brushing requires little in the way of specialized equipment. A soft-bristled brush is all that is required. The downside of this low-tech approach is that the bees are likely to get riled up unless they are smoked before and after being brushed or shaken from their combs. Care must be taken, however, because too much smoke can adversely affect the flavor of the honey extracted. However, because of its simplicity, this method is a good option for the beekeeper with few hives.

Blowing the bees off the frames of honey to be harvested is a quicker method and is more desirable

for those beekeepers with numerous hives. Blower systems are sold by bee supply companies and are similar to leaf blowers, or shop vacuums with the ability to run in reverse. Air is directed between the frames of the honey super as it sits up on end and the bees are blown through the super and out the other side. Although it's fast and efficient, this method definitely makes the bees angry, and a veil is standard operating equipment whenever one is harvesting with a blower. The wrath of the bees can be tempered somewhat by utilizing the blower only when temperatures are below 60°F (16°C).

The bee escape is a small device that acts as a one-way door, and it is placed in the center opening of the standard inner cover that features an oval hole in its center. Supers of honey that are to be harvested but are filled with bees are placed above the inner cover, and as bees move down toward the brood area over a period of days, they pass through the bee escape, which prevents them from reentering the honey super above later on. The escape board works on the same principle and takes the place of an inner cover fitted with a bee escape. The larger opening on the escape board allows the removal of bees from honey supers to proceed at a faster rate than with the bee escape, typically in 24 to 48 hours. The downside of using either the bee escape or the escape board is that it requires a minimum of two trips to the hive to bring in the harvest. In addition, bees are reluctant to leave supers that contain even the smallest amounts of brood, and these frames will require shaking or brushing anyway. Bee escapes and escape boards work best when evening temperatures drop down into the 50s or below, encouraging the bees residing in the honey supers above to leave and gravitate toward the brood nest cluster. When using a bee escape or escape board, it is critical to seal up all other openings—between supers, in the inner or outer cover, and so on—that are large enough to allow bees to exit and enter. Scraps of beeswax, mud, drywall compound, or duct tape all work well in this regard. Once the honey supers have been removed, it is common for the bees to beard up on the front of the hive, because there is suddenly significantly less space to house all those bees!

A key to a successful honey harvest is to wait until the bees have transformed the nectar into ripe honey with a moisture content below about 18 percent. A thin layer of wax is the covering the bees will use to seal combs filled with fully ripened honey. Once capped, the honey in these cells is protected from contamination from dirt and debris, thievery by robbers, and, to some extent, moisture. If frames of honey that contain primarily uncapped cells are harvested, the honey extracted will typically contain too much moisture and it will ferment over time. The general rule of thumb is to be sure that, at minimum, 75 percent of the combs that are extracted are fully capped. This means that on average each frame can contain up to 25 percent of its cells uncapped, or that for every frame of uncapped honey that is harvested, three frames of fully sealed combs must be extracted along with it. Those who want to remove the guesswork should purchase a refractometer, a precision instrument that will accurately measure the moisture level in honey.

• HONEY PROCESSING •

In the old days, hives were killed off and the harvested honeycomb was squeezed to separate the honey from the wax. Today, extracting machines efficiently remove honey using centrifugal force in much the same way that water is forced out of clothing in our washing machines. Removal of honey without destroying the wax allows the bees to reuse the frames of honeycomb, and allows beekeepers to harvest honey without destroying the hive. The comb is the furniture that makes up the honey bees' home. The bees are born in these combs, live on them throughout their lives, and use them for food storage. Drawn comb represents a substantial investment on the part of the hive in terms of time, labor, and resources. It is estimated that worker bees must consume approximately

FIGURE 10-5. Whether you use an old-fashioned handheld model like the one on the left, or a modern digital one like that shown on the right, the refractometer will allow you to remove the guesswork in deciding when your honey has the right moisture content and is ripe and ready for harvest.

7 to 8½ pounds of honey for every single pound of beeswax that they excrete through their wax glands. Aside from sacrificing some of the honey harvest to allow combs to be built, the beekeeper's investment lies in the cost of acquiring foundation and inserting it into each frame.

When a hive is allowed to build natural comb within a frame, the comb will be more fragile than newly drawn-out frames that utilize foundation reinforced with wires or support pins, especially during the initial season. It is advisable when extracting honey from a frame of naturally built comb that was built that year to use a variable speed extractor that can extract honey from the comb at a slow speed initially. Once the majority of the honey has been flung out of the comb, the rotation speed of the extractor can be increased to help remove the rest of the honey in the frame. By initially running the extractor at a slower speed than is usually used, damage to the combs can be minimized. Typically, by the second or third year, such frames of comb have been built up and reinforced by the bees to the point where they can withstand the rigors of the rotating honey extractor running at faster speeds without breaking, cracking, or collapsing.

Before frames of ripe honey sealed neatly within the comb can be extracted, the wax cappings must first be removed so that the honey contained in each cell can flow out freely. Organic production does not allow for the use of heat: approved ways of

FIGURE 10-6. With its offset handle, the uncapping knife is designed to effectively remove the thin wax cappings covering the honeycomb.

uncapping frames of honey range from the use of machines that mechanically uncap each frame to manual methods such as the uncapping knife and the capping scratcher or uncapping fork. An uncapping knife with a serrated edge is the best way to effectively remove cappings by hand and without heat. The handle of the uncapping knife should be offset and the blade long enough so that each end of the knife will ride along the bottom and top bar of each frame, helping to guide the cut and prevent gouging into the delicate combs. Care must be taken to hold the knife at the appropriate angle so that the blade will not cut into the wooden components of the frame. The scratcher comes in handy for removing cappings from the areas of comb that the bees did not draw out past the plane created by the outer edges of the wooden parts of each frame. In large operations, when extremely high numbers of supers need to be extracted, an automated electric-powered uncapping machine helps keep one's wrists from tiring out.

The decision as to whether to use deep supers, shallow supers, or medium supers to collect and harvest the honey is a personal one. In terms of strict efficiency, the choice is simple. Each shallow frame holds less honey than a medium or deep frame, and therefore it will take longer to extract the same amount of honey when using shallow supers for honey collection than it will with the frames of medium or deep supers.

FIGURE 10-7. For safety, be sure to keep fingers well back from the front of the frame, or one little slip can lead to a nasty cut.

On the other hand, the larger the super, the heavier it will be when filled by the bees with the ambrosial gifts of their labor. In recent years, bee supply companies have promoted smaller boxes that hold a maximum of eight frames as an alternative for those who wish to save their backs. Whether you choose to utilize eight-frame or ten-frame hive bodies and supers, I strongly recommend that once you have made your choice you stick with one type or the other. Otherwise, the day will inevitably come when you find yourself needing a hive body or super for an eight-frame hive, and the only empty equipment you have available to use is made for ten frames, or vice versa.

When extracting honey, variables such as temperature and the types of materials that are allowed to come in contact with the honey become factors. As with all good production practices, uncapping, extracting, and bottling equipment must be clean and free of foreign materials and debris prior to use. Stainless steel, food-grade plastic, and other food-grade materials must be utilized to reduce the possibility of compromising the harvest's integrity. In keeping with this train of thought, all organic standards limit the temperature that honey is allowed to be exposed to, with most specifying 95° to 120°F (35° to 49°C) as the maximum range, and some going even further, also limiting the length of time that honey may be held at the maximum temperature. This is because all the enzymes in honey are adversely affected at temperatures approaching 120°F, and high temperatures will affect the honey's color and flavor. These temperature requirements refer to the temperature range of uncapping equipment and bottling equipment. Conventional honey that has been heated will stay liquid on the shelf for a much longer period than unheated honey before crystallizing, and, because it is clear, it is typically filtered to remove the tiny particles of pollen and propolis that may have entered the honey during the extraction process. Not only do these minute particles detract from the clarity of the final product, but they also act as nuclei around which crystals may form, speeding

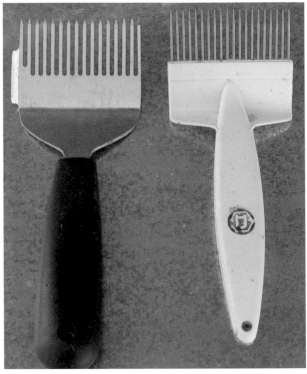

FIGURE 10-8. The cappings scratcher, or uncapping fork, is used to remove the wax cappings from areas of comb that the uncapping knife or machine is unable to reach.

up the crystallization process and shortening the shelf life of the liquid honey. Organic honey standards prohibit the filtration of honey, thereby preserving the nutritional and medicinal benefits that the bits of pollen and propolis impart to the final product. As a result, most natural and organic honey that is available will be crystallized, with the exception being honey that has been freshly harvested or honey from floral sources that naturally resist crystallization, such as tupelo honey or sourwood honey.

To remove the majority of the beeswax, pollen, and propolis that gets into the honey during extraction without using heat, a screen or sieve can be used. However, running honey through a sieve can be a slow process, and when harvesting large amounts of honey, settling tanks are the preferred method of making the final product presentable. As honey sits, foreign debris in the honey will slowly rise to the surface, where it will float. The relatively clear honey at the bottom of the tank can then be drawn off and bottled.

Once the harvest has been bottled, the national organic standards have much to say about the label that goes on the final product. The National Organic Program (NOP) separates organic foods into three categories: 100 percent organic, organic, and made with organic ingredients. The labeling requirements for each are as follows:

(a) Products sold, labeled, or represented as "100 percent organic." A raw or processed agricultural product sold, labeled, or represented as "100 percent organic" must contain (by weight or fluid volume, excluding water and salt) 100 percent organically produced ingredients. If labeled as organically produced, such product must be labeled pursuant to § 205.303.

(b) Products sold, labeled, or represented as "organic." A raw or processed agricultural product sold, labeled, or represented as "organic" must contain (by weight or fluid volume, excluding water and salt) not less than 95 percent organically produced raw or processed agricultural products. Any remaining product ingredients must be organically produced, unless not commercially available in organic form, or must be nonagricultural substances or nonorganically produced agricultural products produced consistent with the National List in subpart G of this part. If labeled as organically produced, such product must be labeled pursuant to § 205.303.

(c) Products sold, labeled, or represented as "made with organic (specific ingredients or food group(s))." Multi-ingredient agricultural product sold, labeled, or represented as "made with organic (specified ingredients or food groups(s))" must contain (by weight or fluid volume, excluding water and salt) at least 70 percent organically produced ingredients which are produced

and handled pursuant to requirements in subpart C of this part. No ingredients may be produced using prohibited practices specified in paragraphs (1), (2), and (3) of § 205.301(f). Nonorganic ingredients may be produced without regard to paragraphs (4), (5), (6), and (7) of § 205.301(f). If labeled as containing organically produced ingredients or food groups, such product must be labeled pursuant to § 205.304.[1]

These labeling requirements were created in 1990 when Congress recognized the growing demand for organic products and passed the Organic Foods Production Act (OFPA). This act called for the creation of a National Organic Standards Board (NOSB), a citizen review panel whose job is to develop a national list of approved substances allowed for use in organic production and to advise the US Secretary of Agriculture on program implementation policies. Prior to the OFPA, organic standards had been developed through the years by numerous certification agencies, all working on their own, and as a result, organic standards had become anything but standard. Due to the wide diversity of certification organizations, many different regulations governing the production of organic crops were developed. Thus, a major part of the reasoning behind the establishment of federal organic standards was the desire to standardize the rules governing organic certification so that they were consistent from state to state and region to region.

The USDA allows the green-and-white organic logo to be used only on products consisting of 95 to 100 percent organic ingredients. Unfortunately, because the controversial issues surrounding honey have not yet been resolved, resulting in the NOP dragging its feet on the establishment of organic honey standards, the use of the USDA organic logo on honey labels has itself become controversial. Some organic certification agencies—typically those that had organic honey standards in place before the NOP standards were released—will allow the use of the term *organic* as long the apiculturist meets the requirements as written in their standards. The honey bee recommendations put forth by the NOSB tend to be referred to heavily when honey regulations are being updated in order to harmonize the standards being written with what are likely to be the final rules once the NOP finalizes its decision. Certification agencies that don't yet have standards in place have tended to put the work of writing standards on hold, the rationale being that it makes little sense to spend time and effort creating a set of standards for honey when the NOP regulations, once released, will trump their own. How long it will be until national standards are finally realized is an open question. Not only is there currently little momentum behind the creation of national organic honey standards, but the controversial issues surrounding the establishment of such standards are so sticky that they suck in the participants and mire down progress, like a bee drowning in a tank full of honey. As with all things, honesty is the best policy when it comes to organic agriculture. The penalties for fraudulently labeling organic products consist of a civil penalty of $10,000 for each occurrence, in addition to criminal penalties under federal law.

• HONEY INVENTORIES •

One of the most important elements for a colony's survival during winter is access to adequate stores of honey to ensure that the hive does not starve before spring rolls around. Beekeepers who are truly concerned with the welfare of the bees will harvest only the excess honey that has accumulated during the course of the year, and they are careful to leave the colony with enough honey to see it through the winter. The amount of honey a hive requires for its winter use will differ from location to location. Here in Vermont, bees typically need a minimum of 60 to 100 pounds of honey per colony. A good rule of thumb is to harvest all the supers that are filled with honey with the exception of the last full super that is situated just above the brood within the hive. During the harvest, as each full

super is lifted off a hive, the super directly below it should be inspected for brood. If there is a significant amount of brood mixed in with the honey stored in the super below, leave the super that is chock-full of honey sitting above it in place. In addition, make sure that any nucleus colony created during the course of the year that was unable to fill at least a shallow super above the brood nest with honey receives a full super harvested from a stronger hive. This way we ensure that every colony goes into winter with a full super of honey above the brood nest. Capped honey above the brood nest is key to the colony's winter survival and is much more effective than storing individual frames of honey in the freezer and trying to feed them back to the bees. During the course of the winter months the cluster will naturally make its way up through the hive, often ending up near or even right up against the inner cover by spring. When a full super of honey is used to feed the hive, there is less chance that the cluster will eat itself into a corner and starve compared to the stopgap measure of trying to position a few frames of stored honey in the right place in the hive at the right time in late winter/early spring. Thus, the anatomy of a hive prepared for winter finds the bees clustered around the brood nest near the bottom of the hive with plenty of honey stored above to provide room for their slow but steady upward winter migration.

This standard operating procedure goes hand in hand with the management technique that utilizes hives consisting of a deep (9⅝-inch) hive body sandwiched between two shallow (5¹¹⁄₁₆-inch) supers, which act as shallow hive bodies, as opposed to the conventional hive consisting of two deep hive bodies. There are a number of benefits to using two shallows (or two mediums) and a single deep as the main body of a hive. The greatest benefit is realized when having to feed colonies. When it comes to feeding a hive, the best food we can offer is honey already sealed in the comb. As long as care has been taken to ensure that honey fed to a hive does not contain American foulbrood or other disease organisms, a full super of honey filled with

FIGURE 10-9. When deep snow covers the entrances to a hive, it is a good idea to dig out the snow so the hive can vent moist air and the bees have the opportunity to leave the hive and make cleansing flights on warm days.

capped frames is first-class bee feed. A weakened hive makes an especially good recipient of such feed, because it will save that hive's bees from expending their energy filling combs and converting sugar syrup or bee tea into honey. For many years beekeepers who have used shallow or medium supers as part of the main hive unit have been able to feed honey in the comb to bees more efficiently than if only deep hive bodies are used for the hive proper. A nuc that has filled its deep hive body with honey, pollen, and brood has historically been provided with all the feed it will require for the winter by adding a full shallow containing about 30 pounds of honey on top of the deep rather than giving the colony another full-depth super that can hold as much as 75 or 80 pounds of honey—especially since there is honey already stored along with the brood in the hive body below. However, things have been changing, and with the shifting climate patterns and milder winters, a single shallow or medium above the brood nest is not always enough, especially when the bees being kept are

Italians. Today, in the northeastern part of America, bees are less likely to starve and tend to be stronger in the spring when a full deep super of honey (or its equivalent—two shallow or two medium supers) are left on the hive above the brood nest as autumn gets underway. In spring, whichever of the supers on top of the deep brood box contains the most brood can be reversed with the shallow underneath the deep hive body, creating a situation where the deep brood chamber is now sandwiched between a medium or shallow below and two mediums or shallows above.

The other big benefit is the fact that shallows full of honey are much easier to lift than full deeps—a benefit that becomes more and more significant the older one gets. Also, if mice get into the hive during the winter, they typically build their nest in the bottom super. In such instances it is preferable to have shallow frames of drawn comb damaged by mice rather than the more valuable deep frames.

If no frames filled with honey are available, extracted honey can be fed to needy colonies.

FIGURE 10-10. Hives that feature shallow or medium supers as part of their year-round hive structure can be fed full supers of honey for winter much more efficiently that hives that consist of all deep hive bodies.

Because of the threat of disease, all extracted honey from unknown or questionable sources should always be mixed with some water and the resulting syrup boiled for 20 to 30 minutes before being fed to a hive, so that any disease spores that the honey contains will be destroyed. However, boiled honey is not as desirable as clean raw honey that has been unheated, because heat destroys the enzymes naturally found in Mother Nature's sweetener while increasing the level of HMF (see "Feeding" in chapter 3 on page 55). If honey is not available, the next best feed to use is white sugar. The reason white sugar is preferred over brown sugar, turbinado sugar, or other types when making syrup for bees is that white sugar has had all its nonsugar components removed and will not contribute to dysentery within the hive during the winter. The same refining process that provides empty calories that prove detrimental for human consumption is a little less harmful when used for bees. A ratio of one part sugar to one part water is typically recommended for spring feeding, whereas a 2:1 ratio is best when mixing up sugar syrup for feed in the fall, a time when hives need to make and store as much honey as possible as quickly as possible. Since I don't feed my bees in order to artificially stimulate them to build up quicker and only feed when starvation is imminent, I always use a 2:1 ratio of sugar to water. The use of hot water will help ensure that all the sugar will dissolve—just make sure the syrup cools sufficiently before feeding it to the bees. Honey produced from sugar syrup is not as desirable as honey produced naturally, because syrup is deficient in the natural minerals and enzymes found in honey produced from fresh nectar. (See the Bee Tea recipe in chapter 3 on page 60.)

Most organic standards require that organic sugar be used when making sugar syrup and restrict the use of high-fructose corn syrup (HFCS). As indicated earlier, this is because feed made from HFCS may contain HMF and will result in honey that is likely to contain genetically modified material, which has the potential to weaken the vitality of the hive, making it more prone to contracting diseases or becoming a target for parasites. However, it is most likely that, should certified organic HFCS become available, organic standards would find it acceptable for use as bee feed. Should this occur, beekeepers concerned about the well-being of their hive should avoid HFCS anyway, for the reasons explained in chapter 3.

When fall feeding is complete, each hive should be given a "heft test" by grabbing the back of the hive by the handhold of the bottom or middle hive body and lifting. If the hive feels like it's filled with lead and you are unable to move it without excessive strain or leaning your whole body into the hive, then the hive is in good shape for winter as far as honey goes. It takes some experience to consistently estimate the amount of honey stored in a hive accurately by hefting alone. Initially it is a good idea to open up a few hives to confirm the level of capped honey they contain and to get a good sense for how heavy a hive should feel when properly provisioned for winter. One final manipulation that should be made prior to winter is to position the center of the brood nest so that it is located near the middle of the hive body, with an equal number of honey frames on either side. This way the colony will have equal access to nourishment no matter which direction they may move during the course of the coming months.

Although winter in the southern United States is not marked by the extreme cold that characterizes the North Country, it is often still characterized by a dearth of nectar sources. Lack of sufficient amounts of stored honey is as serious a problem for the southern beekeeper as it is for those in the North. Depending on location, a colony heading into winter in the South may require almost as much honey as one in the North, and unless it is well supplied, brood rearing will be held in check, the result being a colony that is weak and unfit to take full advantage of the first major flows of honey come spring.

CHAPTER 11

Marketing Products from the Hive

Instead of dirt and poison we have rather chosen to fill our hives with honey and wax; thus furnishing mankind with the two noblest of things, which are sweetness and light.

—JONATHAN SWIFT

Once your honey is harvested, bottled, and labeled, you have to figure out what to do with it. If you have only one or two hives and are lucky, you will produce enough honey to provide for you and your family's needs. In some years, you may even have enough left over to give an occasional jar as a gift to your friends and neighbors. If you produce more honey than you can use, then it's time to think about how to market your excess and help defray some of the expenses you have incurred producing your crop.

I strongly recommend you sell your excess honey (or at least barter with it), rather than find additional people to whom you can give it as a gift. Why? It comes down to supporting your local beekeepers and the beekeeping industry. When people receive free honey, it affects honey sales because folks who would have otherwise purchased honey now do not have to. And if a local beekeeper is giving away honey for free or selling it very cheaply, this has an impact on the perception of honey's value in the area. Why would someone pay top dollar for a jar of honey when it is available at a much lower price or even for free? Stop to consider that even high-end honey sales typically do not earn enough to cover the true costs of production, especially if one were to factor in a living wage in return for all the time that the beekeeper invests in producing that

jar of honey. I would contend that even the most expensive honey is likely underpriced. After all, honey is not just a sweetener and food, it is also a powerful healing medicine. And a jar of honey is the only food that will never spoil in its natural form, as long as it is stored properly.

When it comes to selling your honey you have three primary options: sell it bottled and labeled at retail prices directly to the end user; sell it bottled or bulk at wholesale prices to another business that will sell it to the end user; or sell it as a commodity item in bulk to a packer or processor, who will turn around and sell it either retail or wholesale or both. In the rest of this chapter, I hope to give you an idea of some of the benefits and drawbacks of each marketing option, as well as some tips and suggestions to help you be successful at whatever type of marketing you choose.

One of the first decisions you should make is who your customer will be. This decision will help clarify the type of packaging you will use for your products. If you are going to sell your products in bulk, less attention will need to be given to specialty packaging and labeling. The type of package and label become much more important when you decide to sell directly to the final user, or to another business who will in turn sell your product to the consumer. In our modern marketplace, the quality of the packaging and labeling is just as important as the quality of the product inside the package. The more attractive and unique your packaging, the more your products will stand out among the competition. Needless to say, sticky or dirty jars that contain less than the amount of honey indicated on the label will not win you customers and may get in you trouble with government regulators.

Once you've identified your desired customer base, customize your packaging and labeling design to be most effective in meeting your customer needs and tastes. Do you want to use plastic containers or glass jars? What sizes and shapes? Where will the labels be placed and which information—nutritional data, bar codes, product name—will go where? To be legal, the label on the front of the container must state what's in the jar, how much is in the jar, who you are, and where you are located. For a lot more practical information on honey labeling, visit the National Honey Board website (www.honey.com).

Offering a variety of items for sale will help to increase sales. Rather than packaging your honey in only one size of jar, consider offering, at minimum, one small size and one larger size. If you have enough honey to offer three or four different sizes, even better. The more size offerings you have, the greater the likelihood that your customers will find something that will fill their needs at the price they are looking for. This idea can be expanded to offer different types of honey: liquid, crystallized or creamed, comb, chunk, flavored, seasonal, or varietal. Extrapolate this idea further and you will find yourself offering other related bee products such as wax, candles, soap, salves, and balms. Again, this is all based upon the basic premise that the more items you offer for sale, the more likely a prospective customer will find something they want. Keep in mind, however, the desire to offer a large variety of items for sale needs to be balanced with the amount of money, time, and space you can afford to tie up in all the product development, manufacturing, and storage that each additional item will require.

If you find yourself challenged in a market with buyers who are hesitant to spend as much as you want to charge for a jar of honey, consider additional labeling. Labels that indicate the type of flowers the honey comes from, where the honey was produced, or story behind the honey all can add value to your product.

• THE RETAIL MARKET •

Selling your products primarily or solely through retail outlets is more critical and more doable for small- and medium-scale operations. It requires more time and money invested in packaging and labeling than bulk sales, but when you sell your honey at retail prices directly to the end user, you often have the opportunity to receive the highest return on your investment. The growing interest in locally produced food in many parts of the United States helps to support the small, hand-crafted, artisan business model, which is personified by the small- to medium-scale beekeeper. Interest in local production will only grow stronger as fuel and energy costs continue to rise, which in turn make food production and transportation costs go up, and food prices increase. As energy costs increase, locally produced goods that don't require long-distance transport become more competitive.

Retail sales start at your doorstep. Your family, friends, and neighbors will typically make up your initial customer base. You already know them, they know you, and you already have a relationship with them. However, if you are not able to sell off all your excess by selling to your family, friends, and neighbors, you will have to reach out to your local retail market.

A natural extension of the family and friends model is to set up a farm stand. The ideal farm stand is visible from the road and clearly marked with signs. There should also be plenty of room for folks driving by to pull their vehicles off the road and park while visiting your stand. A nice feature of the farm stand is that one can sometimes rely on the honor system for payment and not have to staff the stand the entire time it is open. Unfortunately, this can be problematic in some communities, because there are folks who will take advantage of your trust. The level of trust a farm stand owner exhibits can range widely, from an unattended stand with a jar of coins for people to make change, to an unstaffed stand with a padlocked money box and security cameras, to a stand staffed 100

FIGURE 11-1. A neat and attractive honey booth has a variety of products all labeled and clearly priced, with an attentive company representative on hand. PHOTO BY ALICE ECKLES.

percent of the time. Design and decorate the stand in a way that enhances the atmosphere you want to create. Lighting, if available, can help to extend your business hours and provide an extra measure of security.

The next step up from a farm stand is to have a vendor booth at your local farmers' market. Farmers' markets nationwide have seen tremendous growth in recent decades as consumer interest in obtaining fresh products directly from the farm has grown, According to the USDA, the number of operating farmers' markets was 6,132 in 2010. This was a 16 percent increase in the number of farmers' markets from 2009 and up almost 350 percent since 1994. Not only do farmers' markets

allow consumers to have access to locally grown, farm-fresh produce, they afford farmers the opportunity to develop a personal relationship with their customers and cultivate consumer loyalty.

Farmers' markets are typically weekly or biweekly affairs and usually require a time commitment of at least three hours on market day. For best results, it is important to attend consistently so your customers will know what to expect and won't be disappointed after traveling to the market only to find that you are not there. Since honey is not an item that most people need to purchase on a weekly basis, you will want to include a mix of bee-related products in order to make renting a booth worth the time and effort. Fairs and festivals are also excellent places to sell your products; these events tend to have many more people in attendance than a farmers' market, but most take place only once a year.

Whether it's a farmers' market, fair, or festival, pre-event planning is critical. First, clarify the primary purposes of your market booth and consider this when planning your booth design. Is your goal to sell products retail, find new wholesale outlets, promote name brand recognition, train your staff on how to run your booth, or some combination of the above? It is important to have a clear vision of the outcome you want to create. You may also need access to money to reserve your space and assemble the needed portable tables, chairs, and attractive display.

Learn the rules governing the booth space at the event you are attending and abide by them. Be sure you have enough assistance and leave plenty of time to set up and break down your display on schedule. The organizers of the event will appreciate this and will be much more likely to invite you back in the future.

Once your booth is set up, look at the booth from all angles to see it as your potential customers will, and make adjustments as needed. If you have time, walk around and check out the other booths to get additional ideas of how to enhance your presentation.

Your product/booth display at a farmers' market, fair, or festival will often be more involved than what is required for a successful farm stand, because you will be competing with other vendors for the attention of the customers. The decor of the booth should reflect your own style and tastes, the image you want your business to have, and the type of customers you hope to attract.

I like to display lots of product without having my display space look too cluttered. Generally, folks are more inclined to purchase an item when there are plenty of choices on the shelf rather than just a few lonely jars. You create the perception of abundance when the display shelf or table is filled with a large number and variety of goods. The space should be open, friendly, warm, inviting, attractive, and clean. Add interest to a display by including shelving or boxes so products are displayed at various height levels. Another way to add height is to stack products in pyramids or other shapes. I find it is best to position small items as close to eye level as possible; I put larger, more prominent items directly on the display table. To prevent small items from being hidden from view, place large and tall items towards the back of the display space and shorter or smaller items up front. A nice tablecloth (perhaps with a honey bee motif) can add color and class. A vase of flowers or some potted plants strategically placed within and around the display area can also help create an inviting, festive, and attractive atmosphere.

The location of your booth can play a role in how successful or unsuccessful you are. It is usually preferable to have a booth on the end of a row rather than to be sandwiched between other booths in the middle of the row of vendors. When you are between two other vendors, you are limited to displaying along the front of your booth space. However, if you are located at the end of a row, you can often enhance your visibility by making use of the display space at the side of the booth as well as that facing the aisle; check first with the event management to be sure they do not object. This increases your presentation area and display options. People may tend to take a shortcut through a booth located at the end of a row, especially if you position your display table in the back of the booth and leave the rest of the space open. This is not a bad thing. You *want* people to come into your booth. The more individuals that pass through your booth, the more likely some of them will stop and buy something. A large sign announcing your company name or the primary products you sell (e.g., "Honey") will help people spot your booth from a distance and draw them in.

Whenever possible I like to get a booth space next to or near the place where people enter and leave the market area. Being the first booth that a person sees upon entering gives you an edge on the other vendors, which can be especially important if competing vendors are selling items similar to yours. As a honey vendor, you can benefit by being the last booth a person passes on their way out, because honey is heavy and many people will not want to carry it around while they shop. Most would rather make their honey purchase last. However, while prospective customers may state such an intention, there is the chance that they may forget to buy honey from you on their way out, they may purchase honey from a competing vendor instead, or they may run out of money before they make it back to your booth. One way to help guard against losing sales due to such cir-cumstances is to offer to make the sale immediately and then keep the purchase behind the counter for the customer to pick up as they are leaving. This approach is especially helpful if your supplies are limited and you expect to sell out before the end of the day. It guarantees that the customer will have honey to take home at the end of the day even if you sell all your stock before they are ready to leave. At events where people will spend many hours walk-ing around, such as at festivals and fairs, small jars that don't weigh too much will tend to be your best sellers. Larger jars of honey sell best in a farmers' market situation where folks are basically grocery shopping and won't have to lug their purchases around too long.

To increase interest in your booth and boost sales, feature an interactive aspect that will draw folks in. The easiest and most obvious interactive approach is offering samples of your honey for people to taste. Some beekeepers like to offer a smear of honey on a cracker or put out little plastic spoons for scooping honey out of the sampling jar. I find that including crackers or other foods in the taste sample can interfere/compete with the flavor of the honey, and plastic spoons create a lot of waste that will not biodegrade. I prefer to offer prospec-tive customers wooden coffee stirrers for sampling the honey; the used stirrers can be composted. I buy 7-inch coffee stirrers and cut them in half. Not only is this option better for the environment,

but it is also much less expensive than buying little plastic spoons. Just keep an eye on the kids (and some adults) when they are sampling so that they don't dip more than once! And be sure to provide a place for folks to dispose of their sample sticks or spoons after trying your honey.

Another great booth design feature that will encourage interaction between you and the public is an observation hive. Be forewarned: when an observation hive is present, about twice as many people as usual will stop to ask you questions about bees. As a result you will want to have assistance at your booth when an observation hive is present. The majority of the questions will tend to be some version of "Where is the queen?" and "How are the bees doing; are they still having problems?" I find that it is good to have one person who handles the questions and another person who can handle sales and make change. Otherwise you may find yourself engrossed in conversation and either ignoring a potential customer, who may end up leaving empty-handed in frustration, or constantly excusing yourself from explaining the intricacies of the hive in order to make a sale.

If you have an observation hive set up with bees in it, make sure it contains a queen; people will be expecting it. One helpful tactic is to mark the queen so that folks can find her for themselves without your help, which frees you up to help other potential customers.

Don't be afraid to use sales aids such as recipes to promote your products. Professionally pro-duced brochures and other aids promoting honey can often be obtained inexpensively from local beekeeping associations or through the National Honey Board (see resources on page 273).

It is a good idea to develop a short introductory greeting to draw people in for more conversation and a sale. Your comment or remark should quickly give folks an idea of what you offer and why your products may be better than what they are used to. Some examples include: "Would you like a taste of raw unheated honey that was harvested just last month?" or "This handmade honey soap won't dry out your skin." It is never a good idea to talk negatively about the vendor selling honey across the aisle; rather, stick to the positive attributes of your products that explain why your stuff is better than what they may find elsewhere. Remember that price isn't the only factor that influences a person to buy.

The most important advice I can offer is to really *listen* to what your prospective customer is telling you. Buyers want to be able to look you in the eye, shake your hand, and walk away feeling that you have earned their trust.

Once you have initiated a conversation, keep in mind that studies have shown that as much as 70 percent of your nonverbal behavior can impact how a person perceives and responds to you. How you or your staff are perceived can also reflect on how your business is perceived. Your body language, facial expressions, cleanliness, professionalism or neatness of clothes, and even your walk and energy level can all influence how a prospective customer will perceive you. When taken together, all these impressions can make or break the potential customer's decision as to whether they want to do business with you. Even your voice, tone, and choice of words can influence how you are perceived. Some examples of desirable nonverbal communications that professionals recommend include:

- Stand, don't sit in your booth.
- Make and maintain eye contact.
- Smile—it is the universal sign of friendship.
- Nod your head while listening.
- Lean toward the person you are speaking with.
- Move at a quick but unhurried pace.
- Be confident and responsive.
- Use the customer's name or title if known, especially when dealing with repeat customers.
- Be polite and professional, no matter how you really feel.
- Remember that the customer is always right, even when they are wrong.
- Don't take complaints or negative comments personally.

- Use courtesy and tact.
- Always be patient and calm.
- If you or your staff make a mistake, admit it, apologize, and fix it as soon as possible.

Before you try to make your first sale, however, figure out what price you want to receive for all your hard work. If you are highly organized, you will have records of how much you have spent on your bees, equipment, and production, as well as how much time you put into working the bees and processing your honey. When you know what your total harvest is, divide the number of pounds of honey you have produced by your costs and the hourly wage you believe you should receive to come up with an accurate amount each pound of honey should sell for. Unfortunately, few of us will manage to take the time to track all our labor hours and expenses accurately in order to make such calculations, and the few who actually do so are likely to discover that the price of their honey would have to be so high that few people would buy it. Thus, the next best thing is to simply find out what comparable products are selling for locally and price your products accordingly.

Other ways to promote and retail your honey and bee products include some type of catalog or brochure. This approach can be very successful, especially if you have access to a large mailing list of folks who are predisposed to purchasing bee products. Don't have access to a mailing list? You can create one by simply offering a free product raffle at every event you attend. Have people register for the raffle by providing you with their name, address, and e-mail address.

Another popular way to promote and sell honey is through the Internet. Listing your apiary on the Internet is a great way to get your message and product information out to a potentially large audience all over the world. There are numerous online companies that offer fairly easy step-by-step instructions on how to use their websites as a platform to purchase your own domain name and set up a website to promote and sell your products.

If you take orders via the Internet or distribute a catalog or brochure you have to be prepared to ship your products. You will need packing and shipping supplies, room to store them, and a work space where you can pack up your orders. A well-run catalog business and/or website can generate large numbers of orders that may push your business to grow significantly. This can be a double-edged sword: history is littered with businesses that failed after growing faster than their management could competently handle.

• THE WHOLESALE MARKET •

If you find that you don't have the time to staff a booth at a farmers' market, or that you have more product than your retail market can absorb, you may want to start selling to wholesalers. In return for freeing up your time by merchandising your products for you, wholesalers expect to pay a lot less for the items they buy—sometimes as low as half the retail price.

A good place to start when trying to break into the wholesale market is your local grocery store. Other businesses that are likely to want to carry local honey include food cooperatives, farm stands, restaurants, and tourist destinations and attractions. Bakeries are usually also looking for good sources of honey, and since their honey is incorporated into baked goods, bakeries will accept lower grades of honey, such as very dark or off-tasting honeys that are not ideal for selling as table-grade.

FIGURE 11-2. Besides meeting all the legal requirements and being attractive, a honey label that conveys a unique story about the product or the way it is produced will tend to be the most effective.

Even though a business may sell honey or bee products regularly, getting them to start buying *your* products can be challenging. The business you are approaching may have a supplier of honey that they have been working with for many years and like a lot, or they may simply not have any room on the shelf for another honey or bee product such as yours.

One of the easiest ways to get your products onto a shelf without stepping on any toes is to make a market connection with a new store that is just opening. Such opportunities unfortunately are fairly rare, and you will more than likely have to approach established businesses. When contacting a prospective wholesale client, it's crucial to make contact with the person who makes the purchasing decisions. Don't waste time speaking with a clerk or an assistant buyer. Provide the buyer, manager, or owner a sample of your product line and let them know what features make your honey special compared to what they may already have on the shelf. If they are not able to make a decision on the day you visit, follow up with them a few days later, after they have had a chance to try your honey or other products.

One approach that can sometimes tip the balance in favor of your products, especially in small mom-and-pop stores, is to offer to not only deliver but to price the product and place it on the shelf. The staff in many stores are often overwhelmed with work, and having a vendor stock the shelves for them can be very enticing. This also gives you some influence over how your product is displayed. Another way to ease your entry into the wholesale market is to offer unique or hard-to-find products such as comb or chunk honey.

Once you have the account, customer service is another way that you can set yourself apart from the competition. Find out how the store buyer likes to order and make it easy to do so. Do they always order restock items on a certain day of the week? Perhaps they would like you to check in occasionally to see whether they would like to place an order for more product. Each store tends to have its own

system, so work with your buyers to try to align your efforts with theirs.

Price is another important factor that goes into a purchaser's decision of whether or not to carry a new product. By selling wholesale, you have already given up a significant amount of profit compared to retail sales, so I advise that you price your honey in comparison to what is already on the shelf. If you can, find out what kind of markup or margin your customer is working with and use this information to price your products accordingly. If you are new to the area and are persuading customers to try your honey for the first time, or you are working to build your market share, you may want your brand to be priced a little lower than the competing brands. Beware of undercutting the competition too much. Not only do you want to cover your costs and not sell yourself short, but it is rarely desirable to develop the product image of the cheapest brand on the shelf, because quality is often reflected by price. If you are positioning your brand as a high-end gourmet item, for example, then you may want to price your products so that they will sell for slightly more or slightly less than the most expensive competitor on the shelf. After you have established an account, consider offering to demo your product in the store. Many stores appreciate vendors who will do in-store promotions by giving out product samples. Customers love to get free samples, and you will have the opportunity to tell folks about your products and what makes them great. Make the most of your demo time by targeting the busiest time of the week at that location. Be sure to bring along an extra case or two of the items you will be promoting so that you can restock the store shelves before you leave for the day.

When I was first starting out and trying to break into the marketplace, selling my honey wholesale to local stores was a terrific way to get my name and my honey known in the community. Placing your products in your local stores acts like a form of free advertising and is especially effective when combined with in-store demos.

One word of caution: do not insist that a store mark your products at a certain price or promote or display them to your specifications. Sure, you can make suggestions, but once you deliver the order to the buyer, it is no longer under your control. The store management and staff can and should be able to do what they like with it. When a producer/vendor tries to tell the staff in a store how they should be doing their job, it rarely goes over well. Imagine if the store staff started telling you how you should be taking care of your bees and insisting how your product packaging should be different. You get the picture?

Bakeries and institutional customers such as hospitals and schools are another segment of the wholesale market. They will use your honey in products they make or meals they serve. These customers are interested in purchasing honey in bulk, which saves you the trouble of providing special packaging and labeling, thus allowing you to pass along the savings to the customer.

• RETAIL CHAIN STORES •

For the most part, dealing with retail chains is not that different from dealing with smaller independent stores. Chain stores do tend to be more institutionalized and less flexible in how they deal with vendors. For example, deliveries often are accepted only on certain days at certain times without exception unless you deliver to a centralized warehouse location. Many chains may also charge you a slotting fee to place your product on the shelf.

The biggest issue regarding chain stores is to not let them put you out of business. This can happen if you allow a single chain, or customer, to become the majority of your business. Certain retail chain stores are renowned for becoming the primary customer of a business and then using their massive purchasing power to demand additional discounts or special deals over time. A chain store can threaten to pull your items from the shelves and stop buying your products if you don't provide

them with the deal they want. Once such a store or retail chain has become 50 percent or more of a business's revenue source, it becomes very difficult to deny such requests. There are stories of companies that have been forced to sell their products for less than the cost of production after finding themselves trapped in the above situation. My recommendation would be to never allow any one customer to account for more than 30 percent of your business revenue. This way, if that customer goes out of business or stops purchasing from you for some reason, your business should be able to weather the shift in the business climate without it having a devastating and potentially fatal impact.

• THE COMMODITY MARKET •

The production of a bulk commodity for sale to processors, packers, and distributors is based upon a model that dates back hundreds of years, to the days when production was geared to producing as much as possible to ship back to the mother country. Commodity production, being large scale by definition, is typically funded by investors, such as the ruling monarchy centuries ago, or the banking and financial industries today. It is primarily focused on maximizing profits, often with large externalized environmental and social costs. Usually accompanying the high costs of commodity production is the acquisition of debt on the part of producers. It is this debt load, and the desperation to meet one's financial obligations, that typically drives producers to use artificially stimulating practices in order to push production, while at the same time externalizing costs through socially destructive, polluting, and unsafe practices. As pointed out in chapter 9, commodity production based upon the tenets of the Green Revolution is proving itself to be based upon failed philosophical concepts that are incapable of taking us through the twenty-first century without further exacerbating our current daunting environmental, social, and economic problems. The future will be dominated not by commodity agriculture, but by human-scale producers who raise food for people (see "Industrialization and Corporate Globilization of Agriculture" in chapter 9 on page 208). In the future most farms will once again maintain habitat and forage for wild pollinators, and/or keep a few hives to handle their pollination needs instead of relying on large migratory beekeeping operations. Farmers and beekeepers are starting to realize that it is more important to grow their operations slowly and stay out of debt than it is to have a good credit rating. Commodity honey production relies on so many hives that chemicals and drugs are required to save on labor costs, and artificial diets are common in such feedlot beekeeping operations. Obviously, none of these approaches is compatible with natural and organic beekeeping.

• AN ALTERNATIVE • BUSINESS MODEL

I have begun practicing an alternative model in how I conduct my business affairs. Why explore an alternative? Simply because a little study of how our money system works indicates that the current model is guaranteed to fail eventually because it is supported by an economic system whose structure is fundamentally flawed.

Since banks are required to keep only 10 percent cash reserves to cover their loans, a bank that has 1 million dollars in cash deposited in its vault can turn around and loan out 10 million dollars. The catch here is that they also charge interest on the money they loan out, even though they didn't have the money to lend out in the first place. This is possible because most people don't use a lot of cash on a regular basis for anything. Most transactions are electronic, through credit and debit cards, or paid via checks. The eventual outcome of such a system is that all monetary wealth is slowly but surely accumulated in the hands of the large banks and those at the top of the economic food chain. Without the implementation of some type of wealth redistribution mechanism that periodically transfers some of the wealth from the wealthy and

places it back in the hands of the working people, the only money that will be available is money that is created as debt.

As if this were not bad enough, our monetary system (the US dollar) is not tied to anything of value other than faith and trust in the United States government. With the elimination of the gold standard, inflation has been a nagging issue that requires constant economic growth at a level that must, at minimum, equal the growth rate of inflation. When economic growth lags behind the rate of inflation, then incomes and profits decline, businesses go out of business, and the standard of living declines. This would not necessarily be a problem except for the fact that constant economic growth is physically impossible on a finite planet with limited amounts of resources that can be exploited for economic gain. Unrestrained growth is not the solution, it is part of the problem. If played out to its logical conclusion, the economic system of Western civilization will eventually collapse, simply because of the way it is designed. Therefore, to continually invest in it and trust in it to reliably provide for our own futures, as well as those of our children and grandchildren, is "irrational exuberance" (to cop a phrase from one former Federal Reserve Bank chairman).

Thus, the first rule I conduct my affairs by is to stay out of debt as much as possible. If debt is unavoidable, then I always choose first to be indebted to someone or an institution other than a bank that will charge interest, or to a credit union that will at least charge a lower rate of interest. My primary way of staying out of debt is keeping the size of my business on a human scale and growing slowly by financing growth entirely with money that is earned rather than borrowed.

One way to be sure the size of my business stays small is to get back to the way agriculture was practiced historically, producing primarily for my family's and my community's needs. I do not ship honey to anyone anymore; instead I encourage out-of-town customers who contact me to purchase their honey from a local beekeeper in their area. The only exceptions to this rule are if the customer is passing through my area or I am passing through their area. By requiring my customers to purchase honey directly from me, I help to limit the rate of growth that my business experiences, and I avoid the waste and inefficiencies of time and energy involved with shipping honey all around the country and the world. By encouraging folks to source honey produced through organic management locally, I help to increase the number of conversations beekeepers have with folks looking for clean honey, and this provides an incentive for conventional beekeepers to at least consider modifying their hive management practices.

Additionally, I sell my products retail only, and I will consider selling wholesale only if my production ever grows to a level that is more than my local market can absorb—something I work on avoiding. This way I always garner top dollar for my work rather than giving away half my profits to a middleman. By restricting my sales to only those that will benefit me the most (highest profit) and cost me the least in additional expenses, I am able to position myself to not only be financially solvent, but also to not become a victim of never-ending growth. Due to the laws of physics, our economy will have to stop growing at some point. We are simply too many people, consuming too much on a finite planet, and at some point those of us in the developed nations will have to learn how to get by with less. By working to slow growth to a more steady state now, we can help this transition be calmer and more orderly, rather than the chaotic and potentially violent transition that is likely to occur for those who do not plan for and implement systems that acknowledge this inevitability.

• SOME FINAL THOUGHTS •
ON MARKETING

A good policy that will help increase customer satisfaction and create consumer loyalty is to stand behind your product 110 percent. Whenever a customer is not completely satisfied, make it right for that person. This means accepting the old adage, "The customer's always right," even when the customer is wrong.

Research has shown that when a customer has a complaint or problem with a product or service and looks to the company to correct the issue, the company in question is given the opportunity to strengthen the business-customer relationship and potentially make a customer for life. If the company fixes the problem not just to the customer's satisfaction, but goes over and above what the customer expects or had hoped for, that customer will become more loyal to that company than if they never had a problem to begin with. Something to keep in mind when you get that call from someone who complains about finding a bee leg in their honey, for example. This is especially important for small- to medium-sized beekeeping operations supplying the local community. Your reputation will tend to be the most valuable part of your business. A reputation takes a long time to develop but, unfortunately, can be destroyed all too quickly.

Get to know your customers, and keep it small so they get to know you too. Provide personal service. If you truly like your work or—even better—love your work, your customers will too, and that's the real secret. A fully satisfied customer is your best promotion, one that all the advertising money in the world can't buy.

CHAPTER 12

Organics and the Evolution of Beekeeping

She imbibes the song of joy evoked by sunlight on petals.
—THE BEE MISTRESS, AS QUOTED IN
THE SHAMANIC WAY OF THE BEE
BY SIMON BUXTON

Agriculture is the backbone of any sustainable economy. By combining the gifts of Earth, Air, Fire, and Water, agricultural activity is one of the few forms of production that creates new wealth for society, wealth that did not previously exist in any form. Most economic activities other than agriculture simply add value to previously created wealth or materials to create a profit. Historically, major economies have always experienced an agricultural revolution before making a transition to an industrial model.

For many years the American market for organic foods has been the fastest-growing segment of the US food industry, achieving yearly gains of 20 percent or more. Since the US economic collapse of 2008, organic sales still lead the industry, but have slowed, with 2010 sales increasing 9.7 percent over 2009 sales according to the Organic Trade Association.

· TRANSITIONING TO · ORGANIC APICULTURE

Although the Organic Foods Production Act (OFPA) was passed in 1990, it wasn't until 2002 that a final set of organic regulations was finally produced. Despite going through many amendments and modifications since then, the National Organic Program (NOP) has yet to specifically address organic honey production, and the only

guidance beekeepers have at the federal level are the recommendations put together by the National Organic Standards Board (NOSB). One assumes that the patchwork of organic standards that are currently being maintained by various certification agencies throughout the United States will sooner or later be replaced by national standards once the NOSB recommendations, or something like them, have been codified by the USDA. A review of current organic apiculture standards and the NOSB recommendations indicates that there will probably be five basic components with which the organic beekeeper must contend once the NOP completes its work.

The first issue to grapple with will be documentation. Everything that goes into or comes out of organic hives will need to be documented. This means that each hive will have to be given a unique identification name or number to differentiate it from all the other colonies in the apiary. If heat is used during any part of the processing and bottling operation, it will have to be monitored and documented. It has always been a good idea to keep a journal of observations, manipulations, treatments, and honey harvests anyway, but organic standards *require* the apiculturist to take this practice to the next level. Once we get into the habit of keeping careful records of each hive, it is easier to track performance over long periods, and this allows us to more easily identify the breeding stock with the highest potential.

Forage type and availability is the second major component of organic honey bee management. As indicated earlier, it will be challenging, to say the least, to locate bee yards in areas where only organically raised crops and unsprayed wildflowers that conform to organic standards are found. Unless future federal standards make a radical shift from what has historically been the case, organic regulations will limit organic apiculture mostly to places where no pesticides or chemical fertilizers have been used, such as wilderness areas and very large organic farms, where an apiary can be established in the middle of the farm to prevent bees from potentially foraging on blossoms growing on plants that have been sprayed or fertilized in ways that would affect organic integrity.

Much has been said, and continues to be said, regarding the necessity of beekeeping and honey bees for the proper pollination of many crop varieties. Research conducted at Cornell University indicates that approximately one-third of all cultivated food crops require the presence of bees for farmers to realize the sizable harvests they have come to rely upon, and the world's hungry have come to need. Eliminate honey bees and the pollination services they provide, and the economic and social upheaval that would result from the massive loss of yields among affected crops would be staggering. What is not often discussed, however, is the important role that agriculture plays in sustaining the beekeeping industry.

The Champlain Valley that encompasses Lake Champlain and lies between New York's Adirondack Mountains and Vermont's Green Mountains is a vast, fertile expanse that supports a strong agricultural industry made up primarily of dairy farms. These farms maintain many acres of clover and alfalfa to feed their cattle, and it is these same fields that make the Champlain Valley such a wonderful place to keep bees. Beekeeping in this little corner of the world relies upon dairy farmers for its viability. Eliminate the farms, and forests would soon replace the fields, reducing the availability of bee forage so drastically that the number of honey bee colonies the valley could support would likely be reduced by an estimated 60 to 70 percent. Thus the interdependence that exists between the world of the honey bee and the plant kingdom directly impacts the relationship between farmers and beekeepers. It is this dependence that is the most limiting factor for certified organic honey production.

Until there is a significant increase in the percentage of total farm acreage in the United States that is certified organic, the limited availability of forage for certified organic hives will severely depress the potential number of organic colonies that can be managed. This means that organic honey will continue to be rare and producible only by those lucky enough to have access to the limited number of areas where certified organic production can take place. Because organic agriculture makes up only about 2 to 3 percent of the overall agricultural acreage in the United States, the majority of organic apiaries will have to be located in and around wooded areas, parks, and other wild places that typically do not provide an abundance of nectar sources and thus do not contribute to large honey surpluses and bumper honey crops. These circumstances will force the organic bee industry to remain small and localized well into the foreseeable future.

However, in many cases apiculturists can adopt and benefit from another key factor of organic production: hive management. Different techniques and materials are now available to replace the hard chemicals upon which many beekeepers have become dependent. The one-size-fits-all silver-bullet approach will no longer suffice due to variations in scale, goals, time limitations, beekeeping skill, and climate. Each beekeeper will have to develop a unique style that will utilize the three categories of disease and pest control measures: mechanical (such as screened bottom boards, traps, secluded locations, and burning), biological (beneficial fungi, *Bacillus thuringiensis*, genetically tolerant queens, et cetera), and soft chemicals (essential oils, organic acids, and other homemade and commercial treatments). In order to add value to their honey and differentiate their finished products from the rest of the pack, savvy marketers will use as a selling point the fact that alternative methods eliminate the use of toxins and antibiotics.

FIGURE 12-1. White clover is widely considered to be a premier honey plant, and its honey is regarded as one of the finest in the world: the standard of comparison for high-quality honey.

Because most standards in effect today do not allow the organic beekeeper to just let a hive die, the use of conventional methods to restore colony health are typically required whenever a hive is in jeopardy. A "hospital yard" will have to be established, with procedures for dealing with sick hives that must be pulled out of organic production. Beekeepers will also have to take measures to prevent any possibility of commingling organic with nonorganic honey.

All aspects of harvesting, processing, and storing of hive products will come under additional scrutiny. Bees will have to be separated from their honey by mechanical means rather than through the use of chemical fume boards. Old galvanized extractors

will fall by the wayside as modern stainless-steel and food-grade processing equipment becomes the order of the day. The beekeeping industry is fortunate that its primary product, honey, will not support the growth of any bacteria, fungi, or mold that is harmful to humans. Honey, in fact, is so safe and stable that samples of honey found in ancient Egyptian tombs were still edible.

"Do not feed honey to infants under one year of age" or some similar warning is common on honey labels. This is because honey may contain spores of the bacterium *Clostridium botulinum*. Botulism spores are similar to seeds in that they will germinate when placed in a favorable environment and grow into their vegetative phase. Infant botulism is caused when enough spores enter their vegetative stage and start growing rapidly in the infant's immature digestive tract, producing a toxin that impacts the child's neurological functions. Newborn babies lack the intestinal microflora that prevent healthy children and adults from getting sick after ingesting *Clostridium botulinum*. About half of reported cases of infant botulism have occurred in babies less than two months old. While it is believed that by six months of age most infants will have developed their intestinal flora to the point where they become resistant to *C. botulinum* (especially if they are breast-fed), an additional six months has been added to the warning by the national Centers for Disease Control (CDC) as a safety factor.

Symptoms of infant botulism include muscle weakness or loss of control such as droopy eyelids, weak cry, feeble sucking, drooling, lethargy, irritability, constipation, and progressive "floppiness," all of which may follow an initial fever. An infant exhibiting these symptoms should receive prompt medical attention. Recovery is almost certain as long as the condition is diagnosed and treated early and the baby has not suffered brain damage. Infant botulism should not be confused with food-borne botulism (resulting from the consumption of preformed botulism toxin), which can sicken older children and adults as well as infants.

Infant botulism often occurs in babies who have not been fed honey. This is because *Clostridium botulinum* spores are ubiquitous in our environment and found in soil and water and on dust floating in the air. It is not known how many spores must be ingested before infant botulism will occur, or why only some babies seem to become sick. However, it does seem that the chances of contracting the illness depend a lot on the immediate environment of the infant and the overall health and susceptibility of the baby. Infants may contract botulism from surfaces in the environment, breathing dust in the air, or from water or food. Most foods will contain *Clostridium botulinum* spores unless they have been processed in a way that has cleaned off or destroyed the spores and bacteria. The states of

Pennsylvania, Utah, Arizona, and California tend to have higher instances of infant botulism than other states. It is believed that these states may have soils that have higher than average levels of *Clostridium botulinum* spores.

In its vegetative stage, *Clostridium botulinum* cannot survive in honey due to honey's antibacterial and antimicrobial properties, which are well documented. In fact, when honey is ripe, with a moisture content below about 18 percent, nothing harmful to humans can grow in it. Raw honey that is ripe is the only natural unprocessed food that, for all intents and purposes, will never spoil when stored properly in an airtight, moisture-proof container. Honey does not need to be refrigerated.

There are several ways that honey controls the growth of bacteria and mold. The higher sugar content and the pH of honey inhibits the growth of molds and other pathogens in much the same way that sugar is used to preserve jams and jellies. Raw honey that has been unheated and unfiltered also contains the enzyme *glucose oxidase*, which converts into hydrogen peroxide and gluconic acid as it breaks down on the skin. Thus, raw honey applied to a wound will be constantly releasing hydrogen peroxide that will help sterilize the wound area. In addition, honey is hygroscopic. This means that it draws moisture to itself. As a result, when honey comes into contact with bacteria, it will suck the moisture out of the bacteria, killing off the microscopic critters. In fact, if you have botulism bacteria growing in a petri dish and add raw honey, the honey will kill the bacteria. Given that honey itself will kill off *Clostridium botulinum* in its vegetative stage and that the spores are prevalent throughout our environment, and thus present in many foods besides honey, why do we focus specifically on the potential botulism risk in feeding honey to infants under the age of one?

To date I have not been able to find any documented evidence of a single case of infant botulism that can be proven to have been caused by honey. This may be because it would be considered immoral to conduct a study where babies were purposely fed honey contaminated with botulism spores in an effort to clearly prove cause and effect. However, this begs the question: why is honey the only food that is singled out for a warning label stating that it should not be fed to infants less than a year old?

When researchers investigate instances of infant botulism, they find that in most cases the child has not consumed honey. However, there have been cases of infant botulism where the baby had been fed honey at some point prior to getting sick. When this information is combined with the fact that about 5 percent of the roughly 2,100 honey samples tested have been found to contain *C. botulinum* spores, and in at least one case an infant that contracted botulism had eaten honey that tested positive, this identifies honey as a risk factor and establishes a correlation between honey and infant botulism. This is the reason why the CDC, the American Academy of Pediatrics (AAP), and Health Canada, along with other public health associations and the National Honey Board all agree that there is enough of a scientific link between honey and infant botulism to warrant the precautionary measure of a warning statement. However, any scientist worth his or her salt will tell you: one of the most basic principles of science is that correlation does not prove causation. This basic truth seems to be what the CDC, AAP, Health Canada, and the National Honey Board want us all to forget.

Now, don't get me wrong, I am not claiming that because it has yet to be proven that honey is a cause of infant botulism, it is impossible for babies under one year of age to contract botulism from contaminated honey. After all, high concentrations of spores have been found in honey at times. What irks me is that the same can likely be said for many other foods. As a result I will admit that while label warnings may be prudently cautious and appropriate in the case of honey, I am at a loss to understand why honey is the only food item singled out to carry a warning statement.

In the rare cases where honey has been found to be severely contaminated with *C. botulinum* spores,

where are these spores coming from? Some scientists believe that high concentrations of botulism spores may enter honey during rare and extreme conditions within the hive, or when dead bees are mixed into the honey. However, since botulism spores have not been shown to grow in nectar that is being processed into honey inside the hive, nor can it grow in ripe honey that is in the comb or in the jar, it seems most likely that contamination occurs sometime during the honey harvesting, extracting, and bottling processes.

Efforts to produce honey free from botulism spores start by thoroughly cleaning your honey processing area and all equipment prior to use for extracting and bottling. A high-pressure hose and extremely hot water are the best thing you can use to clean honey, wax, and propolis from extracting equipment. For smaller jobs, Citra Solv, an essential-oil-based cleaner and degreaser, and a scrubbing sponge works well. Everyone involved in the processing of honey should wash their hands before work. These are commonsense actions that should be done by everyone on a regular basis anyway.

If you want to go the extra mile, keep bees from getting into the honey during extracting and close open windows to prevent dust buildup in the honey house. To be extra cautious, avoid tracking dirt into the processing area by leaving outdoor shoes outside and wearing only clean indoor shoes while processing the honey. As inconvenient as these last suggestions sound, they may be the easiest way to ensure that honey is free from significant amounts of *Clostridium botulinum*, especially if you keep bees in one of the states identified as having a higher than normal risk. Because adding honey to a petri dish containing a botulism colony will kill the bacteria growing there, honey's natural qualities as a powerful antimicrobial agent and preservative have thus far covered up for many a careless operation, but such sloppiness will no longer be tolerated in the world of organic certification.

The fifth component of beekeeping that will change will be the labeling and marketing of the final product. All organic labels must by law state the name of the USDA-approved third-party certification agency responsible for confirming the organic integrity of the finished product. Given that the folks responsible for your certification will have their name on your label, they will want to ensure that you meet all specifications laid out in the standards. Labeling details such as font size and the positioning of certain statements (for example, the name of the certification agency must appear directly under or next to your company contact information) will become paramount. Organic beekeepers will want to have their organic certification agency preapprove their label before sending it off to the printer, or they run the risk of having to modify and reprint labels after the fact if the labels don't meet with official approval.

A sixth and final factor should be mentioned here: the significant investment that beekeepers will need to make to certify their honey as organic. Not only will there be costs associated with the trial and error of transitioning to organically harmonious ways of keeping bees, but there will also be a yearly fee that will have to be paid to cover the cost of third-party inspections. That's right. Your apiaries, your honey house, your books—everything—will be scrutinized to make sure you have met all organic requirements; so be sure to do your homework before applying for certification, and maintain rigorous adherence to the standards. The fees are typically determined by the size of your business as measured by yearly sales figures. An exemption has been made for small operations that gross less than $5,000 a year. These small businesses don't have to pay to be able to call their honey organic, but they still have to meet all the requirements of the standards. Such small producers do not have to undergo third-party scrutiny and are thus taken at their word. In the end, though, they are not allowed to use the organic logo on their label, nor do they receive a certificate of organic authenticity for their products—a document that more and more retailers are asking for these days to meet their customers' expectations.

FIGURE 12-2. Only products that have received USDA-approved third-party inspection and contain at least 95 percent organic ingredients can sport the USDA organic logo.

Once you've established your plans and set up your systems to meet the organic standards, the work of organic beekeeping can begin. First you have to transition your bees into organic management. The transition period for changing honey bee management from conventional to organic, depending on the standards followed, tends to be less than a year. With all the challenges of getting organically certified with honey bees, this is one of the few bright spots. For instance, dairy livestock are required to pass through a full one-year period of organic management before being certified, and agricultural land must undergo a three-year transition period before becoming fully organic. Of course, all bets are off once the national standards are finally in place, because they may call for a very different length transition period.

One might logically assume that those beekeepers who have managed to become certified through one of the sets of standards written and enforced by an individual certification agency prior to the establishment of the NOP organic apiculture regulations will have an easier time making the transition to national standards. Unfortunately, this is not necessarily so. The NOP defines honey bees as livestock, and as such, they are subject to all the rules governing organically raised livestock. For example, organic livestock may be fed only 100 percent organic feed. The OFPA, however, takes the position of previous legal opinions that distinguish between honey bees and traditional livestock, whose foraging activities can be more easily controlled. As a result, the question of whether bees are to be treated as livestock and must be fed exclusively organic feed while under organic management is about as clear as mud. Until there is an answer to this question issued by the NOSB and codified in the NOP, individual certifiers are being left to their own devices to sort out the issue of honey bees' status as livestock.

Some certification agencies allow the use of nonorganic supplemental feeding and the use of the USDA organic logo on the honey label. Other agencies don't allow one or the other, and in some cases, neither practice among beekeepers is certifiable under their standards. Agencies that do not allow use of the USDA organic logo on honey have more leeway in interpreting the standards, since the final product falls under their own organic standards rather than the national regulations. Many of these certification groups were active before the implementation of the NOP and have their own organic logos that beekeepers can use on their honey label, as long as they meet the relevant group's standards.

• THE STATE OF ORGANICS •

The organic movement was started as an alternative to the status quo and "business as usual." It helped to create the not-for-profit-only business model in which quality is more important than quantity, and now organic has proven so successful as a concept that big business is taking notice and wants a piece of the action. Needless to say, all the confusion and indecision surrounding aspects of the NOP serve the interests of conventional mass-market businesses well. Producers that want to take advantage of the marketing strength of an organic label without having to change their conventional practices have benefited the most from the confusion. This has resulted in the organic standards becoming subject to political manipulation, an influence that does not always have the integrity of the regulations in mind. Some companies have even been able to "beat the system" by lobbying their congressional representatives to pass legislation that bends the

rules and allows exemptions to the rule requiring 100 percent organic feed for livestock, or lets chickens or cows be certified organic even if they don't get to graze outside for any significant time in their lives. Because of this shifting playing field, where every interest is jockeying for advantage, individuals and organizations that wish to preserve the integrity of the federal organic standards must be vigilant in defending them against efforts that would weaken them in the eyes of the consumer.

The "good old days" when local producers sat around a table and hashed out the issues until they all came to a clear conclusion that everybody could live with must seem like a real blessing to today's certification administrators, who have to contend with so much uncertainty. The bureaucratic nightmare created by the federalization of organic agriculture is a major hurdle for the conscientious organic certifier. In addition, the delays and foot-dragging create additional challenges and difficulties for certifiers and producers. Within the first few years of implementation of the NOP, the USDA made and proposed changes to the standards, only to backpedal later, after the consuming public expressed widespread opposition. The USDA has also released statements indicating that it has taken a particular position on an issue, only to stall on releasing the official written decision. This apparent indecision creates an air of insecurity among both certification agencies and producers. True organic production requires a degree of certainty that in many ways is missing from the National Organic Program as it has been implemented in its early years. One can only hope that this situation will improve with time.

Whether consciously or unconsciously, many beekeepers, bee researchers, and beekeeping organizations are helping to perpetuate the air of indecision and confusion surrounding the finalization and implementation of organic apiculture standards. This group includes a variety of interests, from beekeepers who think that all apiculture is natural and that there is "no such thing as organic beekeeping" to agricultural schools that promote a chemical mind-set that tends to overlook natural alternatives and that are influenced by research money doled out by chemical/agribusiness companies. Many of these same institutions and individuals will express the feeling that anyone who is claiming that he or she is keeping bees organically is lying and must be cheating somehow. This viewpoint is understandable and to be expected given the high level of pesticide and antibiotic "misuse" that occurs among conventional beekeepers. Far too many beekeepers do not use beekeeping chemicals and drugs as directed by the label or use unapproved pesticides and antibiotics in their hives. Although these activities are defined as federal crimes under the law, weak enforcement and lack of testing provide the opportunity for rampant abuse. Thus, because cheating is all too common among conventional beekeepers, it stands to reason—the thinking goes—that organic beekeepers must be cheating as well. However, unlike conventional beekeepers, organic producers have a third party looking over their shoulder to ensure that they do not cut corners and cheat the system. This is one of the reasons why the consumer is willing to pay a premium for organically certified foods.

It's a good thing consumers have not yet gotten wind of the corner-cutting activities of some conventional beekeepers. It would take only one well-publicized incident of chemical or antibiotic contamination to ruin the sterling reputation that honey enjoys as a pure, natural product. The consumers' best defense in this regard is to establish a personal relationship with a local beekeeper, ask a lot of questions, and determine for themselves whether the production practices of their local supplier meet with their approval and philosophical preferences. It is precisely this desire by many consumers that is part of the driving force behind the current expansion of the organic food industry, as well as the increasing number of farmers' markets and community-supported agriculture (CSA) farms that eliminate the middleman and allow farmers to deal directly with the end user. Beekeepers can substantially increase their profit

margins and gain significant control over their distribution by packaging their honey themselves and marketing directly to wholesale distributors, retailers, and/or consumers, rather than supplying the bulk commodity market (see chapter 11).

Another reason for the resistance to organic apiculture by some beekeepers is that organic honey is perceived as a threat. Not only do these folks believe, either correctly or mistakenly, that they will be unable to meet the standards for organic production, but they worry they will have to answer to customers who will want to know why their honey is not certified organic. The fear is that nonorganic, conventionally produced honey will be viewed as inferior to organic honey. This is disturbing for many of these beekeepers, since they may sincerely be trying to use chemicals and antibiotics appropriately and are doing their best to produce high-quality products. Many of these folks think that meeting the tough requirements for organic production is simply not possible because they have such high colony numbers, are unaware of the organic management practices they could be utilizing, or are established in a location that would preclude the option of organic certification due to the close proximity of nonorganic foraging sources. When faced with questions from their customers, many beekeepers will be faced with the choice of either trying to explain their use of toxic chemicals and antibiotics in the hive or simply lying. As a result, many in this struggling industry—those who are unable to take advantage of the marketing power of the organic label—will find the business climate all the more challenging as organic apiculture gains in popularity.

Unfortunately, the industrial agricultural model emphasizes profit and relegates stewardship of the land to secondary status at best. The organic philosophy, on the other hand, places stewardship of the land and its denizens as a more important goal than the guarantee of a bumper crop. Systems and processes that renew and rejuvenate themselves define organic agriculture that is truly sustainable in the long run. By investing in the health of the hive and following the Hippocratic oath of "First, do no harm," the organic apiculturist profits by working in the honey bee's best interests, rather than profiting through exploitation. The health of a hive is the sum total of a wide variety of large and small details. Such a state of health can be accomplished consistently, not by approaching the craft of beekeeping through the industrial, one-size-fits-all model, but only by recognizing and respecting apiculture as the natural biological activity that it is—an endeavor that touches on the health and well-being of the land and all of its inhabitants.

• THE FUTURE OF • ORGANIC BEEKEEPING

Honey bees and the people who care for them will be an essential component in the development of local and regional food systems, which will be required in the future to provide for our long-term food security and maintain a sound, sustainable basis for economic activity. The reliance on toxic, petrochemical-based pesticides by large agribusinesses in their attempt to "feed the world" has resulted in the contamination of our soil and water and has led to the serious decline of natural pollinators in many areas. Their corporate message is based on the assumption that we must accept this environmental degradation in return for an end to hunger, despite the fact that world hunger is an issue of distribution and geopolitics, not one of production.

Our industrial model of farming has created many farms that consist of acres and acres of a single crop. The use of pesticides on these fields has eliminated most of the natural pollinators, and, as a result, farmers have become totally reliant on the services of migratory beekeepers. As the supply of native colonies for providing pollination services continues to decrease in the face of pests and disease, the costs of renting honey bees will climb. The agricultural industry that already has had to rely on importing bees from other states must now import bees from other countries. The economic and ecological costs associated with transporting hives on tractor-trailer trucks cross-country, or by plane from overseas, to provide pollination services for large farms and orchards will continue to rise significantly unless alternative, sustainable, and ecologically sound fuel sources can be fully developed. Unfortunately, in the United States, government and industry seem more intent on trying to hang on to what's left of the world's declining oil supplies than investing in alternatives that will take us into the future. My sense is that factors such as these will ultimately result in a return to smaller farms being run on a human scale and a new reliance on local apiaries and beekeeping outfits for pollination. These farms will produce a diverse variety of crops to feed the families that work them along with the citizens in the surrounding communities rather than focus on mass production for export or commodity markets.

As part of this shift, more and more farms will elect to keep their own hives for pollination purposes in answer to both the increasing costs associated with renting hives and the challenge created by the dwindling supply of bees available for hire. This will result in the improvement of the overall health of the honey bees, because they will not have to undergo the stress of transportation, risk exposure to disease and parasitic pests in other areas of the country, or suffer from the limitations in diet that occur when hives are transported long distances and placed in the middle of acres and acres of a single crop during the time of bloom. The bees will also benefit from the reduction in overmanipulation and artificial stimulation with pollen substitutes and sugar or corn syrup that many migratory beekeepers utilize in their effort to equalize and build up their hives prior to performing pollination services. We say that hives that have been fed in such a manner are stronger because the feeding stimulates brood production, but what does this artificial feeding, when done on a regular basis, do to the bee's immune system and the overall health of the hive?

FIGURE 12-3. Clean, uncontaminated pollen that is fermented and turned into bee bread in the hive is the best possible source of vitamins, minerals, and protein that a colony can get.

I envision beekeeping following the path that all of agriculture will have to travel to become more profitable and sustainable in the future: small-scale producers will market products locally in their final form at wholesale or retail prices, and they will take advantage of the increased profit margins that can be obtained through value-added products, as opposed to producing a low-margin bulk commodity. By taking back control and responsibility for supplying their own inputs, processing, bottling, and marketing, beekeepers can become much more profitable, and their operations will be more likely to successfully weather times of economic stress. By reducing the reliance on outside inputs and developing management

techniques that increase soil fertility and livestock health over time rather than depleting them, organic and other human-scale ecologically focused forms of agriculture offer the real, and only realistic, hope for the future of farming.[1] Beekeeping and beekeepers, by necessity, must be an integral part of this agricultural shift.

One of the most obvious and contentious issues that impedes progress on national organic honey standards is the establishment of appropriate buffer zones between organically managed hives and things that would compromise organic integrity. The proliferation of the products of genetic engineering throughout the United States alone may make organic honey and pollen production a moot point, depending on the level of GMO contamination that is allowed in products from the hive. However, the quagmire of certification issues extends much deeper. For instance, should organic farms and orchards be required to use certified organic bees for pollination? Such a requirement might make sense when we look at the established precedents that have already been set throughout the organic regulatory process. For example, nonorganic ingredients may not be used in products labeled as organic, or 100 percent organic, if an organic option is available, and synthetic inputs are allowed only if they appear on the national list of approved and regulated synthetics. What will it mean if standards provide for the pollination of blossoms on organic plants by bees that have been exposed to chemicals unapproved for organic use? When we recognize the incredibly large numbers of hives required for agricultural pollination, and the shortages that already exist in the number of conventional hives available to pollinate nonorganic crops, an allowance is clearly going to have to be made for the use of nonorganic hives pollinating organic crops—at least until the number of certified hives increases dramatically.

Such a spectacular growth in the number of colonies that are managed without toxic chemicals and drugs is a realistically attainable goal. If I can keep my bees healthy and productive without resorting to toxic chemotherapy treatments, each of you reading this book can do so, too. As you try out different approaches to keeping bees, it is wise to evaluate each new management protocol on just a few colonies initially to determine how well it will work in your situation. Once you have grown comfortable with the various methods and techniques with which you are experimenting, they can be expanded to the rest of your operation.

The honey bee is an endless source of fascination, variation, and contemplation. So much so that the claims that bees have been written about and studied more than any other insect on Earth are not at all far-fetched. The fact that the little honey bee manifests such incredible healing power

in the world—healing that works on many levels, from plants to humans—is a real blessing. Unfortunately, both the huge managed-colony losses and the disappearance of feral swarms, which has been consistently reported in many areas of the globe during the past decade, indicate that the honey bee—like every single major ecological system and many of the species on our planet—is under profound stress.

The answer, in my opinion, will not be found in a fast and easy magic-bullet solution. Instead, it will require a combination of methods and a shift in our current management techniques that must be utilized with success over the course of many decades, and without developing newly resistant pests or diseases. By adopting a variety of natural, nontoxic approaches to hive management that limit stressful sublethal effects, such as some of the ones described in this book, visionary beekeepers will help move their industry toward a permanent solution—or at the very least help buy more time for the honey bee. Such creativity, focused on helping the beekeeping industry to progress in a sustainable manner, has the potential to carry the ancient craft of keeping bees well into the future.

• CULTIVATING • A GREATER AWARENESS, SPIRITUAL AND PERSONAL RELATIONSHIP WITH THE BEES

Humans have long regarded the honey bee and honey as sacred.[2] From the civilizations of old Europe between 7000 and 3500 BCE through the ancient Sumerian and Babylonian times and the Egyptian, Greek, and Roman empires to the descendents of honey hunters who carry on their traditions in Asia and Africa today, the honey bee and beekeeping have long been closely associated with spiritual ideas and practices. Honey and bees are prominently mentioned in the Bible and Hebrew scriptures. Frequent mentions of bees and honey are found in the sacred writings of India, the Vedas.

Sura 16 of the Koran, the sacred book of Islam, is entitled "The Bee." Even today many beekeepers in North America who may not necessarily be very religious, when pressed, will admit to a feeling of sacred relationship with the honey bee, either in the way that bees came to be in their lives or in how they feel when they are alone with the bees. As described in the original preface to this book, my involvement with beekeeping came about as a result of my personal spiritual searching. It is this foundation upon which my beekeeping adventures began that has served to shape how my relationship with the honey bee has evolved.

As beekeepers, how much time do we spend thinking about or working on our relationship with the bees? If we don't give it a lot of thought, our relationship with the bees is likely limited to looking out for their welfare, doing our best to keep them healthy and well fed, enjoying the benefits of pollination and gathering up products from the hive for our own benefit in return. While there is nothing necessarily wrong with this relationship, it could be so much more.

What is preventing our relationship with the bees from deepening and becoming greater? For each of us, the answer to this question will be different. A good place to begin may be with the apicultural ethic I describe in chapter 2 on page 16. When beekeepers nurture a personal relationship with their bees, such an apicultural ethic comes naturally. Why would one want to develop a personal relationship with a bunch of insects? Well, first of all it feels good. You will feel closer to your bees and not separate or cut off from them. The personal relationship you develop with your hives has the potential to move you as much as a human relationship will. Over time the connection you develop with certain hives can feel similar to your connection to old friends.

However, like all close personal relationships, it takes time and work to develop. Indigenous cultures had a natural connection to the land, its plants, and animals. We unfortunately have to put more effort into making such connections strong

since our society does not offer much support in this area. Building a close personal relationship with your bees can help eliminate fear of the bees, allowing you to feel safer around the hives. After all, when you think about it, humans are the most powerful and dangerous animals on the planet. If anything it should be the bees that are fearful of us!

So let's say that you think that developing a close relationship with your bees, above and beyond the relationship you have now, may be a good idea—just how might you go about it? To start with, when opening up the hive be full of wonder and try to view the hive as if you were looking at it for the first time. To see the hive with the eyes of an innocent child means being enthusiastic about what you see. This is a way to see what is always there but may often be overlooked, perhaps because we don't slow down enough to take the time to see. While it is natural to focus on the activities the bees are engaged in, urge yourself to notice other things, like the approximate age of the bees, or the comb and its condition and use (for instance, the size of cells, the amount and kind of pollen stored in cells, the age of eggs, et cetera).

This process of careful observation can be further enhanced by closely observing a single comb covered with bees for 15 to 30 minutes or longer. What jobs are the bees engaged in? What condition is the brood in? Are the bees reacting to your presence? While an observation hive really comes in handy for this exercise, it can be done out in the field when the weather is favorable and robbing pressure is minimal.

Make an effort to pay attention to your other senses during your visit. What body sensations do you feel? What smells fill your nostrils? What sounds fill your ears? Do you get goose bumps when you are with the bees? What emotions come up for you when you are opening the hive or manipulating the frames of comb? You may be surprised by the range of nonvisual information that you receive during your time with the bees. While children are often very tuned into these nonverbal cues, as adults we may have to make more of a conscious effort to observe them. Pay attention also to how these sensations and emotions may change over time as your relationship with the bees matures.

An essential part of what I refer to as feeling a connection to the bees is being able to feel what is going on inside of ourselves. This is something we are not given much encouragement for in our everyday lives. Being aware of bodily sensations, or emotions that are evoked by what is going on around us, can provide valuable information as to how well the situation, person, or hive resonates with our being. This is because our feelings never lie to us. Our rational, logical minds on the other hand deceive us all the time. We have all had the experience of thinking that a certain course of action was the best course to take only to find out later that we made a mistake. When something feels right, it is right. When something doesn't feel quite right, no matter how right we *think* it is, it is not right.

The ability to form a strong connection with the bees relies on regular interactions. Visit your hives regularly, at least once a week. Those that are interested in forming a strong personal relationship with the bees may even choose to make a date with them. Pick a certain day and time and show up every week, rain or shine. You don't have to open the hive every time you visit. Just spend time with the hive. Regular visits are especially important when you are just starting to develop your relationship with the bees; it will take a few seasons to develop the relationship strongly.

When dating it is common to bring gifts. While the bees may appreciate a gift of flowers, they may not appreciate a gift of chocolate as much as you or I. Honey bees are always giving to us through their work as pollinators and their production of honey, wax, propolis, pollen, and even bee venom (for therapy). It is good to give back both through our concern and care for their welfare and through gifts

of the heart, mind, or prayer. When I approach a hive with a jar and ask the colony, a superorganism, to give up a part of its body by letting me take several bees for bee venom therapy, I will offer part of my body in return by pulling out a few hairs from the back of my neck and leaving the hairs by the hive. Such symbolic gifts act as a physical reminder of the respect and appreciation we may want to cultivate for the bees.

You might even consider making offerings to the bees without taking anything in return. Have you ever experienced how worn down you can become when others are constantly taking from you without offering anything in return? We receive so many blessings from the bees, to give to them without expecting to receive anything ourselves is one way to form a healing relationship with the bees.

Another way to help forge a personal connection with bees is to talk to them or even sing to them. Talking to bees, along with prayer, gift giving, song, and dance, all have a long history in beekeeping and are ways of making a heart connection. Many of us already talk to our cars, televisions, and radios, so why not the bees? Once you have gotten comfortable talking to the bees, feel free to give yourself permission to express to others how you feel toward the bees.

Once you have established a strong connection with the bees, it will carry over to other places.

When we are developing a personal relationship with honey bees, it can help us nurture relations with the rest of life on Earth, as well as help us become better beekeepers. I am not claiming that honey bees share the same kind of feelings that we do, but bees do respond when we take the time to manage our beehives using natural and organic methods and when we express our love, appreciation, and respect for them. The bees can't help themselves. Such a response is shared by all living things.

NOTES

Chapter 1:
Why Organic Beekeeping?

1. Malcolm T. Sanford, "Using Liquid Formic Acid for Mite Control," *Bee Culture* 131 (June 2003): 18.

2. *SEER Cancer Statistics Review, 1975–2008*, table I-14, "Lifetime Risk (Percent) of Being Diagnosed with Cancer by Site, Race/Ethnicity Both Sexes, 17 SEER Areas, 2006–2008," National Cancer Institute. Accessed October 14, 2012 at http://seer.cancer.gov/csr/1975_2008/results_merged/topic_lifetime_risk_diagnosis.pdf.

3. Larry Connor, "More on Drone Biology," *Bee Culture* 133 (September 2005): 19–21; Jeff Pettis, Anita M. Collins, Reg Wilbanks, and Mark Feldlaufer, "Survival and Function of Queens Reared in Beeswax Containing Coumaphos," *American Bee Journal* 146 (April 2006): 341–44.

4. Brian Halweil, *Eat Here: Reclaiming Homegrown Pleasures in a Global Supermarket* (New York: W. W. Norton & Company, Inc., 2004): 63–64.

Chapter 3:
Hive Management

1. M. Beekman and F. L. W. Ratnieks, "Long-range Foraging by the Honey-bee, *Apis Mellifera* L.," British Ecological Society, Department of Animal and Plant Sciences, University of Sheffield, *Functional Ecology* 14 (2000): 490–96.

2. R. R. Sagili and C. Breece, "Effects of Pollen Quality (Diversity) on Honey Bee Physiology, Immunocompetence and Colony Growth," Proceedings of the American Bee Research Conference 2010, *American Bee Journal* (April 2012): 406.

3. Eva Crane, *The World History of Beekeeping and Honey Hunting* (New York: Routledge, 1999): 395–96.

4. Thomas Seeley, *Honeybee Democracy*, (Princeton: Princeton University Press, 2010) 50.

5. Blaise W. LeBlanc et al., "Formation of Hydroxymethylfurfural in Domestic High-Fructose Corn Syrup and Its Toxicity to the Honey Bee (*Apis mellifera*)," *Journal of Agriculture and Food Chemistry* 57 (16) (July, 31, 2009): 7369–76.

6. A. Juarez-Salomo and P. Valle-Vega, "Hydroxymethylfuraldehyde Thermogeneration as Honey Quality Parameter," *Tecnologia-de-Alimentos* 30 (6) 13–17, 17 ref. Accessed at http://www.airborne.co.nz/HMFref.html.

7. Edoardo Capuano and Vincenzo Fogliano, "Acrylamide and 5-hydroxymethylfurfural (HMF): A Review on Metabolism, Toxicity, Occurrence in Food and Mitigation Strategies, " *LWT—Food Science and Technology* 44 (4)(May 2011): 793–810; European

Food Safety Authority (EFSA) Panel on Food Additives and Nutrient Sources Added to Food (ANS), "Scientific Opinion on the Re-evaluation of Caramel Colours (E 150 a,b,c,d) as Food Additives," *EFSA Journal* 9 (3)(2011): 103. Accessed October 2012 at http://www.efsa.europa.eu/en/efsajournal/pub/2004.htm#.

8. A. Thrasyvoulou, "Heating Times for Greek Honeys," *Melissokomiki-Epitheorisi* 11 (2)(1997): 79–80. Accessed at http://www.airborne.co.nz/HMFref.html; I. Kubis and I. Ingr, "Effects Inducing Changes in Hydroxymethylfurfural Content in Honey," *Czech Journal of Animal Science* 43 (8): 379–383, 11 ref. Accessed at http://www.airborne.co.nz/HMFref.html.

9. Ibid.

10. Bob Harrison, "Weslaco Bee Lab and Current Research," *American Bee Journal* 147 (4): 323–26.

11. Blaise W. LeBlanc et al., "Formation of Hydroxymethylfurfural in Domestic High-Fructose Corn Syrup and Its Toxicity to the Honey Bee (*Apis mellifera*)," *Journal Of Agriculture and Food Chemistry* 57 (16) (July, 31, 2009): 7369–76.

12. R. Dufault et al., "Mercury from Chloralkalai Plants: Measured Concentrations in Food Product Sugar," *Environmental Health* 8 (2)(2009). Accessed at http://www.ehjournal.net/content/8/1/2.

13. "The Facts about High Fructose Corn Syrup," Corn Refiners Association, http://www.sweetsurprise.com.

14. National Honey Board: http://www.honey.com/downloads/carb.pdf.

15. Los Altos Health Research Clinic Study conducted by Dr. Gene Spiller; Frederick J. Bates et al., "Polarimetry, Saccharimetry and the Sugars," *Circular C440*, US National Bureau of Standards, (May 1, 1942), from S. Fallon Morell and R. Nagel, "Agave Nectar: Worse Than We Thought,"

Wise Traditions in Food, Farming, and the Healing Arts 10 (2)(Spring2009): 52.

16. Andrew Pollack, "Judge Revokes Approval of Modified Sugar Beets," *New York Times*, (August 13, 2010): B1.

17. Rudolf Steiner, *Beekeeping: Nine Lectures on Bees,* 2nd printing (Blauvelt, NY: Steinerbooks, 1988): 24–25.

18. Jeanne Hansen, "The Nation's Best Bee Ordinance!," *American Bee Journal* 152 (5) (May 2012): 447–49.

Chapter 4:
Genetics and Breeding

1. Jeffery W. Harris and John R. Harbo, "The SMR Trait Explained by Hygienic Behavior of Adult Bees," *American Bee Journal* 145 (May 2005): 430–31.

2. Rudolf Steiner, *Beekeeping: Nine Lectures on Bees,* 2nd printing (Blauvelt, NY: Steinerbooks, 1988): 14–15.

3. Ibid., 40.

4. Ibid., 42–43.

Chapter 5:
Parasitic Mites

1. H. Shimanuki, Nick Calderone, and D. Knox, "Parasitic Mite Syndrome I. The Symptoms," *American Bee Journal* (1994): 827–28.

2. P. A. Macedo, J. Wu, and Marion D.Ellis, "*Using Inert Dusts to Detect and Assess Varroa Infestations in Honey Bee Colonies,*" Faculty Publications: Department of Entomology, University of Nebraska–Lincoln, Paper 174 (2002). Accessed at http://digitalcommons.unl.edu/entomologyfacpub/1743.

3. Joe M. Graham, ed., *The Hive and the Honey Bee* (Hamilton, IL: Dadant & Sons, 1993): 1121.

4. A. I. Root Co., *The ABC & XYZ of Bee Culture* (Medina, OH: A. I. Root Co., 1990): 322.

5. Ibid.

6. Judy Y. Wu, Carol M. Anelli, and Walter S. Sheppard, "Sub-Lethal Effects of Pesticide Residues in Brood Comb on Worker Honey Bee (*Apis mellifera*) Development and Longevity," *PLoS ONE* 6 (2)(2011): e14720. doi:10.1371/journal.pone.0014720.

7. Steve Shepherd, "Research Reviewed," *Bee Culture* 138 (3)(March 2010): 54.

8. US Environmental Protection Agency, "A Summary of the Emissions Characterizations and Noncancer Respiratory Effects of Wood Smoke," EPA-453/R-93-036, from Citizens for Safe Water around Badger, "Fact Sheet: Open Burning at Ravenna Arsenal." Accessed at http://www.cswab.org/ravenna.html.

9. Randy Oliver, "Fighting Varroa-Biotechnical 2," *American Bee Journal* 147 (5) (May 2007): 399–406. Accessed at http://scientificbeekeeping.com/fighting-varroa-biotechnical-tactics-ii.

10. Jerry Hayes, "Sugar Dusting," *American Bee Journal* 145 (January 2005): 25.

11. S. Hart, "Baby Bee Odor Lures Cradle-robbing Mites," *Science News* (August 12, 1989): 103.

12. Ibid.

13. C. C. Miller, "Stray Straws,"*Gleanings in Bee Culture* 35 (September 1907): 1127.

14. E. R. Root, "80 Years among the Bees: The Evolution of Modern Comb Foundation," *Gleanings in Bee Culture* 76 (May 1948): 295.

15. Randy Oliver, "A Trial of HoneySuperCell Small Cell Combs," *American Bee Journal* 148 (2008): 455–58; G. A. Piccirillo and D. de Jong, "The Influence of Brood Comb Cell Size on the Reproductive Behaviour of the Ectoparasite Mite *Varroa destructor* in Africanized Honeybee," *Genetics and Molecular Research* 2 (2003): 36–42; D. Message and L. S. Goncalves, "The Effect of the Size of the Honey Bee Cells on the Rate of Infestation by *Varroa jacobsonii*,"

XXXth International Apicultural Congress of Apimondia, Nagoya, Japan (1985): 250; M. Maggi, N. Damiani, S. Ruffinengo, D. de Jong, J. Principal, and M. Eguaras, "Brood Cell Size of *Apis mellifera* Modifies the Reproductive Behavior of *Varroa destructor*." *Experimental and Applied Acarology* 50 (3)(2010): 269–79.

16. J. A. Berry, W. B. Owens, and K. S. Delaplane, "Small Cell Comb Foundation Does Not Impede Varroa Mite Population Growth in Honey Bee Colonies," *Apidologie* 41 (1)(2010): 40–44; M. A. Ellis, G. W. Hayes, and J. D. Ellis, "The Efficacy of Small Cell Foundation as a Varroa Mite (*Varroa destructor*) Control," *Experimental and Appied Acarology* 47 (4)(2009): 311–16; M. F. Coffey, J. Breen, M. J. F. Brown, and J. B. McMullan. "Brood-Cell Size Has No Influence on the Population Dynamics of *Varroa destructor* Mites in the Native Western Honey Bee, *Apis mellifera mellifera*," *Apidologie* 41 (1)(2010): DOI: 10.1051/apido/2010003; T. D. Seeley and S. R. Griffin, "Small-Cell Comb Does Not Control Varroa Mites in Colonies of Honey Bees of European origin," *Apidologie* 42 (2010): 526–532; M. A. Taylor, R. M. Goodwin, H. M. McBrydie, and H. M. Cox, "The Effect of Honey Bee Worker Brood Cell Size on *Varroa destructor* Infestation and Reproduction," *Journal of Apicultural Research* 47 (4) (2008): 239–242; M. W. Wilson, J. Skinner, L. Chadwell, "Measuring the Effects of Foundation on Honey Bee Colonies," *American Bee Journal* (June 2009).

17. "Natural Cell Size and Its Implications to Beekeeping and Varroa Mites," Michael Bush. Accessed March 22, 2012, at http://www.bushfarms.com/beesnaturalcell.htm.

18. Lambert Kanga, Rosalind R. James, and Walker Jones, "Field Trials Using the Fungal Pathogen *Metarhizium anisopliae* (Deutermycetes: *Hyphomycete*) to Control

the Ectoparasitic Mite, *Varroa destructor* (Acari: *Varroidae*) in Honey Bee Colonies," *Journal of Economic Entomology* 96 (4) (2003): 1091–99.

19. Rosalind R. James, Gerald Hayes, and Jarrod E. Leland, "Field Trials on the Microbial Control of Varroa with the Fungus *Metarhizium anisopliae*," *American Bee Journal* 146 (November 2006): 968–72.

20. R. Rivera, F. A. Eischen, R. H. Graham, and G. M. Acuna, "Varroa Control Trials with the Thymol-based Gel Product," *American Bee Journal* 145 (May 2005): 433.

21. P. J. Elzen, R. L. Cox, and W. A. Jones, "Evaluation of Food Grade Mineral Oil Treatment for Varroa Mite Control," *American Bee Journal* 144 (December 2004): 921–23.

22. Ibid.

23. Pedro P. Rodriguez and C. E. Harris, "Food Grade Mineral Oil—Thymol Widen Alternatives for Honey Bee Parasite Control," *American Bee Journal* 143 (September 2003): 727–30.

24. Department of Entomology, Washington State University, Dandant advertisement in *American Bee Journal* 144 (July 2004): 526.

25. Reed E. Burns, Justin E. Emerson, and Gerald B. McElroy, "An Economical and Effective Spray Device for Applications of Liquids to Beehives," *American Bee Journal* 145 (October 2005): 803–4.

26. Joe M. Graham, ed., *The Hive and the Honey Bee* (Hamilton, IL: Dadant & Sons, 1993): 877–78.

27. Malcolm T. Sanford, "Using Liquid Formic Acid for Mite Control," *Bee Culture* 131 (June 2003): 17–19.

28. Ibid.

29. Ibid.

30. Gunther Hauk, *Toward Saving the Honey Bee* (San Francisco, CA: Biodynamic Farming and Gardening Association, Inc., 2002): 70–71.

31. Ibid.

32. Eva Rademacher and Marika Harz, "Effectiveness of Oxalic Acid for Controlling the Varroa Mite," *Ameican Bee Journal* 146 (July 2006): 614–17.

33. Gunther Hauk, *Toward Saving the Honey Bee* (San Francisco, CA: Biodynamic Farming and Gardening Association, Inc., 2002): 70–71.

Chapter 6:
Insect Pests

1. Peter Neumann and Patti J. Elzan, "The Biology of the Small Hive Beetle (*Aethina tumida*, Coleoptera: *Nitidulidea*): Gaps in Our Knowledge of an Invasive Species," *Apidologie* 35 (3)(2004): 229–247.

2. Doug Somerville, "Study of the Small Hive Beetle in the USA," NSW Agriculture, Rural Industrial Research and Development Corporation, Publication No. 03/050, RIRDC Project No. DAN-213A, (June 2003): 5–7; M. Michael Hood, "The Small Hive Beetle, *Aethina Tumida*: A Review," *Bee World* 85 (3)(2006): 56.

3. Doug Somerville, "Study of the Small Hive Beetle in the USA," NSW Agriculture, Rural Industrial Research and Development Corporation, Publication No. 03/050, RIRDC Project No. DAN-213A, (June 2003): 6.

4. Ibid.

5. M. Michael Hood, "The Small Hive Beetle, *Aethina Tumida*: A Review," *Bee World* 85 (3)(2006): 54.

6. J. D. Ellis, S. Spiewok, K. S. Delaplane, S. Buchholz, P. Neumann, and W. L. Tedders, "Susceptibility of *Aethina tumida* (Coleoptera: Nitidulidae) Larvae and Pupae to Entomopathogenic Nematodes, *Journal of Economic Entomology* 103 (2010): 1–9; D. I. Shapiro-Ilan, J. A. Morales-Ramos, M. G. Rojas, and W. L. Tedders, "Effects of a Novel Entomopathogenic

Nematode-Infected Host Formulation on Cadaver Integrity, Nematode Yield, and Suppression of *Diaprepes abbreviatus* and *Aethina tumida*," *Journal of Invertebrate Pathology* 103(2010): 103–108.

7. M. Michael Hood, "The Small Hive Beetle, *Aethina Tumida*: A Review," *Bee World* 85 (3)(2006): 54.

8. A. Core, C. Runckel, J. Ivers, C. Quock, T. Siapno et al., "A New Threat to Honey Bees, the Parasitic Phorid Fly *Apocephalus borealis*," *PLoS ONE* 7 (1)(2012): e29639. doi:10.1371/journal.pone.0029639.

Chapter 7:
Four-Legged and Feathered Pests

1. Joe M. Graham, ed., *The Hive and the Honey Bee* (Hamilton, IL: Dadant & Sons, 1993): 1136.

Chapter 8:
Environmental and Human Threats

1. Joe M. Graham, ed., *The Hive and the Honey Bee* (Hamilton, IL: Dadant & Sons, 1993): 890.

2. A. I. Root Co., *The ABC & XYZ of Bee Culture* (Medina, OH: A. I. Root Co., 1990): 263.

3. Joe M. Graham, ed., *The Hive and the Honey Bee* (Hamilton, IL: Dadant & Sons, 1993): 1197.

4. E. Laurence Atkins, *Analysis of the Apicultural Industry in Relation to Geothermal Development & Agriculture in the Imperial Valley, Imperial County, California*, University of California, College of Biological Science, (Riverside, CA: April 1979); J. Raloff, "Environment: Pollution May Confuse Pollinators: Smog Dilutes Scents Needed to Guide Floral Foragers," *Science News*, 173 (16)(May 2008): 14. doi:10.1002/scin.2008.5591731617.

5. Louise A. Malone and Minh-Hà Pham-Delègue, "Effects of Transgene Products on Honey Bees (*Apis mellifera*) and Bumblebees (*Bombus* sp.)," *Apidologie* 32 (5)(2001): 1–18. Accessed at http://www.hortresearch.co.nz/files/science/gmimpacts/m1403malone.pdf.

6. Minh-Hà Pham-Delègue et al., "Long-term Effects of Soybean Protease Inhibitors on Digestive Enzymes, Survival and Learning Abilities of Honey Bees," *Entomologia Experimentalis et Applicata* 95 (2000): 21–29.

7. New Zealand Ministry for the Environment, *Report of the Royal Commission on Genetic Modification*, (2002): 52.

8. Michael S. Engel, Ismael A. Hinojosa-Diaz, and Alexandr P. Rasnitsyn, "A Honey Bee from the Miocene of Nevada and the Biogeography of *Apis* (Hymenoptera: Apidae: Apini)," *Proceedings of the California Academy of Sciences*, Series 4 60 (3)(May 7, 2009): 23–38.

9. Eva Crane, *The World History of Beekeeping and Honey Hunting* (New York: Routledge, 1999).

Chapter 9:
Hive Diseases

1. M. D. Simone-Finstrom and M. Spivak, "Increased Resin Collection after Parasite Challenge: A Case of Self-Medication in Honey Bees?" *PLoS ONE* 7 (3)(2012): e34601. doi:10.1371/journal.pone.0034601.

2. A. I. Root Co., *The ABC & XYZ of Bee Culture* (Medina, OH: A. I. Root Co., 1990): 119.

3. Mark Goodwin and Cliff Van Eaton, *Elimination of American Foulbrood without the Use of Drugs: A Practical Manual for Beekeepers* (Napier, New Zealand: National Beekeepers' Association of New Zealand, 1999).

4. S. Fenoy, C. Rueda, M. Higes, R. Martin-Hernandez, and C. del Aguila: "High-level Resistance of *Nosema ceranae*, a Parasite of the Honeybee, to Temperature and

Desiccation," *Applied and Environmental Microbiology* 75 (21)(2009): 6886–89. doi:10.1128/AEM.01025 09.

5. Ibid.

6. Thomas C. Webster, "Sunlight, Water and Nosema Spores: Managed Pollinator Coordinated Agricultural Project (CAP)," *American Bee Journal* 152 (May 2012): 501–02.

7. F. A. Eischen, R. H. Graham, R. Rivera, and R. James, "Controlling Nosema Spores on Stored Honeycomb," American Bee Research Conference, January 2011, *American Bee Journal* 141 (May 2011): 508–09.

8. P. R. Rhoades and J.A. Skinner, "Effects of Treatment with Thymol, Fumagilin, Honey-B-Healthy, and Nozevit on Caged Honey Bees Infected with *Nosema apis* and *N. ceranae*," American Bee Research Conference, January 2011, *American Bee Journal* 141 (May 2011): 514.

9. Ibid; Ivana Tlak, Gajger, "Nozevit Aerosol Application for *Nosema ceranae* Disease Treatment," *American Bee Journal* 151 (November 2011): 1087–89.

10. Renée Johnson, *Honey Bee Colony Collapse Disorder*, Congressional Research Service Report for Congress, (January 7, 2010).

11. Dennis van Engelsdorp, Diana Cox Foster, Nancy Ostiguy, Maryann Frazier, and Jerry Hayes, "'Fall-Dwindle Disease': Investigations into the Causes of Sudden and Alarming Colony Losses Experienced by Beekeepers in the Fall of 2006. A Preliminary Report," (December 15, 2006), January 23, 2007.

12. Ibid.

13. Ibid.

14. Nikos G. Tzortzakis and Costas D. Economakis, "Antifungal Activity of Lemongrass (*Cympopogon citratus* L.) Essential Oil Against Key Postharvest Pathogens," *Innovative Food Science & Emerging Technologies,* 8 (2) (June 2007): 253–58; Cristiane de Bona da Silva, "Antifungal Activity of the Lemongrass Oil and Citral

against *Candida* spp.," *Brazilian Journal of Infectious Diseases* 12 (1) (February 2008).

15. M. Minami, M. Kita, T. Nakaya, et al., "The Inhibitory Effect of Essential Oils on Herpes simplex Virus Type-1 Replication in Vitro," *Microbiological Immunology* 47 (2003): 681–84.

16. D. A. Shearer and R. Boch, "Citral in the Nassanoff Pheromone of the Honey Bee," *Journal of Insect Physiology* 12 (12) (December 1966): 1513–21.

17. S. F. Pernal, D. S. Baird, A. L. Birmingham, H. A. Higo, K. N. Slessor, and M. L. Winston, "Semiochemicals Influencing the Host-finding Behavior of *Varroa destructor*," *Journal of Experimental and Applied Acarology* 37 (1–2)(September/October 2005): 1-26.

18. Personal correspondence (December 2012).

19. Kathy Kellison, "One Colony, One Acre," *Bee Culture* (January 2012): 30.

20. N. M. Schiff and W. S. Sheppard, "Genetic Analysis of Commercial Honey Bees (Hymenoptera: Apidae) from the Southern United States," *Journal of Economic Entomology* 88 (5)(1995): 1216–20.

21. D. A. Delaney, N. Meixner, M. Schiff, and W. S. Sheppard, "Genetic Characterization of Commercial Honey Bee (Hymenoptera: Apidae) Populations in the United States by Using Mitochondrial and Microsatellite Markers," *Annals of the Entomological Society of America* 102 (2009): 666–73.

22. H. R. Mattila, D. Rios, V. E. Walker-Sperling, G. Roeselers, and I. L. G. Newton, "Characterization of the Active Microbiotas Associated with Honey Bees Reveals Healthier and Broader Communities when Colonies are Genetically Diverse," *PLoS ONE* 7 (3): e32962. doi:10.1371./journal .pone.0032962

23. D. J. Hawthorne and G. P. Dively, "Killing Them with Kindness? In-Hive Medications May Inhibit Xenobiotic

Efflux Transporters and Endanger Honey Bees," *PLoS ONE* 6 (11)(2011): e26796. doi:10.1371/journal.pone.0026796.

24. L. P. Dahlgren, R. M. Johnson, M. D. Ellis, and B. D. Siegfried, "Varroacide Toxicity to Honey Bee Queens," *American Bee Journal* 151 (May 2001): 508.

25. J. L. Kirschvink, M. Winklhofer, and M. M. Walker, "Biophysics of Magnetic Orientation: Strengthening the Interface between Theory and Experimental Design," *Journal of the Royal Society Interface* 7 (2010): S179-S191. doi:10.1098/rsif.2009.0491.

26. T. E. Ferrari and A.B. Cobb, "Honey Bees, Correlations between Geomagnetic Storms and Colony Collapse Disorder," *American Bee Journal* 151 (May 2011): 509.

27. Robert O. Becker, *Cross Currents: The Perils of Electropollution, the Promise of Electromedicine,* (New York: Tarcher, 1990).

28. Daniel Favre, "Mobile Phone-Induced Honeybee Worker Piping," *Apidologie* 42 (2011): 270–79.

29. S. Sainudeen Sahib, "Impact of Mobile Phones on the Density of Honey Bees," *Munis Entomology & Zoology* 6 (1) (2011): 396–99.

30. Vandana Shiva, *Earth Democracy: Justice, Sustainability, and Peace* (Cambridge, MA: South End Press, 2005).

31. Vandana Shiva, *Yoked to Death: Globalism and Corporate Control of Agriculture,* Research Foundation for Science, Technology and Ecology, (New Delhi, India: 2001): 6.

32. Oliver DeSchutter, *Agroecology and the Right to Food,* United Nations Human Rights Office of the High Commissioner, Report presented at the 16th Session of the United Nations Human Rights Council, (Geneva: UN, March 8, 2011); Jonathan A. Foley et al., "Solutions for a Cultivated Planet," *Nature* 478 (October 20, 2011): 337–42; Natural Research Council, *Toward Sustainable Agricultural Systems in the*

21st Century, (Washington, DC: National Academies Press, 2010); Food and Agriculture Organization (FAO) of the United Nations, *Toward the Future We Want: End Hunger and Make the Transition to Sustainable Agriculture and Food Systems* (Rome: FAO, 2012).

33. Maryann Frazier, Chris Mullin, Jim Frazier, and Sara Ashcroft, "What Have Pesticides Got to Do With It?" *American Bee Journal* 148 (June 2008): 521–23.

34. Dennis van Engelsdorp et al., "'Entombed Pollen': A New Condition in Honey Bee Colonies Associated with Increased Risk of Colony Mortality," *Journal of Invertebrate Pathology* 101 (2)(June 2009): 147–49.

35. Tom Theobald, "Do We Have a Pesticide Blowout?" *Bee Culture* 138 (July 2010): 66–69.

36. Consensus Statement from the work session on *Chemically-Induced Alterations in Sexual Development: The Wildlife/ Human Connection,* (Wisconsin, July 1991). Accessed at http://www.endocrine disruption.com/files/wingspread _consensus_statement.pdf.

37. Philippe Grandjean et al., "The Faroes Statement: Human Health Effects of Developmental Exposure to Chemicals in Our Environment," Journal Compilation, *Basic and Clinical Pharmacology & Toxicology* 102 (2007): 73–75. Accessed at http:// www.precaution.org/lib/faroes_statement _pub.070801.pdf.

38. "Paracelsus Revisited," Peter Montague, Environmental Research Foundation. Acessed October 16, 2002 at http://rachel .org/en/node/5579.

39. Bijal Trivedi, "Toxic Cocktail," *New Scientist* (September 2007), 44. Accessed at http://www.precaution.org/lib/07/prn _toxic_cocktail.070903.htm.

40. A. V. Adrianov, "The Sixth Wave of Extinction," *ScienceDaily,* (August 23, 2004).

Accessed at http://www.sciencedaily.com
/releases/2004/08/040816001443.htm;
American Museum of Natural History,
"National Survey Reveals Biodiversity
Crisis: Scientific Experts Believe We Are
in the Midst of Fastest Mass Extinction
in Earth's History," Press release, (April
20, 1998); Bradley J. Cardinale, J. Emmett
Duffy et al., "Biodiversity Loss and Its
Impact on Humanity," *Nature* 486 (June 7,
2012): 59–67. doi:10.1038/nature11148.

41. James Lovelock, *Gaia: A New Look at
Life on Earth* (Oxford: Oxford University
Press, 1979).

42. Curtis C. Travis and Sheri T. Hester,
"Global Chemical Pollution," *Environmental Science and Technology*, Oak Ridge
National Laboratory 25 (5), (1991):
http://www.precaution.org/lib
/travis_and_hester.1991.pdf.

Chapter 10:
The Honey Harvest

1. *Vermont Organic Farmers 2004 Certification
Guidelines and Applicant Information,*
(Richmond, VT: NOFA, 2004): 70–71.
Accessed at www.nofavt.org.

Chapter 12:
Organics and the Evolution
of Beekeeping

1. Oliver DeSchutter, *Agroecology and the Right
to Food*, United Nations Human Rights
Office of the High Commissioner, Report
presented at the 16th Session of the United
Nations Human Rights Council, (March 8,
2011); Jonathan A. Foley et al., "Solutions for
a Cultivated Planet," *Nature* 478 (October
20, 2011): 337–42; Natural Research Council, *Toward Sustainable Agricultural Systems
in the 21st Century*, (The National Academies
Press, Washington, D.C., 2010); *Toward the
Future We Want: End Hunger and Make the
Transition to Sustainable Agriculture and Food
Systems*, Food and Agriculture Organization
(FAO) of the United Nations, (Rome: 2012).

2. Hilda M. Ransome, *The Sacred Bee in
Ancient Times and Folklore*, (Mineola,
NY: Dover Publications, Inc., 2004); Eva
Crane, *The World History of Beekeeping
and Honey Hunting* (New York: Routledge,
1999); Simon Buxton, *The Shamanic Way
of the Bee—Ancient Wisdom and Healing
Practices of the Bee Masters* (Rochester,
VT: Destiny Books, 2004).

GLOSSARY

ABSCOND: when an entire colony of honey bees abandons a hive and settles in a new location

ACARICIDE: a pesticide that kills mites

ACHROIA GRISELLA: see *wax moth*

AFB: abbreviation for American foulbrood

AFRICANIZED HONEY BEE: the strain of honey bee that is descended from the African bees brought to Brazil in the 1950s. The Africanized honey bee migrated up through Central America and has spread into the southern United States. This bee is characterized by its aggressive defensive behavior that is a result of its historical exposure to an environment that provided short and sporadic honey flows coupled with a large number of natural predators.

AHB: abbreviation for Africanized honey bee

AMERICAN BEE JOURNAL: monthly periodical on beekeeping published by Dadant & Sons of Hamilton, Illinois.

AMERICAN FOULBROOD: bacterial disease that is highly contagious among honey bee colonies, produces a foul odor in advanced stages, and usually leads to the death of a colony unless treated with antibiotics, the burning of infected combs, or the removal of all AFB spores from the colony

APIARY: location where one or more beehives are kept

APICULTURE: things of or relating to beekeeping

APICULTURIST: one who practices apiculture

APITHERAPY: use of honey bee and hive products (honey, propolis, pollen, royal jelly, and bee venom) for health and the treatment of disease conditions

ARTIFICIAL SWARM: see *nucleus colony*

BACILLUS THURINGIENSIS: bacterium that, when ingested by certain insect larvae, produces crystals and spores that kill the host

BEE BREAD: pollen that has been fermented by the bees and stored in comb cells, usually near the brood nest

BEE CULTURE: monthly publication on beekeeping published by the A. I. Root Co. of Medina, Ohio; formerly called *Gleanings in Bee Culture*

BEE ESCAPE: device that fits into the inner cover of a hive and allows for the one-way passage of honey bees; used primarily for removing bees from honey supers prior to harvest

BEE GLUE: see *propolis*

BEE SPACE: the area between objects in a hive that measures approximately 5/16 of an inch (between 1/4 and 3/8 of an inch) and within which the bees tend not to build comb, preferring to keep the space free and open for use as a passageway within the hive

BEE TEA: a sugar syrup fortified with sea salt and herbal tea to make it a little healthier for the bees to consume

BEESWAX: a substance produced by the wax-secreting glands in young honey bees' abdomens that is molded and shaped by worker bees into combs

BIRTHING CELL: the cell within a comb into which the queen lays an egg and the resulting bee emerges once it has developed to the point where it is ready to join the rest of the colony as a mature bee

BOTTOM BAR: the lower or bottom piece that makes up a modern beekeeping frame

BOTTOM BOARD: the piece of hive equipment that makes up the bottom or floor of the modern beehive

BRACE COMB: wax comb built between frames within a hive

BROOD: young honey bee larvae, pupae, and immature bees

BROOD CHAMBER: the area where the queen bee lays eggs and the hive raises brood

BT: abbreviation for *Bacillus thuringiensis*

BURR COMB: the wax comb built on the top, bottom, and ends of frames within a hive

CAPPINGS: the thin wax covering bees build over cells to cover them. Cells of honey and developing brood of a certain age are capped in such a way. The cappings used to seal cells full of honey must be removed as part of the modern honey extraction process.

CAPPINGS SCRATCHER: see *uncapping fork*

CHALKBROOD: a disease state created by the presence of the fungus *Ascosphaera apis* that affects honey bee larvae and pupae, resulting in their "mummification" and taking on the appearance of little pieces of chalk

CHILLED BROOD: brood that is killed due to exposure to cold temperatures

CHUNK HONEY: a piece of bottled comb honey completely submerged in liquid honey

CLUSTER: the formation of honey bees within a hive that occurs in cold weather of about 57°F (14°C) or below and characteristically consists of a dense ball of bees with the queen and brood at its center

COLONY: a family of honey bees occupying a single cavity

COLONY COLLAPSE DISORDER: a term developed during the winter of 2006–2007 to describe the death of honey bee colonies that exhibit the following symptoms: The rapid decline of a hive's population to the point where few or no adult bees are found dead or alive in the hive, either on the combs or on the bottom board. If any living bees are present they are only the queen accompanied by a handful of very young workers. Brood in various stages are present in larger quantities than any remaining cluster can maintain. There is a noted delay of two to three weeks after the initial collapse before there is any robbing/scavenging activity by other bees, wax moths, small hive beetles, wasps, or hornets within the dead hive.

COMB: a structure manufactured by honey bees out of beeswax and consisting of hexagon-shaped cells fitted side by side and used by bees to raise brood and store honey and pollen; also called honeycomb

COMB HONEY: honey in the comb prepared for human consumption as stored by honey bees and found naturally occurring within the hive

DEARTH: a lack or absence. In beekeeping, dearth typically refers to the lack of nectar or pollen available for foraging honey bees.

DEEP SUPER: boxes 9⅝ inches tall that are filled with frames of comb and used to provide room for a colony to store excess honey intended for harvest by the beekeeper. When used to house the brood nest of a hive, it is called a hive body.

DIVISION BOARD FEEDER: type of feeding device that is shaped like a frame and takes the place of one or more frames in the hive, thereby providing feed close to the brood nest

DRAWN COMB: comb that has been built up from a sheet of foundation and consists of thousands of individual wax cells ready for the bees to fill with nectar, pollen, or eggs

DRONE: the male of the honey bee species characterized by a wider body mass and larger eyes as compared to the female honey bee. The drone is not endowed with a stinger and does not participate in any of the work activities taken up by the hive.

DRONE BROOD: the immature stage of the male honey bee

DRONE LAYER: an infertile queen bee that is capable only of laying unfertilized eggs that develop into drones. May also be used to describe a hive that has no queen but is instead populated with one or more laying worker bees whose unfertilized eggs develop into drones.

ECONOMIC INJURY LEVEL: the level of varroa mites within a hive that will negatively affect the the hive's honey production to the extent that it will adversely impact the income of beekeepers who rely on honey sales for some portion of their income. This level is typically considered to be around 3,000 mites.

EIL: abbreviation for *economic injury level*

END BARS: the two pieces of a frame that fit between the top bar and the bottom bar

ESCAPE BOARD: a flat device shaped like an inner cover that is placed between two honey supers and allows for the one-way passage of honey bees; used primarily for removing bees from honey supers prior to harvest

EUROPEAN FOULBROOD: a bacterial disease that is highly contagious and is similar to American foulbrood in that it may lead to the death of a colony unless treated with antibiotics or the burning of infected combs

FONDANT: a toffeelike substance made primarily from sugar and used to plug up entrances in queen cages and as a source of feed for hives, usually used during cold winter months when the additional moisture from sugar syrup can be harmful to a colony

FORAGE RANGE: the distance covered by worker bees searching for items to bring back to the hive

FORAGING: the gathering activities of the worker bee. Honey bees typically forage for nectar, pollen, plant resins to make propolis, and water.

FOUNDATION: a thin sheet, typically made of wax or plastic, that has the outline of hexagonal-shaped honeycomb cells embossed onto both sides of its surface

FRAME: a man-made contraption typically made of wood or plastic consisting of a top bar, a bottom bar, and two end bars, and usually fitted with foundation to encourage bees to build comb within the confines of the frame. Frames are designed to hold and support honeycomb and allow for easy removal from a hive for inspection and harvesting without damaging the comb.

FRAME GRIP: a mechanical tool used for gripping and moving frames within a beehive

FUME BOARD: a metal outer cover fitted on the inside with an absorbent material that is soaked in a liquid chemical that gives off a strong odor. When placed on top of a hive on a sunny day, the metal cover heats up, increasing evaporation of the liquid and driving the bees out of the top super of the hive so the honey can be harvested.

GENE: a piece of DNA (deoxyribonucleic acid) that carries information for the biosynthesis of a specific product within the cell. A gene is the unit by which inheritable characteristics are transferred to succeeding generations in all living things. Genes are contained within, and arranged along the length of, the chromosome. Each chromosome of a species has a specific number and arrangement of genes, which govern both the structure and the metabolic functions of the cells and thus the entire organism. Genes provide information for the synthesis of enzymes and other proteins and specify when these substances are to be manufactured. A change in either gene number or gene arrangement can result in mutation (a change in inheritable traits).

GENETICALLY MODIFIED: the genetic material (DNA) of cells or organisms that has been altered or manipulated to make them capable of making new substances or performing new functions. This gene-splicing technique transports selected genes from one species to a different, often totally unrelated species. These genes or portions of DNA molecules are removed from the donor (insect, plant, animal, or other life-form) and inserted into the genetic material of a virus. The virus is then allowed to infect the recipient organism so that the organism ends up absorbing both the virus and the genetic material. When the virus replicates within the organism, large amounts of the foreign, as well as the viral material are manufactured.

GENETICALLY MODIFIED ORGANISM: a life-form that is the product of genetic engineering. GMOs are considered to be intellectual property and thus these life-forms can be patented under current laws.

GM: abbreviation for *genetically modified*

GMO: abbreviation for *genetically modified organism*

HIVE: a home for bees. Hives manufactured by humans typically consist of a bottom board, an inner cover, an outer cover, and one or more boxes stacked directly one over the other and filled with frames of comb or foundation.

HIVE, STRONG: a colony of bees characterized as having an abundant population of worker bees, enough to cover all the combs in the hive, for example

HIVE, WEAK: a colony of bees characterized as having a smaller than normal population of worker bees that is not able to cover all the comb surfaces within the hive

HIVE BEETLE: see *small hive beetle*

HIVE BODY: the body of a hive typically consisting of a box 9⅝ inches deep and containing frames of comb within which can be found the majority of the brood within the hive

HIVE TOOL: an essential part of beekeeping equipment consisting of a piece of flattened metal that resembles a minicrowbar and is used for prying, scraping, and manipulating the parts of a hive

HMF: see *hydroxymethylfurfural*

HONEY: the nectar from flower blossoms that has been gathered, processed, and concentrated by honey bees into a sweet, viscous material composed primarily of glucose and fructose, along with numerous trace minerals, and a small amount of vitamins, proteins, and enzymes

HONEY BOUND: the state in which a queen bee is unable to lay a significant number of eggs because the cells in the brood area are filled with honey due to the lack of available space in the rest of the hive to store processed nectar

HONEYCOMB: see *comb*

HONEY FLOW: when flowers are producing enough nectar to allow significant honey storage that results from a concentrated period of nectar foraging usually over a relatively short period; sometimes referred to as a *nectar flow*

HONEY HOUSE: a building usually in or near an apiary where beekeeping equipment and supplies are stored, where honey is extracted and stored, and which serves as the beekeeper's workshop

HONEY SUPER: a box used for honey production that can consist of a shallow, medium, or deep super

HYDROXYMETHYLFURFURAL: a compound produced when high-fructose corn syrup or honey is exposed to high temperatures. Small amounts of HMF will occur naturally in honey as it ages during storage. Hydroxymethylfurfural is toxic to honey bees and mildly toxic to humans.

INNER COVER: a cover that sits on top of the uppermost super on a hive, underneath the outer cover, and contains one or more openings to allow for ventilation, feeding, and the use of bee escapes

KENYAN TOP BAR HIVE: A style of top bar hive developed in Kenya.

LAYING WORKER: an infertile female honey bee (worker) that has been stimulated to develop her egg-laying capacity due to the absence of a fertile queen and the accompanying queen pheromones that inhibit the development of the ovaries of worker bees found within the hive. The eggs laid by a laying worker are infertile and will result only in the production of drones.

MANDIBLES: the mouth or jaws of the honey bee

MEDIUM SUPER: a box that is 6⅝ inches deep and typically contains frames of comb in which the bees can store honey or raise brood

MIGRATORY BEEKEEPERS: apiculturists who move hives to one or more locations during a season to pollinate a specific crop or to take advantage of a honey flow

NASONOV GLAND: a gland that gives off a pheromone with an odor that attracts bees to food and nest sites. Sometimes called the scent gland.

NASONOV PHEROMONE: the scent given off by the Nasonov gland.

NATURAL COMB: honeycomb that is built by bees without the aid of foundation

NECTAR FLOW: see *honey flow*

NEONICOTINOIDS: a family of systemic pesticides that are based on nicotine

NOP: abbreviation for National Organic Program

NOSB: abbreviation for National Organic Standards Board

NOSEMA: a disease caused by the microorganisms *Nosema apis* and *Nosema ceranae* that affects the midgut of the honey bee

NUC: abbreviation for nucleus colony

NUCLEUS COLONY: a new hive of bees created by the beekeeper and typically containing three to five frames of bees, brood, honey, and pollen. A nucleus colony is a miniature version of a regular beehive and often is used to expand an apiary or replace dead colonies. Sometimes referred to as an artificial swarm or split.

OFPA: abbreviation for Organic Foods Production Act of 1990

OUTER COVER: the outermost cover of a modern beehive that protects the hive from precipitation

PARTHENOGENESIS: the process by which unfertilized bee eggs are able to develop into queens

PIPING: the term used to describe the sound that a worker or a queen bee will make when they press their thorax against the comb or another bee and vibrate their wing muscles at a rate of 200–250 Hz. The piping sound is emitted by a queen that is in distress, but is most often believed to represent a challenge to all other queens that may be in the hive, while the piping sound of the worker bee is associated with preparing a swarm to take off in flight.

POLLEN: dust-sized particles produced by the anthers of plants for the purpose of procreation. Pollen is picked up by the hairs on the honey bee's body and then formed into a compact clump by the bee, who brings it back to the hive, where it is stored in fermented form in honeycomb cells. Pollen is necessary for the rearing of brood.

PROBOSCIS: the tongue of the bee

PROPOLIS: a sticky substance with powerful antimicrobial properties that is composed mainly of plant resins. The bees use propolis both as a building material to anchor parts of the hive and fill in cracks and holes and as a varnish over the interior of the hive to provide a sterile environment and prevent the growth of bacteria, mold, and disease.

PUSH-IN CAGE: a queen cage made of hardware cloth that allows the queen to lay eggs in a small patch of comb while still being separated from the rest of the hive

QUEEN: a completely developed female honey bee that is capable of laying both fertilized and unfertilized eggs and usually has a life span of approximately four to six years

QUEEN CAGE: a small box containing a screen used to hold and transport a queen bee (also see *push-in cage*)

QUEEN CANDY: a form of *fondant*

QUEEN CELL: the peanut-shaped, vertically oriented cell in which a queen bee develops from egg to adult. A queen cell is considered "ripe" a couple days before hatching.

QUEEN EXCLUDER: a device used to prevent the queen from moving from one part of the hive to another. While there are many styles of excluder, they all feature openings large enough for a worker bee to pass, but too small for a queen. Queen excluders are typically used to prevent the queen from laying eggs in the honey supers.

QUEENLESS: having no queen bee present (as in a hive)

QUEEN-RIGHT: in a state of containing a fertile laying queen (as in a queen-right hive)

RABBET: a recessed area located along the upper inside edge of the ends of supers and hive bodies for the purpose of providing a place for the frame to rest when hanging within the hive

RENDERING: the act of melting down beeswax, combs, and cappings and removing the impurities

REVERSING: a hive manipulation that expands the area used for brood rearing. Reversing a hive typically involves moving a hive body full of brood and honey from the top of the hive down to the bottom board and placing the mostly empty hive body that had occupied the space on the bottom board to the top of the hive to provide room for the queen to lay as she naturally moves upward during the course of the season.

ROBBING: the removal of stored honey from a hive by members of another nest or hive. This will tend to occur in a colony when it is weak and the hive entrance opening is too large for the population of guard bees to adequately defend the hive.

SACBROOD: a virus-induced disease of bee brood that is contagious but not too serious

SEALED BROOD: young bees in the pupa stage whose cells have been sealed over with a thin layer of mostly beeswax. Sealed drone brood will jut out from the comb and have a bulletlike shape, whereas sealed worker brood will appear flatter and more in line with the surface of the comb.

SHALLOW SUPER: a box that is 5¹¹⁄₁₆ inches deep and typically contains frames of comb within which the bees will store excess honey

SHB: abbreviation for *small hive beetle*

SMALL HIVE BEETLE: a parasitic beetle whose larvae feed on the honey, pollen, and brood of weak or dying hives. Feces from the feeding larvae cause honey in combs to ferment. Bees finding themselves in such a situation will tend to abscond.

SMOKER: a metal apparatus equipped with bellows into which a burning material is introduced with the intention of producing smoke that can be directed toward honey bees in an effort to render them docile and easy to work with

SMR: abbreviation for *suppressed mite reproduction*

SPERMATHECA: a structure within the queen bee's reproductive system that is used to store the sperm collected from drones during mating

SPLIT: to divide a colony of bees for the purpose of increasing colony numbers. A split will typically be made up of 8–10 frames from the original colony. A split with a laying queen is also called a nucleus colony.

STARTER STRIPS: narrow strips (typically 1 to 2 inches wide) of foundation that are used in place of a full sheet of foundation, allowing the bees to build the majority of their comb freely without the additional guidance provided by the foundation.

SUGAR SYRUP SPRAY: a mixture of sugar and water that is sprayed on the bees during hive manipulations and replaces the use of smoke to prevent the bees from becoming defensive

SUPER: a box filled with frames of comb or foundation that is placed above the brood chamber of a hive to provide the colony with the space to store excess honey typically intended for harvest by the beekeeper

SUPERING: the act of adding supers filled with frames to a hive for the purpose of providing the colony with additional honey storage space

SUPERSEDURE: the process by which the worker bees raise a new queen from a fertilized egg to replace the mother queen, typically because she is old and failing

SUPPRESSED MITE REPRODUCTION: see *varroa-sensitive hygiene*

SWARM: a group of worker bees of various age groups and a queen bee that have left behind a mature colony to start a new one. The swarm will usually leave behind a hive that is full of honey, pollen, brood, worker and drone bees, and one or more virgin queens that have recently hatched, or are about to hatch, out of their birthing cells.

SWARM CELL: a queen cell that is found along the outer edge of the comb or at the edge of damaged comb and contains a queen being raised by a colony that has intentions of swarming

TOP BAR: the uppermost cross-member of a frame. The ends of the top bar will rest in the rabbets of the supers or hive bodies to suspend the frame and make it accessible to the bees.

TOP BAR HIVE: a hive that utilizes top bars instead of full fames, upon which the bees build their comb naturally without the aid of a sheet of foundation

TRACHEAL MITE: a microscopic mite that lives in the trachea of the honey bee. This mite is very difficult to detect without the aid of a microscope.

UNCAPPING FORK: a multipronged instrument used to scratch the cappings off honey sealed in cells prior to its extraction from the comb

VARROA: a mite from Asia that has spread around the world parasitizing European honey bee colonies. Because the European honey bee is not varroa's native host and has not developed methods of naturally keeping the mite's population under control, varroa typically overwhelms European honey bee colonies within one to two years if left untreated, causing colony death primarily from the transmission of viruses from mite to bee.

VARROA-SENSITIVE HYGIENE: a genetic characteristic expressed within a hive of honey bees by the tendency of the workers to identify developing brood that is sharing its cell with actively reproducing varroa mites and to remove the infested brood from the colony, thereby breaking the reproductive cycle of the mite. Formerly known as suppressed mite reproduction.

VIRGIN QUEEN: a queen bee that has not gone on a mating flight and is unable to lay fertilized eggs

VSH: abbreviation for *varroa-sensitive hygiene*

WAX MOTH: a moth that parasitizes weak or dead colonies, laying its egg masses in cracks and crevices within the hive

WAX WORM: the larvae stage of the wax moth. Wax worms bore their way through combs spinning silken tunnels and eating beeswax combs, pollen, and honey as they go.

WORKER BEE: a female honey bee with an underdeveloped reproductive system that carries out the majority of the work necessary for the proper functioning of a colony and has a life span of approximately five to six weeks during the summer and four to six months during the winter

WORKER BROOD: young worker bees in the larval and pupal stages

The following resources are not meant to be a complete list of beekeeping resources, but are provided to get you started. The following companies and organizations are simply ones that I personally have had dealings with or am most familiar with.

General

A. I. Root Co.
623 W. Liberty St.
Medina, OH 44256
800-289-7668
e-mail: info@beeculture.com
website: www.beeculture.com
Publishers of Bee Culture magazine

American Apitherapy Society
818-501-0446
e-mail: aasoffice@apitherapy.org
website: www.apitherapy.org
Information on the use of honey bee products for the treatment of disease conditions

Bee Alert Technology, Inc.
1620 Rodgers St., Suite 1
Missoula, MT 59802
406-541-3160
Anti-theft technology

Better Way Wax Melter Honey Processors Ltd.
116 11th Str. SE
Altoona, IA 50009
515-967-4952
"The Cyclone" lactic acid vaporizer

Brushy Mountain Bee Farm
610 Bethany Church Rd.
Moravian Falls, NC 28654
800-233-7929
e-mail: info@brushymountainbeefarm.com
Source for small-cell foundation of 5.1 and 4.9 millimeters

Dadant & Sons
51 South 2nd St.
Hamilton, IL 62341
888-922-1293
e-mail: dadant@dadant.com
website: www.dadant.com
Source for small-cell foundation of 5.1 and 4.9 millimeters
Publishers of the *American Bee Journal*

Golden Bee Products
344½ Aris Ave.
Metairie, LA 70005
504-456-8805
Ventilated, sting-resistant bee suit

Heilyser Technology Ltd.
685 Dalkeith Ave.
Sidney, BC V8L 5G7
Canada
250-656-8727
e-mail: orioleln@shaw.ca
website: www.members.shaw.ca/orioleln
Oxalic acid vaporizer

Honey-B-Healthy
website: www.honeybhealthy.com/DealerLocator.html

National Honey Board
11409 Business Park Circle, Suite 210
Firestone, CO 80504
303-776-2337
website: www.honey.com

Beekeeping Supply Companies

Betterbee
8 Meader Rd.
Greenwich, NY 12834
800-632-3379
website: www.betterbee.com

Brushy Mountain Bee Farm
610 Bethany Church Rd.
Moravian Falls, NC 28654
800-233-7929
e-mail: info@brushymountainbeefarm.com

Dadant & Sons
51 South 2nd St.
Hamilton, IL 62341
888-922-1293
e-mail: dadant@dadant.com
website: www.dadant.com

F. W. Jones and Sons
Box 1230
Bedford, Quebec J0J 1A0
Canada

450-248-3323
e-mail: info@fwjones.com

Humble Abodes
636 Coopers Mills Rd.
Windsor, ME 04363
877-423-3269
e-mail: humbleabodes@prexar.com

Mann Lake Ltd.
501 1st St. South
Hackensack, MN 56452
800-880-7694
website: www.mannlakeltd.com

Maxant Industries
P.O. Box 454
Ayer, MA 01432
978-772-0576
e-mail: sales@maxantindustries.com

Rossman Apiaries
P.O. Box 909
Moultrie, GA 31776
800-333-7677
website: www.gabees.com

Walter T. Kelley Co.
P.O. Box 240
807 W. Main St.
Clarkson, KY 42726
800-233-2899
website: www.kelleybees.com

Breeders of Bees with Mite-Resistant Genetics

B. Weaver Apiaries
6301 Highland Hills
Austin, TX 78731
866-547-3376
e-mail: worker@beeweaver.com
website: www.beeweaver.com

Hardeman Apiaries
P.O. Box 214
906 South Railroad Ext.
Mt. Vernon, GA 30445
912-583-2710

Harper's Honey Farm
421 Louveteau Rd.
Carencro, LA 70520
337-298-6261
e-mail: labeeman@russianbreeder.com

Kirk Webster
Champlain Valley Bees and Queens
P.O. Box 381
Middlebury, VT 05753
802-989-5895

Pendell Apiaries
P.O. Box 40
Stonyford, CA 95979
530-963-3062

Organic Certification Agencies with Apiculture Programs

MOFGA Certification Services, LLC
P.O. Box 170
294 Crosby Brook Rd.
Unity, ME 04988
207-568-4142
e-mail: mofga@mofga.org

Oregon Tilth, Inc.
260 SW Madison Ave., Suite 106
Corvallis, OR 97333
503-378-0690
e-mail: organic@tilth.org
website: www.tilth.org

Organic Crop Improvement Association (OCIA)
1340 North Cotner Blvd.
Lincoln, NE 68505
402-477-2323
e-mail: info@ocia.org
website: www.ocia.org

Vermont Organic Farmers (VOF)
P.O. Box 697
14 Pleasant St.
Richmond, VT 05477
802-434-4122
e-mail: info@nofavt.org
website: www.nofavt.org

Pesticide Testing and Reporting

Environmental Protection Agency (EPA)
Report non-target pesticide damage:
beekill@epa.gov.

Maryann Frazier
Penn State, Department of Entomology
501 ASI Building
University Park, PA 16802
e-mail: mxt15@psu.edu

National Pesticide Information Center (NPIC)
Ecological Pesticide Incident Reporting portal:
http://pi.ace.orst.edu/erep

Roger Simonds, Laboratory Manager
USDA-AMS-National Science Laboratory
801 Summit Crossing Pl., Suite B
Gastonia, NC 28054
704-867-3873
e-mail: roger.simonds@ams.usda.gov

INDEX

Note: Page numbers followed by f refer to Figures

About the Author

Ross Conrad learned his craft from world-renowned beekeeper and founder of Champlain Valley Apiaries Charles Mraz, and his son Bill. Former president of the Vermont Beekeepers Association, Conrad is a regular contributor to *Bee Culture: The Magazine of American Beekeeping*. Ross has led bee-related presentations and taught organic beekeeping workshops and classes throughout North America for many years. His small beekeeping business, Dancing Bee Gardens, supplies friends, neighbors, and local stores with honey and candles, among other bee-related products, and provides bees for Vermont apple pollination in spring.

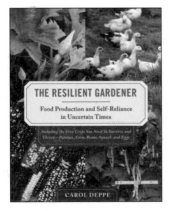
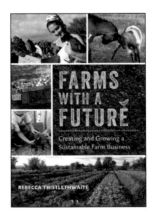